Deep-Fat Frying

Fundamentals
and Applications

Food Engineering Series

Series Editor
Gustavo V. Barbosa-Cánovas
Washington State University

Deep-Fat Frying

Fundamentals and Applications

Rosana G. Moreira, PhD, PE
Associate Professor
Department of Agricultural Engineering
Texas A&M University
College Station, Texas

M. Elena Castell-Perez, PhD, PE
Assistant Professor
Department of Agricultural Engineering
Texas A&M University
College Station, Texas

Maria A. Barrufet, PhD, PE
Associate Professor
Department of Petroleum Engineering
Texas A&M University
College Station, Texas

A Chapman & Hall Food Science Book

AN ASPEN PUBLICATION®
Aspen Publishers, Inc.
Gaithersburg, Maryland
1999

The author has made every effort to ensure the accuracy of the information herein. However, appropriate information sources should be consulted, especially for new or unfamiliar procedures. It is the responsibility of every practitioner to evaluate the appropriateness of a particular opinion in the context of actual clinical situations and with due considerations to new developments. The author, editors, and the publisher cannot be held responsible for any typographical or other errors found in this book.

Aspen Publishers, Inc., is not affiliated with the American Society of Parenteral and Enteral Nutrition.

Library of Congress Cataloging-in-Publication Data

Moreira, Rosana G.
Deep-fat frying: fundamentals and applications/Rosana G. Moreira, M. Elena Castell-Perez, Maria A. Barrufet.
p. cm.
Includes bibliographical references and index.
ISBN 0-8342-1321-4
1. Oils and fats, Edible. 2. Deep frying. I. Castell-Perez, M. Elena. II. Barrufet, Maria A. III. Title.
TP670.M67 1999
664'.3—dc21
99-20349
CIP

About Aspen Publishers • For more than 35 years, [...] publisher in a variety of disciplines. Aspen's vast in[...] print and electronic formats. We are committed to pro[...] available in the most appropriate format for our customers. Visit Aspen's Internet site for more information resources, directories, articles, and a searchable version of Aspen's full catalog, including the most recent publications: **http://www.aspenpublishers.com**
Aspen Publishers, Inc. • The hallmark of quality in publishing
Member of the worldwide Wolters Kluwer group.

Editorial Services: Denise Hawkins Coursey
Library of Congress Catalog Card Number: 99-20349
ISBN: 0-8342-1321-4

Printed in the United States of America

1 2 3 4 5

This book is dedicated to our parents
and to all the students who made this possible.

Table of Contents

Preface

I shall be telling this with a sigh
Somewhere ages and ages hence:
Two roads diverged in a wood, and I—
I took the one less traveled by,
And that has made all the difference.
—Robert Frost, "The Road Not Taken"

This book is the result of a need to cover a specific subject in a concise and deep format to facilitate the exchange of scientific information to people from both industry and academia. Its subject matter, deep-fat frying, is a particular aspect of food engineering and one of the oldest and most common food processing unit operations. This book is aimed at everyone interested in understanding the theoretical principles of deep-fat frying as they are applied in practice. We hope readers will find this book useful as a reference text for a specific area of interest and for use in selected courses addressing unit operations in the food industry.

Deep-Fat Frying of Foods is the outcome of several years of intensive research and experimentation on the effect of deep-fat frying on the food properties, process optimization, process control, and mathematical modeling of several aspects of the process. Throughout the book readers will learn about the importance of deep-fat frying worldwide, the impact of food type on frying design, and what makes oil a suitable medium for food processing. In addition, we present an in-depth look at the effects of frying on quality, the standards for production of good quality fried foods, the basic mechanisms responsible for cooking (heat and mass transfer), standard frying equipment, and packaging requirements.

Snack foods are one of the most significant groups of deep-fat fried foods. However, low fat content and texture are desirable quality attributes in these products. Low-fat snacks will be the driving force of the snack food industry in the next five years. As we investigated this topic, we realized there was still much more to be learned. Therefore, readers will also be exposed to a novel attempt to describe the oil absorption mechanism, the development of low-fat products, and the use of control theory to optimize the fryer operation.

Those wishing to expand further on various specific areas of deep-fat frying will find an excellent compilation of related works presented here. We originally aimed at providing a comprehensive review of the scientific developments in deep-fat frying. However, such an extensive literature search soon became an impossible task. Therefore, the major available works are presented here and we apologize for works that may have been excluded.

Acknowledgments

Our sincere thanks go to those who provided the facilities, resources, and expertise to conduct experiments, analyze data, and interpret results. Special thanks to Texas A&M University's Cereal Quality Laboratory (Dr. Lloyd Rooney's group), the Drying Research Center, and Frito-Lay Inc.'s Research and Development Division.

CHAPTER 1

Introduction

Deep-fat frying of foods is considered everywhere in the world to be the most common unit operation used in food preparation. Therefore, to produce, preserve, and market fried foods optimally, it is important to understand the frying mechanism. This chapter briefly introduces the importance of frying; the equipment, oil, and fats used in frying; oil and food quality characteristics; snack food production; and snack marketing trends of the previous 30 years.

The term *frying* is interpreted broadly in this book and includes only the process of deep-fat frying, defined as a process in which the food is cooked by immersion in hot oil.

IMPORTANCE OF FRYING

Deep-fat frying is one of the oldest processes of food preparation. For decades, consumers have desired deep-fat fried products because of their unique flavor-texture combination. Some of these are potato chips, french fries, doughnuts, extruded snacks, fish sticks, and the traditional fried chicken products. The technology was originated and developed around the Mediterranean area, due to the influence of olive oil (Varela et al., 1988). Today, deep-fat fried foods are found in many countries of Europe, Asia, and North and South America.

The aim of deep-fat frying is to seal the food by immersing it in hot oil so that all the flavors and the juices are retained in a crisp crust. Many foods may be deep-fat fried. If they have a soft, rich texture that contrasts with the crispy outside (for example, scallops and eggplant), so much the better. Deep-fat fried fish, chicken, and vegetables—especially potatoes—are particularly successful. An entire range of deep-fat fried savory croquettes and fritters is very popular, and fruit fritters and deep-fat fried pastries are favorite desserts the world over.

The amount of oil in the pan is important; there should be enough to cover the food to be deep-fat fried by at least 2 cm. Food with high starch content, such as potatoes, can be deep-fat fried without a coating, but others need protection from the hot oil. A coating also prevents oil from penetrating the food and stops the food from flavoring the oil, so that it can be used again. Coating with bread crumbs gives the best protection, whereas a batter is good for less tender foods. Small pieces, such as strips of fish, can be simply coated with flour. Cornmeal and oatmeal are popular alternatives.

The frying technology is important to many sectors of the food industry: suppliers of oils and ingredients, food service (hotel, restaurant, and institution) operators, food industries (snack, fully fried, and par-fried foods), and manufacturers of frying equipment. The amounts of food fried and oils used at both the industrial and commercial levels are vast. The U.S. produces more than 2.5 million metric tons (MMT) (5×10^9 pound) of snack food per year, the majority of which are fried (Snack Food Association, 1997). There are more than 500,000 institutional and commercial restaurants in the U.S. that use approximately 1 MMT (2×10^9 pound) of frying fats and oils annually (O'Brien, 1993).

Even though frying is an old process, it is poorly understood (Levine, 1990). Today, manufacturers of frying equipment still lack sufficient knowledge about what happens with a product during frying. Design of most fryers continues to depend extensively on pilot tests, coupled with vast experience that is often reduced to various "rules of thumb."

Deep-fat frying is a process of cooking and drying through contact with hot oil, and it involves simultaneous heat and mass transfer. The quality of the products from deep-fat frying depends not only on the frying conditions but also on the types of oil and food used during the process. Oils play a dual role in the preparation of fried foods because they serve as a heat transfer medium between the food and the fryer. They also contribute to the food's texture and flavor characteristics.

In deep-fat frying of foods, the temperature of the heated oil, the frying time, and the fryer type (batch or continuous) are factors that affect the process. The chemical composition of the frying oils, the physical and physicochemical constants, and the presence of additives and contaminants also influence the frying process. Additives or contaminants can have a marked effect on the palatability, digestibility, and metabolic utilization of a fried food (Varela et al., 1988). The food weight/frying oil volume and surface area/volume ratios determine how much fat penetrates the food.

FRYING EQUIPMENT

A deep-fat fryer consists of a chamber where heated oil and a food product are placed. The size of the fryer may differ from small oil baths (used in food services) to large continuous industrial baths. In a batch fryer, the product is placed in small net cages that are lowered into the batch. In a continuous fryer, a conveyor belt transports the product through the bath. The product is often pushed through the bath by means of a screen and/or paddles.

The speed and efficiency of the frying process depend on the temperature and quality of the frying oil. The temperature of the oil is usually between 150°C and 190°C. Oil turnover time (mass of used oil/oil usage rate) is generally about 10 hours. It is important to understand what happens to the temperature and the moisture and oil content of the product during the frying process to determine safe temperatures and turnover times of the frying oil for a given fryer type.

THE FRYING OIL

Fats and oils include vegetable oils, palm oils, and animal fats. The annual world production of major oils and fats from 1992 to 1996 is shown in Table 1–1 (U.S. Department of Agriculture, 1997). The total for 1995–1996 was 82.24 MMT (16.45×10^9 pound), about 3% higher than the previous year. Production of canola oil increased the most in 1995 and 1996, followed by sunflower oil and cottonseed oil. Production of olive oil has been reducing since 1992. The 1995–1996 period

Table 1–1 World Production of Major Oils and Fats, in 1,000 metric tons (MT), 1992–1993/1995–1996

Commodity	1992–1993	1993–1994	1994–1995	1995–1996
Edible vegetable oils				
Cottonseed	3,632	3,353	3,731	3,971
Olive	1,777	1,725	1,665	1,455
Peanut	3,597	3,625	4,031	3,957
Canola	8,393	9,036	10,063	11,144
Soybean	17,533	18,307	20,157	19,860
Sunflower	7,326	6,914	7,994	8,736
Palm oils				
Coconut	3,095	3,087	3,441	3,172
Palm	13,006	13,686	14,753	15,576
Palm kernel	1,741	1,880	1,992	2,079
Animal fats				
Butter	4,394	4,248	4,258	4,298
Tallow and grease	7,511	7,572	7,723	7,995
TOTAL	72,005	73,433	79,808	82,243

Source: Data from *Agricultural Statistics, 1997*, United States Department of Agriculture.

was not good for coconut oil either, showing an 8% decrease in production compared with the 1994–1995 period. Production of palm oil, on the other hand, has been steadily increasing since 1992, as is observed for tallow and grease. Butter production increased slightly in 1995–1996.

Grains and oil seeds are raised on every continent except Antarctica. Table 1–2 lists the annual worldwide production of the major oils and fats in 1996; the total was about 70 MMT (14×10^{10} pound) (Food and Agricultural Organization, 1997). Soybean and palm oils are the major oils in terms of tonnage. The U.S. produces almost 90% of soybean oil in the world. Asia produces the most cottonseed, peanut, canola, palm, and coconut oils; Europe produces the most olive and sunflower oils and butter. Soybean and sunflower are the most popular oils in South America; palm oil is the most popular in Africa and Oceania.

In 1995, the U.S. imported about 1.33 MMT (2.66×10^9 pound) of vegetable oils such as cocoa butter, coconut, olive, palm, peanut, and canola oils. Canola and olive oils accounted for 42% of the total vegetable oils imported in 1995 (USDA, 1997). In terms of export, the U.S. sold 1.27 MMT (2.54×10^9 pound) of vegetable oils (cocoa butter, coconut, cottonseed, linseed, peanut, and soybeans oils, and margarine); soybean oil accounted for 82% of the total U.S. vegetable oil export in 1995.

The annual per capita U.S. oils and fats consumption between 1986 and 1995 is presented in Table 1–3 (USDA, 1997). It fell below 28 kg in 1989 for the first time since 1986 but reached a high of 31.71 kg in 1994.

Fats and oils play important functional and sensory roles in food products. They are responsible for carrying, enhancing, and releasing the flavor of other ingredients as well as for interacting with other ingredients to develop the texture and "mouthfeel" characteristics of fried foods. In addition, fats and oils are sources of energy, providing 9 kcal/g—more than twice that supplied by proteins or carbohydrates—fat-soluble vitamins (A, D, E, and K), and the essential linoleic and linolenic acids responsible for growth (Giese, 1996). Fats are considered to be solid at ambient temperature, whereas oils remain liquid at environmental conditions.

More than 95% of fats and oils consist of triglycerides, a compound formed from glycerol and free fatty acids. The type and location of the fatty acids on the glycerol backbone define the physical and chemical characteristics of the various fats (McGrady, 1993). Fatty acids may be either saturated or unsaturated. Color Plate 1 is a comparison of the percentage of saturated and unsaturated fatty acid levels for common edible oils.

There are many types of oils and fats available for frying. Until 1986, animal-origin fats were the primary fats used by the food services. The food industry tended to use animal/vegetable blends, as well as partially hydrogenated vegetable oils (Brooks, 1991). The vegetable oils used include soybean, cottonseed, corn, peanut, olive, canola, safflower, and sunflower. Soybean, safflower, sunflower,

Table 1–2 World Production of Major Oils and Fats by Region, in 1,000 MT, in 1996

Commodity	Africa	Asia	Europe	NC America	Oceania	S America	World
Cottonseed oil	310	1,789	79	594	77	248	3,097
Olive oil	342	329	1,411	2	0.09	9	2,093.09
Peanut oil	869	3,382	82	157	2	68	4,560
Soybean oil	98	3,287	2,714	7,631	20	5,876	19,626
Canola oil	76	4,852	2,986	1,359	81	26	9,380
Sunflower oil	491	1,456	3,231	394	34	2,096	7,702
Palm oil	1,756	13,998	0	204	276	812	17,046
Coconut oil	122	2,476	70	134	71	22	2,895
Animal fat (butter)	98	432	2,102	677	422	181	3,912
TOTAL	4,162	32,001	12,675	11,152	983.09	9,338	70,311.09

Source: Data from *Agricultural Statistics, 1997*, United States Department of Agriculture.

and canola oils are always partially hydrogenated before being used for frying to increase their stability. Cottonseed, corn, peanut, and olive oils are used as a stable source of polyunsaturated fatty acids because of their low linolenic acid content.

THE OIL AND FOOD QUALITY

For fried snack foods, surface appearance and texture are the most significant factors for acceptability. A desirable frying oil must be low in free fatty acids and polar compounds and must have a high breakdown resistance during continuous use. Hence, a thorough understanding of oil degradation and the effects of degraded oil on the quality of final products is important.

Frying oils degrade with continued use. In frying, the food is submerged in oil heated in the presence of air. The oil is exposed to the action of four agents that cause drastic changes in its structure: (1) moisture from the food, giving rise to oxidative alteration; (2) atmospheric oxygen entering the oil from the surface of the container, giving rise to oxidative alteration; (3) the high temperature at which the operation takes place ($\cong 190°C$), which results in thermal alteration; and (4) contamination by food ingredients.

The type of oil used and the length of time that the oil has been used for frying affect the desired flavor of fried foods. The method still most often used in different countries to discard frying oils is sensory evaluation (Melton et al., 1994). In general, the frying industry monitors product quality by how it looks, tastes, and smells. The appearance of the fried product is monitored by color charts and taste panels. Food processors also incorporate shelf life tests to determine flavor stability by using sensory panels to evaluate foods that have been stored for a period of time; others even apply gas liquid chromatography (Brooks, 1991).

THE SNACK FOOD INDUSTRY: PRODUCTION AND MARKETING

In the previous 40 years, there has been a great increase in the use of deep-fat frying processes in the U.S. and in Europe. Table 1–4 shows the variety of deep-fat fried products that have been successfully developed in the recent years. Most of these products are snack foods with an oil content varying from 6% (roasted nuts) to 40% (potato chips). In this book, *snack food* is any fried snack, including extruded, sheeted, and puffed foods, such as potato chips, tortilla chips, french fries, or corn chips.

The snack food industry in the U.S. represents more than $15 billion annually (Snack Food Association, 1997). The annual sale in dollar and pound volume between 1987 and 1996 is shown in Figure 1–1. From 1987 to 1994, sales increased steadily from 2.12 MMT (4.23×10^9 pound) to 2.85 MMT (5.69×10^9 pound). In 1995, after standing substantial losses during the 1990s, Eagle Snack and Kleber

Table 1–3 Fats and Oils—U.S. Use in Products for Civilian Consumption per Capita, 1986–1995

Year	Butter (kg)	Lard and Tallow (kg)	Margarine (kg)	Baking and Frying Fats (kg)	Salad and Cooking Oils (kg)	Other Edible Use (kg)	All Food Products (fat content) (kg)
1986	2.09	1.59	5.17	10.02	10.98	0.77	29.21
1987	2.13	1.22	4.76	9.71	11.52	0.59	28.53
1988	2.04	1.18	4.67	9.75	11.70	0.59	28.58
1989	2.00	0.95	4.63	9.75	10.89	0.59	27.40
1990	2.00	1.13	4.94	10.07	10.98	0.54	28.21
1991	2.00	1.41	4.81	10.16	11.43	0.59	28.99
1992	2.00	1.86	4.99	10.16	11.62	0.64	29.80
1993	2.13	1.77	5.03	11.39	11.39	0.82	31.03
1994	2.18	2.31	4.49	10.93	11.08	0.73	31.71
1995	2.04	2.22	4.17	11.20	11.20	0.73	30.57

Source: Data from *Agricultural Statistics, 1997,* United States Department of Agriculture.

Table 1–4 Deep-Fat Fried Products in the U.S.

Product	Oil Content (%)
Potato chips	33–38
Tortilla chips	23–30
Corn chips	30–38
Expanded snack products	20–40
Roasted nuts	5–6
French fries	10–15
Doughnuts	20–25
Frozen food (fish, chicken, pancakes)	10–15

Source: Data from L.M. Smith et al., Lipid Content and Fatty Acid Profiles of Various Deep-fat Fried Foods, *Journal of the American Oil Chemists' Society,* Vol. 62, pp. 996–999, © 1985, American Oil Chemists' Society, and I.D. Morton and J.E. Chidley, Methods and Equipment in Frying, in *Frying of Foods, Principles, Changes, New Approaches*, G. Varela, A.E. Bender, and I.D. Morton, eds., © 1988, VCH Publishers.

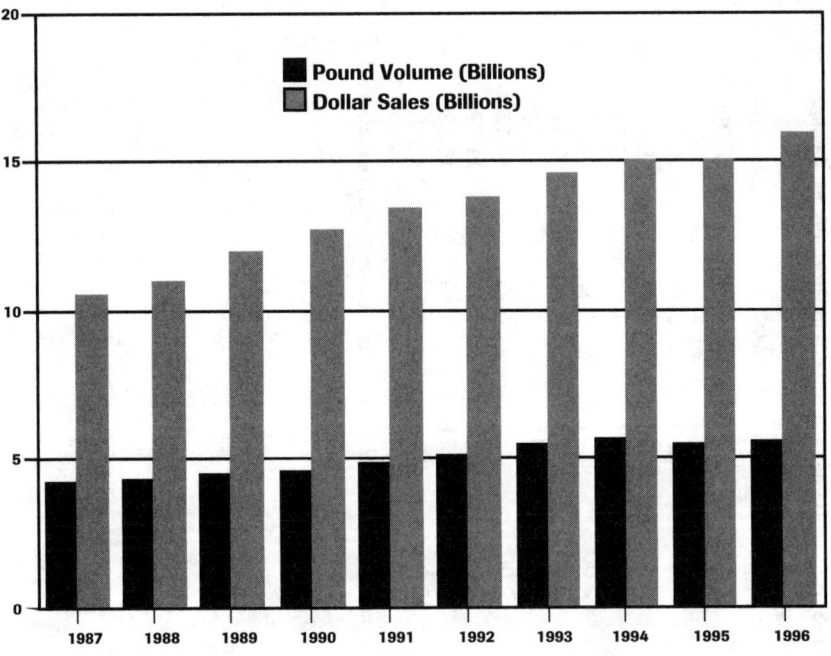

Figure 1–1 Snack Sales in the U.S. between 1987 and 1996. *Source:* Adapted with permission from *State of the Snack Food Industry Report*, p. 2, © 1997, Snack Food Association.

Company left the market. After a 2.6% loss in pound volume in 1995, a positive 1.1% gain was observed in 1996, due to reduced-fat potato chip varieties (2.2%), increased popularity of meat snacks (3.2%) and extruded snacks (9%), renewed consumer interests for corn chips (12.5%), and the low-fat snack mix (5.2%). In the "other snacks" segment of the market, onion rings and potato sticks increased in pound volume by 13.8% and 2.8%, respectively.

In the U.S., snack foods include potato chips, tortilla chips, corn chips, pretzels, popcorn, meat snacks, snack nuts, extruded snacks, pork rinds, party mix, multigrain snacks, and other snacks (onion rings, bagel chips, potato sticks, etc.). Color Plate 2 presents the snack segment shares of dollar sales in 1996.

Potato chips are the leading salted snack in the U.S. Although there is an increased demand for reducing fat consumption, reduced-, low-, and no-fat (known as "better for you") varieties represent only 11% of potato chip supermarket sales. The continued availability and marketing of potato chips made using fat substitutes (i.e., Procter & Gamble Company's Olean brand Olestra) could shift the popularity of reduced-, low-, and no-fat varieties in the future (SFA, 1997).

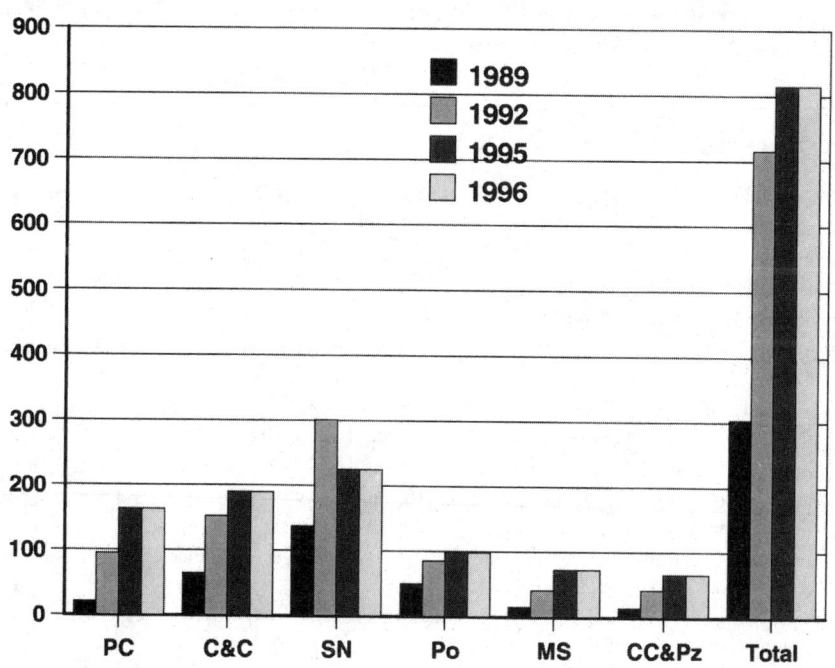

Figure 1–2 U.S. Exports of Snack Food Products, in Millions of Dollars. *Source:* Data from Snack Food Association, 1997. *Note:* PC, potato chips; C&C, cookies & crackers; SN, snack nuts; Po, popcorn; MS, meat snacks; CC&Pz, corn chips & pretzels.

Tortilla chips rank second in the market shares but outpace potato chips by 20% in the Pacific region, which includes California, Washington, and Oregon. Tortilla chip volume is also strong in all other states west of the Mississippi River, averaging about 25% of the market in those areas, compared with an average potato chip market share of about 30%. Consumers have still preferred the regular tortilla chips to the low- or reduced-fat varieties. Pretzels, popcorn, nuts, meat snacks, extruded snacks, and corn chips are the most sold snacks following potato and tortilla chips and represent a market share ranging from 4% to 8%.

The snack food industry expects a changing marketplace in the next five years. Competitive pricing and consumer demand for low-fat and fat-free products are the two main challenges that will affect the snack industry over that time period.

In terms of exports, about 54% of all companies exported snack products in 1996 (SFA, 1997). Canada imports the most U.S. snacks, followed by the Asia/Pacific Rim, Mexico, Eastern Europe, the Middle East, Latin America, and Africa. Figure 1–2 shows the U.S. export of snacks between 1989 and 1996. Total exports increased by almost 170% since 1989. Snack nuts have been the preferred snack for exporting, followed by cookies, crackers, and potato chips. The top five U.S. export markets include Canada (24.3% of total), Japan (10.6%), Belgium (4.7%), Russia (3.5%), and Mexico (3.2%).

REFERENCES

Brooks, D.D. 1991. Some perspectives on deep-fat frying. *INFORM*, 2(12):1091–1095.

Food and Agricultural Organization. 1997. *Agricultural Statistics*. Rome: Food and Agricultural Organization of the United States.

Giese, J. 1996. Fats, oils, and fat replacers. *Food Technol,* 50(4):78–84.

Lenne, L. 1990. Understanding frying operations. *Cereal Foods World,* 35 (2): 272–273.

McGrady, J. 1993. Fats and oils. Presented at the 1993 IFT Short Course *Ingredients Technology.* Chicago, IL.

Melton, S.L.; Jafar, S.; Sykes, D.; and Tigriano, M.K. 1994. Review of stability measurements for frying oils and fried food flavor. *J Am Oil Chem Soc,* 71:1301–1308.

O'Brien, R. 1993. Foodservice use of fat and oils. *INFORM,* 4(8):913–921.

Smith, L.M.; Clifford, A.J.; Creveling, R.K.; and Hablin, C.L. 1985. Lipid content and fatty acid profiles of various deep-fat fried foods. *JAOS,* 62:996–999.

Snack Food Association. 1997. *State of the snack food industry report.* Alexandria, VA: Snack Food Association.

U.S. Department of Agriculture. 1979. Agricultural handbook no. 8–4. Washington, DC: Human Nutrition Information Service.

U.S. Department of Agriculture. 1997. *Agricultural Statistics.* National Agricultural Statistics Service. Washington, DC: U.S. Government Printing Office.

Varela, G.; Bender, A.E.; and Morton, I.D. 1988. In *Frying of foods: principles, changes, new approaches,* ed. Varela, Bender, and Morton. New York: VCH Publishers.

CHAPTER 2

Fried Product Processing and Characteristics

Knowledge of the structure, composition, and properties of food products is useful in understanding their frying characteristics. The product structure may affect oil content; for example, french fries have about 30–50% less oil content than do potato chips. The product composition may influence the oil absorption and structure characteristics; for instance, low-moisture tortilla dough results in tortilla chips with lower oil content and higher pore size distribution. The physical properties of the product control the rate of temperature increase; an example is the effect of crust thickness (low thermal conductivity) that will cause a barrier for heat transfer from the surface to the center of the product.

According to Blumenthal (1991), fried products can be classified according to their surface-to-volume ratios, i.e., products that have: (1) large interior (crumb) volume and no crust differentiation, such as chicken meat below the breading crust; (2) large interior volume and large surface area, with a crust differentiating the surface from the crumb, such as french fries; and (3) small interior volume, large surface area, and all crust (no crumb); an example is potato chips.

The manufacturing, structure, composition, and properties of the four major fried products—potato-based products, corn-based products, fried bread and fritters, and breaded or battered products (chicken, seafood, onion rings, etc.)—are presented in this chapter. Descriptions of the process conditions are also included. For detailed information on snack processing, see Matz (1993); for information about other fried products, see Desrosier (1977), Luh and Woodroof (1988), and Lawson (1985).

POTATO-BASED PRODUCTS

Two types of potato products are described in this chapter—potato chips and french fries. Potato chips can be made from raw material (traditional) or from dehydrated potato (fabricated); emphasis will be given to the traditional potato chips. Potato chips utilize 31% of potatoes processed in the U.S. and french fries (mostly frozen) utilize 44%. Table 2–1 shows U.S. potato utilization sales in 1,000 hundredweight (cwt) from 1988 to 1995 (USDA, 1997).

Potato Chips

The first potato chip was created in the nineteenth century by a New York chef who was trying to develop a thinner version of fried potatoes. Traditional potato chips are very thin pieces (1.27–1.78 mm thick) of sliced raw potatoes that are fried to a final oil content of 33–38% wet basis (w.b.). Generally, in the industry, 2–4 tons of raw potatoes are fried per hour in continuous fryers, with an oil temperature ranging from 177°C to 190°C at the inlet side and 160°C to 174°C at the outlet side of the fryer. The residence time of the product in the fryer varies from 3 to 5 minutes. During the process, the starch in the potato is gelatinized, and the water content is reduced from 75–85% (w.b.) to 1–2% (w.b.), resulting in a crispy product all way through. The product shrinks during frying, resulting in a thickness of about 65% of that of raw chips (Mottur, 1989).

Potato chips prepared in batch fryers tend to be harder than do chips prepared in the industrial type. The difference between these two processes is that the batch-style chips are not washed before frying, resulting in the slices being covered with a thin layer of starch, which may contribute to the hard texture (Matz, 1993).

The processing of making potato chips varies among countries and even from manufacturer to manufacturer. Figure 2–1 shows a typical layout of the process. It consists of several unit operations, including receiving, storing, cleaning, peeling, spotting, slicing, washing, frying, inspecting, seasoning, and packaging.

The most important quality characteristics of potatoes selected for chipping include: (a) high specific gravity (SG) (ideally, SG = 1.085, equal to 21% dry matter) that leads to superior yield of chips and lower oil content; (b) low reducing sugars (below 0.25% and preferably below 0.10%)—potatoes containing low reducing sugars usually develop the desired golden brown color when fried; and (c) free of spouting, fungal damage, greening or shrinkage, and softening (Ranken and Kill, 1993).

After the selection of the potatoes is made, they are first washed to separate residual soil and stones. The skins are then removed by tumbling the potatoes in rotating cylinders coated with carborundum abrasive, then the potatoes are visually inspected for defects (e.g., green surfaces). Next, the potatoes are sliced in a horizontally rotating impeller that moves them continuously around a stationary ring containing eight vertical blades. After slicing, the pieces are washed to remove the starch accumulated at the surface that can cause them to stick together

Table 2–1 Potato Utilization Sales

Year	Product (1,000 cwt)		
	Chips and shoestrings	French fries	Processed, total
1988	44,539	95,466	192,737
1989	43,071	100,459	200,726
1990	44,489	108,455	221,997
1991	45,953	111,128	226,560
1992	48,455	112,496	228,922
1993	48,987	121,087	242,087
1994	49,299	136,531	261,258
1995	47,908	129,029	255,548

Source: Reprinted from *Agricultural Statistics*, 1997, United States Department of Agriculture.

during frying. The washed slices are then blanched in steam or hot water (71–93°C), dewatered with air blast, then fried in a continuous fryer. As the chips emerge from the frying oil, they may be deoiled by running the chips through a hot air tunnel or by using a radiant heater (Matz, 1993). Following this, the chips are inspected for defects (dark color, discolored portions, etc.), seasoned, and packaged.

Fabricated potato chips are made from dough containing dehydrated potato and other ingredients. These are uniform but readily modifiable chips resembling conventional potato chips in most of their features. The first successful fabricated potato chip was Pringles. The process consists of forming the dough (35% w.b.), then sheeting and cutting into predetermined shapes. The cut dough is then placed between a pair of molds and fried in deep-fat fryers. After this, the fried pieces are released from the mold, seasoned, and packaged (Liepa, 1971).

French Fries

French fries were created in Belgium in the seventeenth century by peasants who, during a severe winter, were looking for a food substitute for their unavailable staple fried fish. The French call fries *pommes frites*, meaning "fried potatoes." In America, of course, they are called *french fries*, probably because they gained real popularity after World War I, when soldiers enjoyed the fries they had in France.

The structure of french fries is characterized by two regions: a crispy crust (surface layer) of about 1–2 mm thick where most of the oil is located, and a soft

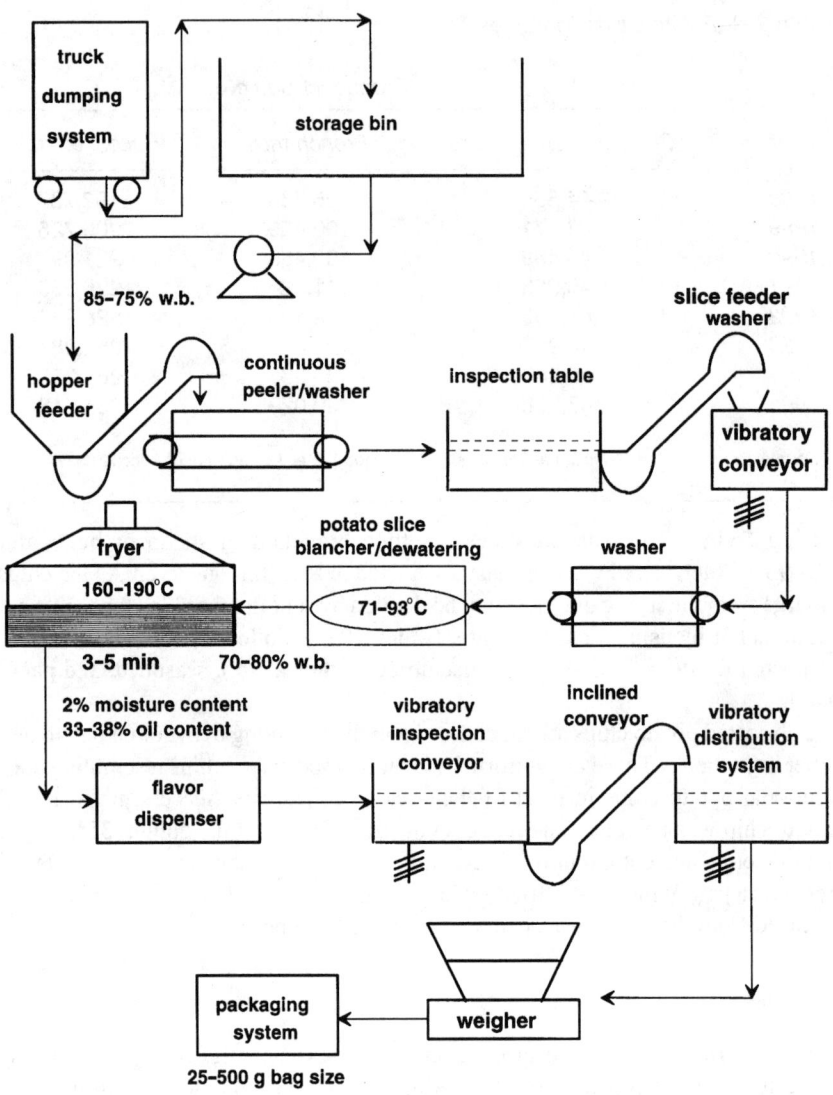

Figure 2–1 Layout of a Potato Chip Processing Line

crumb (interior). Oil content in french fries ranges between 10% and 15% w.b. Traditionally, french fries are prepared by batch frying the potatoes twice—once to cook, once to crisp. The batch process consists of: (1) selecting good, starchy potatoes (russets); (2) peeling and washing; (3) slicing evenly lengthwise (6.4–12.7 mm thick) and washing; (4) dewatering to reduce splashing and uneven cooking during frying; (5) frying (par-frying) at 160°C for 1–2.5 minutes only to cook them; (6) draining and cooling the fries completely at room temperature (at this point, they can be covered and refrigerated for later finishing); and (7) frying for the second time at 190°C to produce crispy potatoes.

In Europe, french fries are also fried in vacuum fryers so that the necessary degree of dehydration can be achieved without excessive darkening or scorching of the product; this product does not require blanching before frying. The oil temperature in a vacuum fryer is much lower than that in an atmospheric fryer (Ranken and Kill, 1993).

Most American french fries are frozen. America's frozen fry industry is a $3-billion business, producing about 4.5 MMT (9×10^9 pound) of french fries in 1995 (USDA, 1997). Use of frozen french fries requires only one frying step. Frozen potatoes are believed to give the best quality of fries; they are crispier and hold their crispness longer after preparation than do those prepared from fresh potatoes (Talburt et al., 1987). About 50% of all American frozen fries are exported annually to Japan. McDonald's alone serves about 3.5 MMT (7×10^9 pound) of french fries to approximately 33 million customers worldwide daily.

Generally, in the industry, frozen french fries are prepared by first washing and peeling the potatoes. Peeled potatoes are then trimmed, inspected, and sorted (small potatoes are usually removed). Potatoes are then cut along the long axis to produce strips with a 0.95–1.27 cm^2 cross-section. Crinkle or corrugated cuts are also produced in great volume today. The strips are blanched before frying. The objective of blanching is to make the product color more uniform, to reduce oil absorption by gelatinizing the surface starch, to reduce frying time, and to improve the fries texture; blanching may be done at 90°C in water for 10 minutes. The potato strips are then dewatered (using screens or warm-air blowers), then conveyed to the fryer. The strips are next par-fried at 121–191°C for 1–3 minutes, then air-cooled on a wire-mesh conveyor belt connecting the fryer to the freezing tunnel. Dark, overfried, or other defective pieces are removed at this point. French fries should be drained immediately after emerging from the fryer by, for example, passing the fries over vibrating conveyors or spraying them with hot water to eliminate the oil from the surface (Lesińska and Leszczyński, 1989). Oil content of these strips varies from 5% to 6% w.b.

French fries can be frozen as loose pieces on a continuous freezer for 12 minutes at −40°C or in packages on air-blast or multiple-plate freezers. It will take about 2.5 hours for packaged french fries to reach a temperature of −18°C in a plate freezer. The size of the packages varies from 0.3- to 0.5-kg cartons or polyethylene bags holding 0.6–2.3 kg for retail to packs ranging from 2.0 to 22.7 kg for the food service industry (Luh and Lorenzo, 1988).

The frozen potatoes are then finished-fried in deep-fat fryers at temperatures ranging from 185°C to 195°C for 2–3 minutes. French fries will absorb more oil if the slices are thawed prior to frying than will those fried frozen (Burr, 1971).

CORN-BASED PRODUCTS

Two major kinds of fried products are made from corn: corn chips and tortilla chips. Both products are made from alkaline-cooked corn. The difference between the two products is that tortilla chips are baked before frying and corn chips are fried without prior baking. This results in products with different oil contents: tortilla chip oil content ranges from 23% to 30% w.b., whereas that of corn chips ranges from 30% to 38% w.b. Because both processes are similar, more emphasis is given to tortilla chips in this chapter.

The Aztec Indians developed the traditional method used to process corn into tortillas in the central region of Mexico. The process of transforming corn into masa is called *nixtamalization*. Nixtamalization typically involves boiling, quenching, and steeping of corn in lime solution (CaO, calcium oxide). The liquor obtained after steeping is discarded, and the cooked corn (nixtamal) is washed to remove the excess alkali and loose pericarp. The nixtamal is then stone ground to produce masa. Modern commercial plants still use many of the same principles from the original process (Serna-Saldivar et al., 1990).

Masa is the main raw material from which tortillas and other products are made. In pre-Columbian times, the Aztecs produced *totopochtli* by roasting tortillas on a hot flat grill. It became a common practice between the households to fry the left-over tortillas in hot oil to improve their flavor and extend storage time. The fried tortillas were called *tostadas* (with a round shape), and *totopos* (tortilla chips) when they were cut into smaller pieces (Serna-Saldivar et al., 1990).

Cooking and steeping the corn in lime solution causes hydration and softening of the endosperm and germ, partial starch gelatinization, flavoring, and disintegration of the pericarp (Gomez et al., 1987). Quenching is a common practice used by processors of corn tortilla chip masa to reduce the temperature of the solution before steeping. This is important to prevent excess water uptake after cooking. Masa with lower moisture content produces chips with reduced oil content (Serna-Saldivar et al., 1990).

An alternative method of making tortilla chips involves production of nixtamalized corn flour (NCF) as an intermediate product. NCF is made by drying and grinding masa into flour. This flour is sieved and blended from streams of different particle sizes to produce NCF with optimum properties for various applications (Serna-Saldivar et al., 1990). Products processed from NCF are more expensive than those made from freshly ground masa; they stale faster and are blander in flavor and poorer in texture (Gomez et al., 1987). However, popularity of NCF continues to increase because it is more convenient and ensures more consistent quality and uniformity of the product than does the traditional method (Serna-Saldivar et al., 1990).

The manufacturing of tortilla chips differs from company to company. An example of an industrial operation for producing tortilla chips is presented in Figure 2–2. Several unit operations are involved: corn cooking, steeping, washing, grinding, masa preparation, sheeting, cutting, baking, tempering, frying, cooling, salting, seasoning, inspecting, and packaging. The corn selected for tortilla chips must be clean and free of broken kernels or foreign materials. The following is a detailed description of an industrial operation (Grant and Jones, 1992).

The first step in the process consists of cooking the corn. The main objectives are: (1) to hydrate the corn kernel and loosen the pericarp; (2) to gelatinize the starch; (3) to increase corn moisture content; and (4) to contribute to the corn flavor (by the lime absorption). Most corn is cooked in steam-jacketed stainless steel kettles. Mixing, cooking, cooling, and agitation all take place in the kettle. Cooking of the corn is made in a solution of lime (normally 1% of the total corn batch weight) and water. The corn is cooked by heat supplied by the steam jacket (steam pressure at 241–276 kPa). Cooking time will vary according to the water temperature and corn variety. The cooking temperature is about 88–93°C and takes 5–10 minutes. Cooling takes 3–9 minutes with water at 66°C. Agitation is made at 60°C for 5 minutes.

The second step is the soaking process. The main objectives are: (1) to increase the corn moisture content and loosen and soften the pericarp; (2) to cool the corn; and (3) to allow for the gelatinized starch to set. Each batch of cooked corn soaks for 10–16 hours. The temperature of the cooked corn is reduced from 88–93°C to below 60°C. During the soaking process, each vat is stirred at least every 2 hours to ensure maximum pericarp removal and to avoid the formation of "hot spots" or pockets of corn that bind tightly together, thus creating heat buildup.

Corn washing is the third step in the process of making tortilla chips. The main goals of this step are: (1) to remove loosened pericarp and excess of lime solution; (2) to separate and remove broken kernels and foreign materials; and (3) to cool the corn. The corn is washed in a tumbler that rotates at 37–40 rpm. The temperature of the washed corn should be below 35°C. The corn then travels through the draining conveyor for about 3–10 minutes, allowing for the surface water to drain off.

Next, the corn milling process starts. During this process, a masa with the correct moisture content (between 51% and 54% w.b.) and uniform particle size is produced. If the masa is too wet (sticky), the finished chip moisture content will be lower, and the chip thickness will decrease. A masa that is too dry (flaky) will

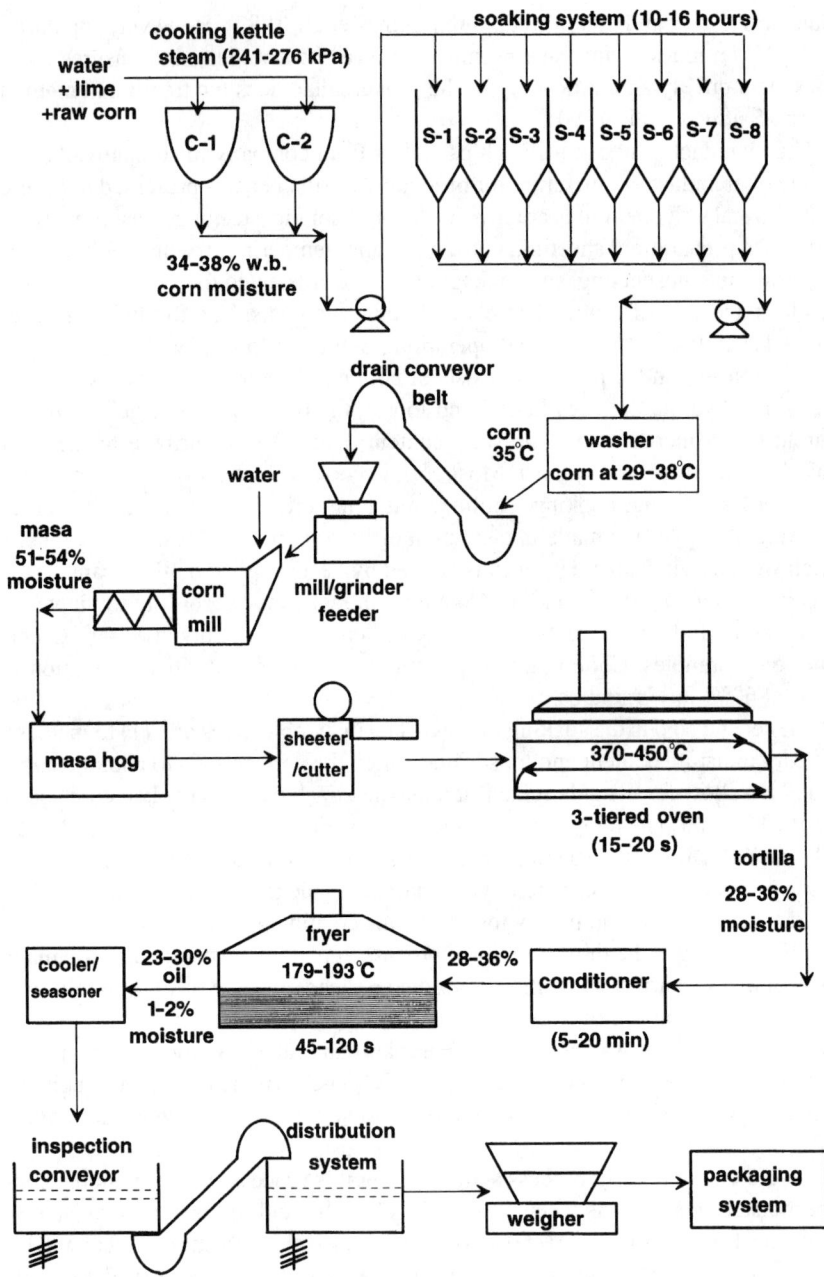

Figure 2–2 Layout of a Tortilla Chip Processing Line

cause the raw chips to fall off the rollers, resulting in a thicker finished chip with higher moisture content. On the other hand, if the masa texture is too fine, the chips will puff and become hard and oily. If the texture is too coarse, the chips will develop holes with no blister at all.

The fifth step consists of sheeting and cutting the masa. The objectives are: (1) to roll the masa into thin sheet of uniform thickness, (2) to cut the sheeted masa into raw chip shapes, and (3) to convey the chips to the oven. The raw chip thickness varies between 0.89 and 1.27 mm.

During the baking process, the raw chip surfaces are toasted, moisture content is reduced from 38% to 40% w.b., the structure strength is increased, and flavor is developed. The baking process is done in a 3-tiered oven for 15–20 seconds at a temperature ranging from 370°C to 450°C to produce a chip that is moist inside but has a dry surface. Generally, higher temperature is used in the top tier than in the middle or bottom tiers.

Once the baked chip leaves the oven, it is conditioned to first cool down the chip, thus equilibrating the moisture throughout the product. This process can be done either in an open belt system or by the use of a closed belt system. Most conditioning systems have a residence time between 5 and 20 minutes.

The baked tortilla chip is then fried in a continuous fryer for about 45–120 seconds at a temperature ranging from 179°C to 193°C. During this stage, the moisture content is reduced to 1–2% w.b., and the chip texture becomes crispy all over. The thickness increases by about 40% of that of a raw chip (Moreira et al., 1995). Oil content after frying will increase to 23–30% w.b. Finally, the tortilla chips are inspected for defects (dark color, discolored portions, etc.), seasoned, and packaged.

FRIED BREAD AND FRITTERS

The most famous deep-fat fried breads are doughnuts. Doughnuts may be eaten warm or cold, and simply sprinkled with sugar, glazed with sugar syrup or chocolate, or rolled in shredded coconut. Some are shaped as buns instead of rings, holding a surprise filling of jam or pastry cream. The richest of all are doughnuts of brioche dough. There are many others, such as the Spanish *buñuelos*, flavored with sugar, anise, and cinnamon; and the Dutch *oliebollen*, flavored with citrus peel, currants, and apples.

At least one type of Chinese *dim sum* dumplings, the *bao*, filled with barbecued pork or sweet bean paste, is deep-fat fried rather than steamed or baked. *Dal puri* are savory fried breads enjoyed in India. They are usually stuffed with cooked split peas or lentils and flavored with spices such as cumin.

When the dough is fried until crisp, the bread is often called a *fritter*. Typical are American crullers, made of yeast or baking powder dough, twisted into an elongated wand, and fried until crisp outside with an airy center. American corn pones are robust little balls of corn meal dough flavored with bacon fat and deep-fat fried; hushpuppies are similar. One popular fritter, the French *beignet*, consists simply of deep-fat fried cream "choux" pastry that puffs to a luscious hollow ball and is served warm, sprinkled with confectioners' sugar. In Spain, *churros* are famous fluted sugar fritters.

Yeast breads are not difficult to deep-fat fry, but both the dough and the oil must be at the correct temperature. The oil should be at 180°C; if it is too hot, the dough tends to scorch, and if it is too cool, the dough soaks up too much fat. The proper temperature for the dough itself when frying is around 24°C, or cool room temperature for baking powder or unleavened dough. Thick shapes should brown on the outside but remain soft in the center, whereas fatter shapes should be crisp right through. Crullers and fancy shapes should be stirred lightly in the hot oil so that they brown evenly, but doughnuts are turned only once.

Doughnuts

The doughnut, also called *fried cake*, has been known, made, and eaten in most countries of Europe from antiquity. It was brought to the U.S. by the first settlers and today accounts for approximately 20% of the total pastry market. About $430 million of doughnuts is sold annually through supermarkets in the U.S. (*Baking and Snack*, 1996). There are two types of doughnuts—cake doughnuts and yeast-raised doughnuts. Both types are deep-fat fried and are made with similar ingredients. Their texture and oil content are different, however; yeast-raised doughnuts have a breadlike crumb and texture, with an oil content around 25% w.b., whereas cake doughnuts have a finer cell structure and softer crumb, like a cake, with an oil content around 20% w.b. (Desrosier, 1977). Cake doughnuts dominate yeast-raised styles in commercial wholesale production by a ratio of 2:1 (*Baking and Snack*, 1996).

Cake doughnuts are made from sweetened dough that is leavened with baking powder. A great variety of cake doughnuts can be made by using different recipes, shapes, flavors, filling materials, and surface finishes. Cake doughnuts are fried for about 90 seconds—45 seconds on each side—at a temperature ranging from 190°C to 198°C.

Yeast-raised doughnuts are generally made from sweet dough fermented with yeast. After fermenting (about 20 minutes), the dough is sheeted to the desired thickness and cut. Then they are allowed to rise (about 20 minutes) in a proof box before frying. Yeast-raised doughnuts should be fried from 182°C to 190°C for about 75 seconds on each side.

There are two ways of forming doughnuts, the sheet-and-cut method and the extrusion method. Soft dough and batter are handled on extrusion equipment. The sheet-and-cut method is generally used for raised doughnuts.

Figure 2–3 shows a schematic of an automatic production line for yeast-raised doughnuts. The process consists of several basic operations: scaling the raw materials; mixing; forming and cutting; proofing; frying; finishing; and packaging. The reader is referred to Matz (1988) for a detailed description of equipment for doughnut processing.

After the raw materials are mixed together to form the dough, it passes through sets of sheeting rollers, gauge rollers, and die cylinders (cutters). The raw doughnuts then are allowed to rise in a proofing system for a period of 25–30 minutes. The raised doughnuts are then fried in a continuous fryer made with conveyors that automatically turn over the doughnut pieces at the right time. The proper control of the turning time is necessary for good-quality doughnuts; they will have a low volume if they are turned over too soon, but will be oily, misshapen, and too dry if they are turned over too late. The doughnuts are then decorated, cooled, and packaged.

Generally, cake doughnuts absorb 7 g of fat/doughnut. Lima et al. (1995) found that frying oil temperature affects the depth of oil penetration in doughnuts. Round doughnuts (15 g) fried at 190°C absorbed oil to a depth of 1.74 mm in comparison with a depth of 4.36 mm for samples fried at 170°C. The most typical sizes of cake doughnuts range from 24 g to 28 g; yeast-raised doughnut sizes average 38 g.

Beignet

An example of a fritter is the French *beignet* that consists simply of deep-fat fried cream puff pastry. Choux, or cream puff pastry, is unlike any other because it is cooked twice. First, butter is melted in water and brought to boil, then flour is mixed. The warmth of the butter water mixture cooks the flour to a ball of dough, which is usually dried over the heat for half a minute. Next, eggs are beaten, one by one, into the dough, which should be warm enough to cook them slightly.

The dough is then deep-fat fried at 180°C until it gets crispy and golden brown. The dough will puff, forming a hollow ball during frying. After removing from the fryer, it is dried in absorbent paper towels and served warm, sprinkled with confectioners' sugar.

BREADED OR BATTERED PRODUCTS

There is a great variety of frozen breaded or battered products available today. These products, fresh or fabricated, are generally par-fried before freezing, ranging from vegetable sticks, cheese sticks, french toast sticks, seafood, hushpuppies, and onion rings to the traditional fried-chicken-on-the-bone products. They are sold mainly to food services and retailers. Among these products, fried chicken products exceeded $8.2 billion in sales in the U.S. in 1996.

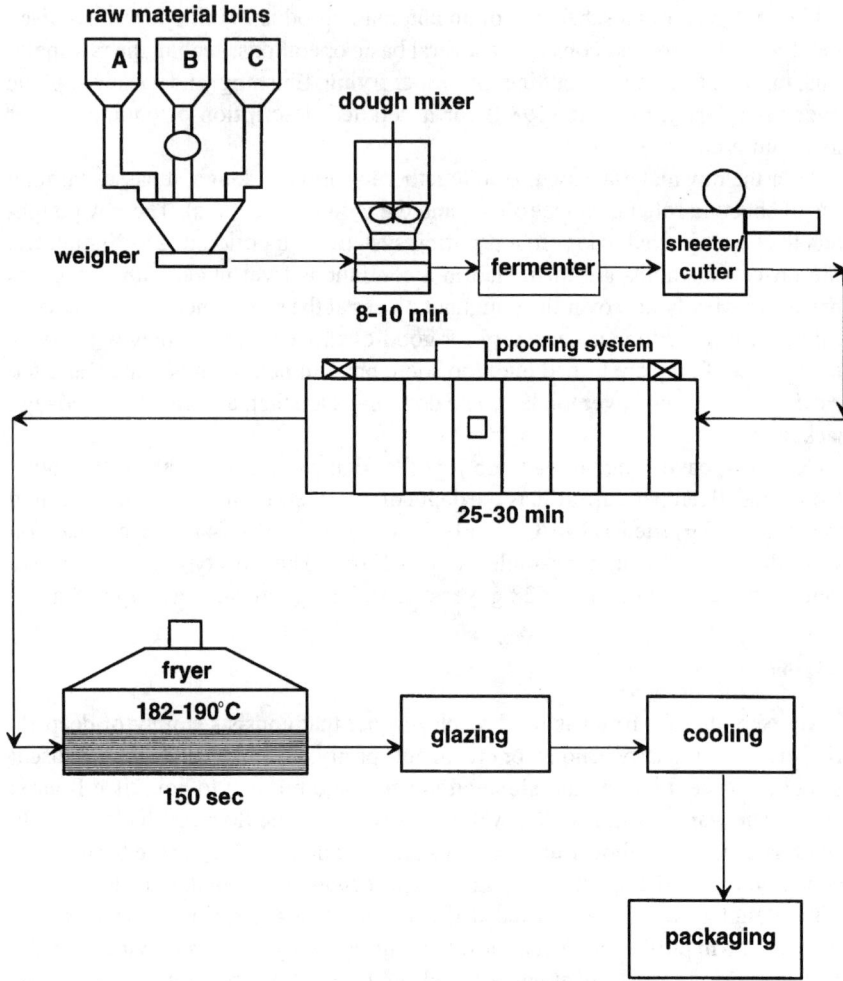

Figure 2–3 Layout of a Yeast-Raised Doughnut Processing Line

Chicken-on-the-bone products used for frying can be fresh or frozen. Generally, fresh chicken is cut up for frying into portions suitable for serving, then seasoned. The chicken pieces are then coated (battered and breaded). The coating reduces dehydration, aids browning, and gives a crisp texture to the fried parts.

The frozen chicken should be thawed before breading and frying. Otherwise, the bones will turn black, due to the blood trapped in the marrow of the bones

(Lawson, 1985). Frozen, prebreaded chicken is available in 10 pieces per chicken cut. This product is marinated before prebreading and freezing, allowing for the product to be fried frozen without the bones turning black. Also available are pre-cooked chicken products that are prebreaded and frozen. These products can be deep-fat fried or cooked in an oven, obtaining attributes similar to a fried product (Peters, 1980).

The breading materials commonly consist of bread crumbs, corn meal, cracker meal, and various prepared breading mixes. The batter, generally having an egg-milk base, is used to moisten the product so that the breading will adhere properly. The use of high-protein batter coating containing egg albumen improves the texture of breaded chicken. Also, batters containing pregelatinized corn flour produce coatings with high moisture content and little oil absorption, resulting in a crispier product (Baker and Scott-Kline, 1988).

A batter is a thick but pourable mixture, usually bound with eggs and containing flour and other starches to give it body. Batters are used to coat food and protect it from the searing heat of deep fat. Water makes a batter light; milk gives it smoothness and makes it brown more quickly, and beer is popular in savory batter because it adds flavor as well as air. Some batters include raising agents; other, more robust batters for fish and vegetables may be raised with yeast. A small amount of oil or butter enriches batters and helps prevent them from sticking to the pan. Salt and pepper are the most common seasonings, though flavorings such as chili pepper may be included, or liqueurs for sweet batters. Sugar should be used only in small quantities because it burns easily in the hot oil, as do delicate ingredients such as herbs.

Yeast or beer batters are appropriate for fish fillets, whole mushrooms, and other raw ingredients that are fried for a relatively long time. For fragile foods such as shrimp or vegetable sticks, or for precooked ingredients, the batter should be thinner. Japanese tempura batter is the lightest of all.

Loewe (1993) classifies batter systems into two categories: interface/adhesion and puff/tempura. The interface/adhesion batters are used with breading, serving primarily as an adhesive layer between the product's surface and the breading; chemical leavening normally is not used. Puff/tempura batters use leavening agents and are used as an outside coating for the food. The batter uniformity and thickness that is related to the batter viscosity determine acceptability of the finished product. A more viscous batter will pick up more breading than will a less viscous product.

The frying processes used in the food services for frying chicken-on-the-bone are classified into pressure frying and atmospheric frying (open kettle). The time and temperature of a fryer depend on the manufacturer's recommendations. When using a pressure fryer, the food service recommends larger models, with frying time ranging from 8 to 10 minutes. Pressure-fried chicken is generally more uniform in color and appearance than is chicken fried in open fryers. Pressure fryers cook food faster than do open fryers, and the food is moister. However, oil degradation is faster in pressure fryers and requires more frequent oil filtration than do open fryers.

The temperature recommended to fry chicken in open fryers varies from 149°C to 163°C; depending on the size of the chicken, raw chicken pieces can take 12–18 minutes to cook to doneness. Lane et al. (1980) found that a minimum of 14.5 minutes of deep-fat frying at 163°C was necessary for breaded chicken thighs to reach an internal temperature of 93°C, at which a trained panel judged the product as done.

CHARACTERISTICS OF RAW AND FRIED PRODUCTS

Dimensions and Compositions

Data on dimensions and weights of raw and processed potatoes are presented in Tables 2–2 and 2–3. Raw potato size varies from 60 mm to 140 mm in length and 50–350 g weight, depending on the variety. The size of potatoes for chip production should average 40–60 mm in width (smaller diameter) and never less than 60 mm in length (largest diameter) (Lesińska and Leszczyński, 1989). The length of raw french fries ranges from 60 mm to 140 mm, and the cross-section from 9.5 × 9.5 mm to 12.7 × 12.7 mm; potatoes for shoestrings have a cross-section smaller than 6.4 × 6.4 mm. After frying, potato chip thickness can increase to about 65% of that of a raw slice (Mottur, 1989); the product also becomes lighter, due mainly to an increase in porosity after frying.

The chemical compositions of raw potato, potato chips, and french fries are given in Table 2–4. The oil content in french fries is about 33% lower than in potato chips, but water content is 25 times higher; ash content is also higher in potato chips.

Table 2–5 presents the composition of raw and processed corn products. In general, tortillas have higher moisture content than does raw corn; on a dry-matter basis, tortillas have less fiber and oil content than does the original grain but comparable amounts of protein, ash, and carbohydrates. Braham and Bressani (1966) found that the tortilla has about 10–20 times as much calcium as raw corn. The oil content in tortilla chips is between 23% and 30% w.b. and 30–38% w.b. in corn chips. Due to oil uptake, tortilla and corn chips have less protein, fiber, and starch by weight than do tortillas.

Table 2–6 presents an example of cake and yeast-raised doughnuts composition. They can be machine or hand made. Fat content and water is higher in yeast-

Table 2–2 Dimensions and Weights of Raw Potato, Raw Potato Chips, and French Fries

Product	Length (mm)	Width (mm)	Weight (g)
Potato	60.0–140.0	60.0–70.0	50.0–350.0
Potato chip	60.0–90.0	1.3–1.8	3.9–12.2
French fry			
Regular	60.0–140.0	9.5–12.7	6.0–25.0
Shoestring	60.0–140.0	≤6.4	2.6–6.2

Source: Data from G. Lisińska and W. Leszczyński, *Potato Science and Technology,* © 1989, Elsevier Applied Science.

raised batter. Table 2–7 shows an example of the changes in composition that occur when cake doughnuts are deep-fat fried; fat content increases from 5% to 25% w.b.

The typical compositions of meat and fish muscle are presented in Table 2–8. Most of the meat product compositions are similar, with the exception of pork meat that has higher fat content. No exact recipes exist for batter systems; they depend on the food product and the desired coating appearance. The optimum formulation of batter systems depends on stresses found during the process, including freezing and final reconstitution. Table 2–9 gives an example of a typical formulation of batter systems.

Physical Properties

It is important to know the physical properties of fried materials because they change significantly during frying. This affects the rate at which heat and mass are transferred. Some physical properties of raw and processed potatoes are presented in Table 2–10. These values were calculated based on composition (see Chapter 4).

Table 2–3 Dimensions and Weights of Finished Potato Chips and French Fries

Product	Length (mm)	Width (mm)	Weight (g)
Potato chip	60.0–90.0	2.0–2.9	3.5–11.0
French fry			
Regular	60.0–140.0	9.5–12.7	5.7–23.0
Shoestring	60.0–140.0	≤6.4	2.6–6.0

Table 2–4 Typical Composition of Raw Potatoes, Potato Chips, and French Fries

Composition	Raw Potato	Potato Chips	French Fries
Water (%)	79.8	1.8	44.7
Protein (%)	2.1	5.3	4.3
Fat (%)	0.1	39.8	13.2
Carbohydrate (%)	17.1	50.0	36.0
Ash (%)	0.9	3.1	1.8

Source: Data from W.F. Talburt et al., Frozen French Fries and Other Frozen Products, in *Potato Processing*, W.F. Talburt and Smith, eds., © 1987, Van Nostrand Reinhold.

Table 2–5 Typical Composition of Raw Corn, Nixtamal, Tortilla, Tortilla Chips, and Corn Chips

Product	Water (%)	Protein (%)	Fat (%)	Carbohydrate (%)	Ash (%)	Fiber (%)
Raw corn	12.2	8.4	4.5	73.8	1.1	1.3
Nixtamal	49.1	5.3	2.0	42.9	0.8	0.7
Tortilla	47.1	5.5	1.5	44.5	0.8	0.6
Tortilla chips	2.0	7.1	25.2	62.3	0.6	0.9
Corn chips	0.9	6.3	36.6	53.0	1.9	1.0

Source: Data from S.O. Serna-Saldivar, M.H. Gomez, and L.W. Rooney, The Chemistry, Technology, and Nutritional Value of Alkaline-Cooked Corn Products, *Advances in Cereal Science and Technology*, Vol. X, © 1990.

Table 2–6 Typical Ingredients of Cake and Yeast-Raised Doughnuts

Ingredients	Cake Doughnut*	Yeast-Raised Doughnut
Sugar (%)	15.5	5.4
Vegetable shortening (%)	2.6	9.2
Salt (%)	0.6	0.8
Nonfat dry milk (%)	3.2	3.1
Soda/other (%)	0.2	0.2
Pastry flour (%)	21.0	21.5
Bread flour (%)	21.0	24.5
Baking powder (%)	1.9	1.0
Egg (%)	12.0	6.0
Water (%)	22.0	24.5
Yeast (%)	0.0	3.8

*Set up for machine-made doughnuts.

Source: H.W. Lawson, Standards for Fats and Oils, *The L.J. Minor Foodservice Standards Series*, Vol. 5, © 1985, Van Nostrand Reinhold.

Table 2–7 Composition of Cake Doughnuts (Batter and Finished Product)

Composition	Doughnut Batter	Finished Doughnut
Moisture (%)	37.5	21.0
Solids (%)	57.5	53.5
Fat (%)	5.0	25.5

Source: H.W. Lawson, Standards for Fats and Oils, *The L.J. Minor Foodservice Standards Series,* Vol. 5, © 1985, Van Nostrand Reinhold.

Table 2–8 Typical Composition of Meat and Fish Muscle[a]

Product	Water (%)	Protein (%)	Fat (%)	Ash (%)
Beef[b]				
Raw	75.6	21.5	2.0	0.7
Fried	39.9	20.0	4.5	0.5
Chicken, raw	73.0–76.0	20.0–23.0	4.7	1.0
Lamb, raw	73.0	20.0	5.0–6.0	1.4
Pork, raw	68.0–70.0	19.0–20.0	9.0–11.0	1.4
Fish (red snapper)[c]				
Raw	77.0	19.3	1.5	1.2
Fried	68.5	20.3	5.5	1.5

Source: Chemical compositions from (a) Fennema (1985); (b) Moreiras-Valera et al. (1988); and (c) Gall et al. (1983).

The bulk density of fried products determines the bag volume required to package a certain mass of product. This property also affects the product speed in a continuous fryer at a certain capacity. The density of potato chips is the property that changes the most during frying; it is lower than those of french fries and raw potatoes (Table 2–10). A potato chip will puff during frying, thus increasing its porosity and resulting in a lighter product. The specific heat expresses the energy required by a unit mass of product to increase in temperature by 1 degree. Raw potatoes require more energy (about 3.6 kJ/kg K) than do frozen, par-fried french fries (around 0.44 kJ/kg K) to reach the desired temperature at which the optimal evaporation rate of water occurs from the product during frying. The thermal conductivity is a measure of the resistance to the conduction of heat within the product. In products with high thermal conductivity, the thermal gradients in the product disappear faster during frying than in an equally sized product with lower thermal conductivity. Frozen french fries' (par-fried) thermal conductivity is about 2 times higher than that of raw potatoes. The thermal conductivity of fin-

Table 2–9 Typical Ingredients of Batter Systems

Ingredients	Batter (%)
Critical	
Wheat flour	30–50
Corn flour	30–50
Sodium bicarbonate	≤3
Acid phosphate	adjusted
Optional	
Flours from rice, soybean, barley	0–5
Shortening, oil	0–10
Dairy powders	0–3
Starches	0–5
Gums, emulsifiers, colors	<1
Salt	≤5
Sugars, dextrins	0–3
Flavoring, seasonings, breadings	desired

Source: Adapted with permission from R. Loewe, Role of Ingredients in Batter Systems, *Cereal Foods World*, Vol. 38, No. 9, pp. 673–677, © 1993, American Association of Cereal Chemists.

ished potato chips (0.11 W/m°C) is very low because of the high porosity of the product after frying.

Moreira and Barrufet (1995) classified the tortilla chip process into three stages: masa, tortilla, and tortilla chips. The masa is described as a dough that contains about 54% (w.b.) water, 16% free starch granules, and 2% gelled starch. Tortillas are baked thin pieces of masa containing 35–40% (w.b.) water, 11% free starch, and 13% gelled starch. Porosity at this stage is on the order of 0.3. Tortilla chips are fried tortillas having a moisture content of about 1–2% (w.b.), free starch granules of 5%; porosity of 0.6, gelled starch of 25%, and oil content of about 23–30%. The bulk density decreases from 880 kg/m^3 before frying to 595 kg/m^3 as a result of an increase in porosity. The specific heat decreases from 3.36 kJ/kg°C to 2.31 kJ/kg°C, due to water loss, and the thermal conductivity is reduced from 2.31 W/m°C to 0.11 W/m°C as porosity increases (see Table 2–10).

The physical properties of doughnuts were calculated based on the composition presented in Tables 2–6 and 2–7. Density of doughnuts decreases the most during frying (porosity increases) from around 1231 kg/m^3 to about 533–822 kg/m^3, depending on the porosity (Table 2–10).

The properties of raw meat products are also shown in Table 2–10. The thermal conductivity value in meat products depends on whether the property was mea-

Table 2-10 Typical Physical Properties of Raw and Fried Products

Product	Density (kg/m³)	Specific Heat (kJ/kg°C)	Thermal Conductivity (W/m°C)
Potato[a]			
Raw	1074	3.65	0.55
Chips[b]	611	1.54	0.11
French fries[c]	1058	2.67	0.37
Corn			
Raw	745	2.01	0.16
Tortilla[d]	880	3.36	2.31
Tortilla chips[e]	595	2.31	0.11
Doughnuts (cake)[f]			
Batter	1230	2.50	0.37
Fried[g]	533–820	2.09	0.13–0.20
Meat/fish[h]			
Beef	1040–1052	3.40–3.60	0.45–0.49
Chicken	1040–1052	3.40–3.60	0.41
Lamb	1040–1052	3.40–3.60	0.42–0.48
Pork	1040–1052	3.40–3.60	0.54
Fish	1052	3.60	0.56

a. Properties were calculated based on composition (Table 2–3 and 2–4)
b. Assuming porosity = 0.5
c. Assuming porosity = 0.09
d. Porosity = 0.3
e. Porosity = 0.6
f. Properties based on composition (Tables 2–6 and 2–7)
g. For porosity in the range of 0.3 to 0.55
h. Cp and density based on composition

Source: Adapted from Moreira et al. (1995), Brooker et al. (1992), and Singh and Heldman (1993).

sured perpendicularly or parallel to the fibers, with thermal conductivity values being higher in the direction of the fibers than across the fibers. Based on product composition, density of meat products may vary from 1040 kg/m³ to 1052 kg/m³, specific heat from about 3.40 kJ/kg K to 3.60 kJ/kg K, and thermal conductivity from 0.412 W/m°C to 0.557 W/m°C.

REFERENCES

Anon. 1996. Doughnut baking, what's new? *Baking and Snack*, May:34–38.

Baker, R.C., and Scott-Kline, D. 1988. Development of high protein coating using egg albumen. *Poultry Sci*, 67:557–564.

Blumenthal, M.M. 1991. A new look at the chemistry and physics of deep-fat frying. *Food Technol*, 45(2):68–71, 94.

Braham, J.F., and Bressani, R. 1966. Utilizacion del calcio del maiz tratado con cal. *Nutr Bromatol Toxicol*, 6:14–19.

Brooker, D.B.; Bakker-Arkema, F.W.; and Hall, C.W. 1992. *Drying and storage of grains and oilseeds*. New York: Van Nostrand Reinhold.

Burr, H. 1971. Frozen french-fried potatoes: Effects of thawing and holding before finish-frying and their no-relation to starch retrogradation. *J Food Sci*, 36:392–394.

Desrosier, N.W. 1977. *Elements of food technology*. New York: Van Nostrand Reinhold/AVI.

Fennema, O.R. 1985. *Food chemistry*. New York: Marcel Dekker.

Gall, K.L.; Otwell, W.S.; Koburger, J.A.; and Appleddorf, H. 1983. Effects of four cooking methods on the proximate mineral and fatty acid composition of fish fillets. *J Food Sci*, 48:1068–1071.

Gomez, M.H.; Rooney, L.W.; Waniska, R.D.; and Pflugfelder, R.L. 1987. Dry corn masa flours for tortilla and snack foods. *Cereal Foods World*, 32(5):372–377.

Grant, R., and Jones, D. 1992. *Making corn and tortilla chips*. Paper presented at SFA Corn and Tortilla Chip Seminar. Louisville, KY.

Lane, R.H.; Muir, W.M.; and Mullins, S.G. 1980. Correlation of sensory doneness with internal temperature of deep fat fried chicken thighs. *Poultry Sci*, 59:719–723.

Lawson, H.W. 1985. Standards for fats and oils. *The L.J. Minor foodservice standards series*, Volume 5. New York: Van Nostrand Reinhold/AVI.

Liepa, A.L. 1971. *Preparation of chip-type products*. U.S. Patent Office, no. 3,576,647.

Lima, M.I.; Vijayan, J.; and Singh, R.P. 1995. *Oil uptake by doughnuts during frying*. IFT Annual Meeting. Anaheim, CA.

Lisińska, G. and Leszczyński, W. 1989. *Potato science and technology*. New York: Elsevier Applied Science.

Loewe, R. 1993. Role of ingredients in batter systems. *Cereal Foods World*, 38(9):673–677.

Luh, B.S., and Woodroof, J.G. 1988. *Commercial vegetable processing*. New York: Van Nostrand Reinhold/AVI.

Luh, B.S., and Lorenzo M.C. 1988. Freezing of vegetables. In *Commercial vegetable processing*. New York: Van Nostrand Reinhold/AVI.

Matz, S.A. 1993. *Snack food technology*. New York: Van Nostrand Reinhold/AVI.

Matz, S.A. 1988. *Equipment for bakers*. McAllen, TX: Pan-Tech International.

Moreira, R.G.; Palau, J.; Sweat, V.; and Sun, X. 1995. Thermal and physical properties of tortilla chips as a function of frying time. *J Food Process Preserv*, 19:175–189.

Moreira, R.G., and Barrufet, M.A. 1995. Spatial distribution of oil after deep-fat frying from a stochastic model. *J Food Engineering*, 27(2):205–220.

Moreiras-Valera, O.; Ruiz-Roso, B.; and Varela, G. 1988. Effects of frying on the nutritive value of food. In *Frying of food—principles, changes, new approaches*, ed. Varela, Bender, and Morton. New York: VCG Publisher.

Mottur, G.P. 1989. A scientific look at potato chips—the original savory snack. *Cereal Foods World*, 34(8):620–626.

Peters, J.W. 1980. Flexible fried chicken forms a base for banquet foods' growing foodservice role. *Food Product Dev*, 14(2):36–40.

Ranken, M.D., and Kill, R.C. 1993. *Food industries manual*. New York: Blackie Academic & Professional.

Serna-Saldivar, S.O.; Gomez, M.H.; and Rooney, L.W. 1990. The chemistry, technology and nutritional value of alkaline-cooked corn products. In *Advances in cereal science and technology*. Vol. X. Minneapolis, MN: AACC.

Singh, R.P. and Heldman, D.R. 1993. *Introduction to food engineering*. New York: Academic Press.

Talburt, W.F.; Weaver, M.L.; Reeve, R.M.; and Kueneman, R.W. 1987. Frozen french fries and other frozen products. In *Potato processing*, ed. Talburt and Smith. New York: Van Nostrand Reinhold/ AVI.

U.S. Department of Agriculture. 1997. *Agricultural statistics*. National Agricultural Statistics Service. Washington, DC: U.S. Government Printing Office.

CHAPTER 3

Frying Oil Characteristics

During deep-frying, fats and oils are repeatedly used at elevated temperatures in the presence of atmospheric oxygen and receive maximum oxidative and thermal abuse. Heating in the presence of air causes partial conversion of fats and oils to volatile chain-scission products, nonvolatile oxidized derivatives, and dimeric, polymeric, or cyclic substances. By affecting the properties of the heat transfer medium (cooking oil), deep-fat frying in turn affects the quality of the fried food. Chemical and physical changes in the oil can prolong the process time, increase the total amount of oil in the product, induce toxicity, and lower food nutritional value. This chapter deals with the effect of heat on the frying oil and the ways to prevent detrimental quality reactions. For more detailed information on oils and fats the reader is referred to Swern (1964) and Giese (1996).

WHAT ARE FATS AND OILS?

The term *fat* is used to refer to animal-origin fat, whereas *oil* is used for plant-origin fat. Fat is one of the body's basic nutrients, providing energy by furnishing calories. It is composed primarily of triglycerides, which are three fatty acid molecules joined by a *glycerol* backbone (Stockwell, 1988):

CH_2OH

|

$CHOH$

|

CH_2OH

There are several fatty acids, but only about 11 are common. An example is *oleic acid*, which has the following chemical formula:

$$HOOC(CH_2)_7CH = CH(CH_2)_7CH_3$$

Glycerol can combine with these fatty acids through esterification to form *mono-*, *di-*, and *triglycerides*:

CH_2OCOR

|

$CHOCOR$

|

CH_2COR

To a large extent, the characteristics of a particular fat or oil and, therefore, its frying properties are dictated by the actual fatty acids that are present in the individual triglyceride molecules. Some of these component fatty acids are longer or shorter in chain length than others:

$CH_2OCO(CH_2)_{14}CH_3$

|

$CHOCO(CH_2)_{10}CH_3$

|

$CH_2CO(CH_2)_{16}CH_3$

The various fatty acids are placed in groups to reflect their functionality in products (and in the human body). If a fatty acid is composed of a chain of carbon atoms with no double bonds, it is said to be *saturated*. These fatty acids are generally solid at room temperature. An example is *palmitic acid, $HOOC(CH_2)_{15}CH_3$.* Fatty acids with one double bond are called *monounsaturated* (such as oleic acid); those with more than one double bond between carbons are called *polyunsaturated,* such as *linoleic acid, $HOOC(CH_2)_7CH = CH(CH_2)CH = CH(CH_2)_4CH_3$.* It is possible for the double bonds to occur in different places. In the case of fatty acid with two double bonds, they can be sited close to each other or far apart. The closer they are together, the more unstable the molecule will be. The position of the double bonds seems to be more important than the number of them. A molecule with two double bonds close to each other is likely to be more unstable than one with three positioned well apart from each other (Stockwell, 1988).

From these descriptions, it should be clear that even with only a few types of fatty acids, a larger number of different triglycerides are possible. With 11 distinct fatty acids that occur commonly in nature, more than 1,300 different triglycerides can be formed. A particular fat is a mixture of many of these different triglycerides. The specific properties of, for example, lard, are dictated by the types of triglycerides and, therefore, the types of fatty acids present. A comparison of percentages of monosaturated, polysaturated, and saturated fatty acid levels for common edible oils is shown in Color Plate 1.

Unsaturated fats are derived primarily from plants and are liquid (in the form of an oil) at room temperature. Generally speaking, oils are composed (in varying percentages) of both monounsaturated and polyunsaturated fats. In general, the more double bonds in a fatty acid and the longer its carbon chain length, the lower the temperature at which an oil remains liquid. Also, the more double bonds, the more unstable the oil is and the more likely it is to undergo various forms of degradation (Stockwell, 1988). Thus, there is a chemical basis for the trade-off between stability (polyunsaturated oils are less stable) and health benefits (polyunsaturated oils are considered to be more healthy).

TYPES OF FRYING OILS

There are many types of oils and fats available for frying (Table 3–1):

1. *Refined, bleached, deodorized* oil performs satisfactorily if the fried food volume is extremely high and an oily product appearance is desired.
2. *Salad oil* provides an oily fried product appearance.
3. Until 1986, *animal-origin fats* were the primary fats used by the food services. Examples are butter, margarine, and lard. Lard is richer than many other fats and, therefore, makes extremely tender, flaky biscuits and pastries. It is a flavorful fat for frying and is widely used throughout South America and in many European countries (Brooks, 1991). These oils have low initial cost and good frying stability. However, there are nutritional concerns and pressure from special interest groups. Later, the food industry tended to use animal/vegetable blends, as well as partially hydrogenated vegetable oils.
4. *Liquid frying shortenings* provide longer frying stability. Products have a semi-oily appearance but are still relatively high in polyunsaturates, which may be important for nutritional considerations.
5. *All-purpose shortenings* were the first plastic vegetable products used for frying. They are preferred for doughnuts and other specialty products with high fat absorption and which require a dry fried appearance.
6. *All-vegetable oils* are among the first shortenings specifically formulated for a definite function. They were developed for maximum frying stability through the reduction of polyunsaturates via hydrogenation. However, they present the inconvenience of solid consistency (O'Brien, 1993).

The three most widely used oils that are high in monounsaturates are olive oil, canola oil, and peanut oil. Polyunsaturated fats are also considered relatively healthy and include the following (ranked in order—most to least—of polyunsaturates): safflower oil, soybean oil, corn oil, and sesame oil (Stockwell, 1988). Soybean, safflower, sunflower, and canola oils are always partially hydro-

Table 3–1 Types of Frying Oils

Frying Oil	RBD Oils	Salad Oils	Liquid Shortenings	All-Purpose Shortenings	Animal Blends	All-Vegetable or Frying Shortening
Consistency	clear liquid	clear liquid	opaque pourable	plastic solid	solid	solid
Typical melting point	liquid	liquid	33–37°C	46–49°C	46–52°C	41–43°C
AOM stability (h)	10–25	16–20	35	40	75	200
Polyunsaturates (%)	34–61	54–60	35–40	15–20	2–4	4–8
Saturates (%)	13–27	14–21	15–20	20–30	45–50	22–25
Selection criteria	pourable, oily fry, nutrition, price	pourable, oily fry, nutrition	pourable, semi-oily fry, nutrition, stability	all-purpose, dry-fry, high melt point	stability, dry-fry, price, meat flavor	stability, dry-fry, nutrition, economy

Note: RBD, refined, bleached, deodorized; AOM, active oxygen method.

Source: Adapted with permission from R. O'Brien, Foodservice Use of Fat and Oils, *INFORM*, Vol. 4, No. 8, p. 914, © 1993, American Oil Chemists' Society.

genated before being used for frying to increase their stability. Cottonseed, corn, peanut, and olive oils are used as a stable source of polyunsaturated fatty acids because of their low linolenic acid content (Pavel, 1985). Other oils, such as jojoba oil, have also been tested for frying (Clarke and Yermanos, 1980; Saguy et al., 1996).

In 1987, *tropical oils* became "suspect" food fats, due to their relatively high saturated fatty acid content. They are coconut, palm kernel, and palm oils. Coconut and palm kernel oils are somewhat similar in that both (a) have high lauric fatty acid contents, (b) are very high in total saturates (coconut, approximately 92%; palm kernel, 82%), (c) hydrolyze to soapy flavors easily, (d) have sharp melting points, and (e) foam immediately in the fryer when blended with a nonlauric oil. Palm oil is less saturated, has a wider melting range, and has only trace amounts of lauric fatty acid, a high palmitic content, and an orange or blood-red color due to its β-carotene content. Palm oil had been used exclusively for prefrying of frozen french fries supplied to the food service industry (Clark and Serbia, 1991). Tropical oils (palm oil, in particular), with their moderate linoleic acid content, very small linolenic acid content, and high level of natural antioxidants are quite suitable for direct use in many frying applications (Ranhotra, 1993).

HOW AND WHEN OIL DETERIORATES

During frying, the oil is continuously and repeatedly used at elevated temperatures (160–180°C) in the presence of air and moisture. This causes both thermal and oxidative decomposition of the oil. These reactions cause formation of both volatile and nonvolatile decomposition products. They also cause foaming when moist foods are deep-fat fried in the oil (Perkins, 1988; Paul and Mittal, 1996).

Frying oils are subjected to three different types of environmental conditions, and each environment has its effect on the oil.

1. *The storage period* starts as soon as the oil is produced and ends when it is placed in the fryer. During this time, the oil is exposed to air at room temperature.
2. *The standby period* occurs when the oil is heated in the fryer. It includes the time taken to bring the oil up to frying temperature and the time taken for it to cool when frying is finished. During this period, the oil is exposed to air at high temperature.
3. *The frying period* is that time during which the oil is actually being used for frying. The oil is exposed to air, steam, heat, and the food being fried.

The main reactions which can occur are shown in Table 3–2. The rate of formation of decomposition products—and indeed of the products themselves—varies

Table 3–2 Degradative Reactions of Oil

Period	Factor	Temperature	Reaction	Speed
Storage	air	ambient	oxidation	slow
Standby	air	hot	oxidation	fast
			isomerization	fast
			polymerization	slow
			pyrolysis	slow
Frying	air, water, food	hot	oxidation	fast
			isomerization	
			polymerization	
			pyrolysis	
			hydrolysis	

Source: Adapted with permission from E.A. McGill, The Chemistry of Frying, *Bakers Digest*, Vol. 6, No. 62, p. 38, © 1980, Sosland Publishing Company.

with the food being fried, the fat being used, the choice of the fryer design, and the nature of the operating conditions (Stevenson et al., 1984). Also, there is a number of factors that influences the deterioration of frying oil: turnover rate, type of the frying process, temperature, intermittent heating and cooling, degree of unsaturation of frying oils, type of food material, design and maintenance of fryer, light, and use of filters (Paul and Mittal, 1996).

The Chemistry of Oil Degradation

Frying oils undergo extensive degradation and complex chemical transformations when they are heated and as a result of interactions between the oil and the food being fried in it. Figure 3–1 shows the different reactions responsible for the changes in frying oil quality. It presents the principal pathways for the formation of degradation compounds in frying oils. These are *volatile compounds*—which, by their nature, are partially eliminated during frying, and whose importance is intimately related to the organoleptic characteristics of the oil and the fried products—and *nonvolatile compounds* (Stevenson et al., 1984).

Oxidation

The only chemical spoilage reaction that normally takes place during the storage period is oxidation. The atmospheric oxygen reacts with the oil at its surface (attacks the double bonds), causing oxidative alterations. Oxidation produces hydroperoxides (flavorless), which can further undergo three major types of degradation: (1) fission, which produces alcohols, aldehydes, acids, and hydrocarbons; (2) dehydration, which produces ketones; and (3) free radical formation, which

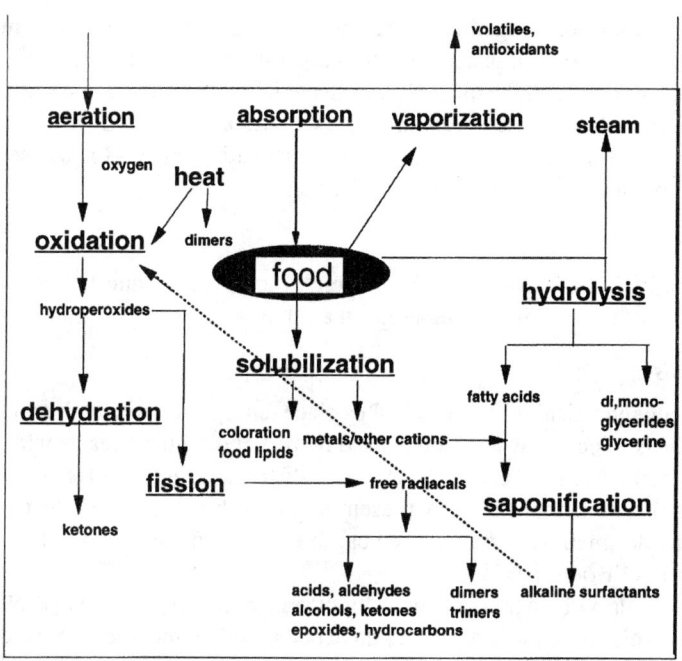

Figure 3–1 Reactions Taking Place during Deep-Fat Frying. *Source:* Adapted with permission from S. Stevenson, M. Vaisey-Genser, and N.A.H. Eskin, Quality Control in the Use of Deep Frying Oils, *Journal of the American Oil Chemists' Society*, Vol. 61, No. 6, p. 1103, © 1984, American Oil Chemists' Society.

produces oxidized monomers, oxidative dimers and polymers, trimers, epoxides, alcohols, hydrocarbons, nonpolar dimers, and polymers (Perkins, 1988). These by-products of oxidation can produce off flavors in the oils (Miller, 1993). A properly refined frying oil will have a life of at least 12 weeks if storage conditions are reasonable.

During the standby and frying periods, the oil is heated in the presence of air, and the process of oxidation becomes much faster. As a result, the double bonds in the oil molecules are broken, and new substances begin to be formed in the fryer. Unfortunately, most of these substances have distinctly unpleasant odors and are responsible in part for the characteristic odor in a well-used oil and for off tastes in the fried food. Also, certain metals, such as iron and copper, accelerate the oxidation of fats and should be avoided (Stevenson et al., 1984).

Polymerization

Thermal alteration results in the formation of cyclic monomers, dimers, and polymers through polymerization. The molecules rearrange, and the double bonds can often end up closer together as a result. Isomerization can, therefore, make the oil more unstable and more sensitive to oxidation. Another effect of polymerization is the thickening of the oil when it is heated. It also results in the formation of a brown, resin-like residue (foaming, caused when the oil will not release the moisture but keeps it trapped while also incorporating air) on the surface of the fryer and on other surfaces exposed to the hot oil.

Pyrolysis

The formation of lower molecular weight compounds, due to the extensive breakdown of the chemical structure of the oil.

Hydrolysis

The major chemical reaction that takes place during commercial deep-fat frying is hydrolysis. It occurs when food is fried in the hot oil. Steam reacts with triglycerides to form free fatty acids (FFA), monoglycerides, diglycerides, and glycerol (glycerine). The amount of FFA present and their breakdown products have an objectionable smell. Eventually, they cause the oil and the fried food to develop off flavors (O'Brien, 1993).

Caustic soda and other alkalis used for cleaning tend to promote hydrolysis and should be well rinsed away after they have been used. Sometimes there are chemicals, such as baking powder, in the food being fried that promote hydrolysis. Moisture in the foods also tends to accelerate this type of deterioration. The polymers formed as a result of oxidative and thermal alterations cause foaming, which traps the steam bubbles longer in the oil to accelerate hydrolysis. The hydrolysis process is also catalyzed by enzymes known as *lipases* (Miller, 1993).

Physical Effects of Chemical Degradation

The first sign of oil deterioration is the formation of distinctive *odors and flavors*. Most of the decomposition products are low in molecular weight and are driven off by the steam as they are formed. They can be smelled in the atmosphere or tasted in the first pieces of fried food but eventually can become more pronounced (Melton et al., 1994).

The oil then begins to *smoke*. The amount of smoking coming from a fryer is proportional to the temperature and to the amount of low molecular weight decomposition products present. In fresh frying oil, temperatures must be over 204°C before enough volatile material is present to actually appear as smoke, but as oxidation and hydrolysis continue, the breakdown products begin to concentrate, and smoke appears at lower and lower temperatures. Actual oil losses due to smoking are not large enough to have any economic significance. Some operators use the appearance of smoke or the appearance of a blue haze over the surface of the oil as an indication of the temperature of the oil. However, it is more an indication that the oil is too hot and is breaking down.

The *smoke point* is the temperature at which an oil begins to smoke continuously and can be seen as a bluish smoke. It is an indication of the chemical breakdown of the fat to glycerol and free fatty acids. The glycerol is then further broken down to acrolein, which is a component of the bluish smoke. It is the presence of the acrolein that causes the smoke to be extremely irritating to the eyes and throat (McGill, 1980):

First step:

$$triglyceride + H_2O \rightarrow FFA + glycerol$$

Second step:

$$\begin{array}{ccc}
CH_2OCOR & & \\
| & & CH_2 \\
CHOCOR & & | \\
| & & CH \\
| & & | \\
CH_2COR & \xrightarrow{heat} & HC = O + 2H_2O \\
glycerol & & acrolein
\end{array}$$

The higher the smoke point, the better suited a fat is for frying. Though processing affects an individual fat's smoke point slightly, the ranges for some of the more common fats are: butter (177°C); lard (183–205°C); vegetable shortenings (180–188°C); vegetable oils (227–232°C). Corn, grapeseed, peanut, and safflower oils all have high smoke points, whereas that of olive oil is relatively low (about 191°C). Fats with smoke points lower than 200°C are not suitable for deep-fat frying. Eventually, the smoke point of an oil may become lower than the normal frying temperature; the operator may then fry at too low a temperature to avoid the production of smoke and, in doing so, encounter all the faults associated with low-temperature frying (McGill, 1980).

An oil reaches its *flash point* (about 320°C for most oils) when tiny wisps of fire begin to leap from its surface. If the oil is heated to its *fire point* (slightly under 400°C for most oils), its surface will be ablaze.

The oil may begin to *foam*. Some compounds act at the surface of the oil by allowing a stable foam to develop. Impurities can accelerate the formation of these foaming agents, particularly the presence of alkaline materials (such as caustic soda), which might have been used to clean the fryers. Foods fried in oils with a foaming tendency are often greasy and less crispy (Chang et al., 1978; Paul and Mittal, 1996). The *color* may also darken. This is mainly the result of oxidative reactions, highly colored compounds being formed by a mechanism that is still not fully understood. Color changes in the frying food can also dissolve in the oil and will tend to darken the frying oil (Melton et al., 1994).

The oil may thicken and become more *viscous* as it is heated. This seems to be due to the process of polymerization and also to oxidation, hydrolysis, and isomerization. It is a serious performance fault because it causes a number of problems. In frying, the oil is really a heat-transfer medium. Thickening reduces the rate of heat transfer, and it takes longer to cook and color the food. It also increases oil absorption (McGill, 1980). This effect will carry over into the finished product, especially as oil content can comprise 30–40% of the product by weight (Paradis, 1993).

The food materials leaching into the oil, breakdown of the oil itself, and oxygen absorption at the oil-food interface all contribute to change the oil from a medium that is almost pure triglyceride to a mixture of hundreds of compounds. Those materials that affect the heat transfer at the oil-food interface must act to reduce the surface tension between the two immiscible materials. These materials act as wetting agents and are regarded as *surfactants*. As the oil degrades, more surfactants are formed, causing increased contact between food and oil. This causes excessive oil absorption and an increased rate of heat transfer to the surface of the food. Eventually, excessive darkening and drying of the product's surface occur, while conduction to the interior is constant (Blumenthal, 1991).

When Should the Frying Oil Be Discarded?

Industrial plants will normally use frying oils with much less than 25% of products of decomposition (total polar materials) because, otherwise, the food would become rancid by the time it is consumed, due to oxidation of the FFA. However, restaurants and food services may use oil containing up to 40% of total polar materials (40% of oil is decomposed) (Blumenthal, 1991).

Frying should be done while the oil is in the *fresh* to *optimum* phases to obtain good-quality foods (Paul and Mittal, 1996). An oil should be discarded when it matches a certain color or when visibility is impaired at a defined distance. An oil should also be discarded if it has a rancid or off odor or if it smells like the foods cooked in it.

The temperature of the oil is the most important factor contributing to the quality of the final product. If the temperature is too low, the food must remain in the oil longer to brown. The longer the food is in the oil, the more oil is absorbed, and the less palatable the food becomes. In terms of oil absorption, oils that have been used repeatedly or not cared for properly may contain polymerized fatty acids, which increase fat absorption (McGill, 1980).

At the parts per million (ppm) level, *surfactants* are important to the ability of hot frying oils to transfer heat to food surfaces against the pressure of steam coming from the food that is frying. Too little surfactant, and an oil produces undercooked food; too much, and an oil produces overcooked food, excessive oil foaming (which promotes oil oxidation), and oil soakage (excess cling) onto/into the fried food. A refined oil usually contains less than 1 ppm of alkaline surfactant materials. Total alkaline materials should be present at less than 10 ppm in an unused oil intended for frying. At concentrations of 20–30 ppm, the effects of alkaline surfactant materials on fried food are noticeable, as evidenced by proper cooking (good heat transfer). When concentrations exceed 60 ppm, quality may be severely compromised (oil-soaked, dark, dried out food).

How to Measure Oil Degradation (Due to Heat Abuse)

Determining the time necessary for frying oils to reach their maximum safe levels of deterioration is a challenging task. Measurements of degradation in frying oils and factors that affect fried food flavor have been reviewed by Melton et al. (1994). However, the method most used by different countries to determine when to discard a frying oil is still sensory evaluation.

No satisfactory and easy method of sensing the frying oil quality has been developed so far (Tyagi and Vasishtha, 1996). The harmful compounds forming in the deteriorated oil can be identified and their quantity measured by using chemical analysis. More than 400 different chemical compounds, including 220 volatile products, have been identified in deteriorated frying oil (Gere, 1982). Degradation products other than nonpolar fraction are collectively called *polar fraction*. The polar fraction is broadly grouped into polymers and decomposition products. The term *polymers* refers to the group of all the degradation products with molecular weight higher than that of triglycerides. The term *decomposition products* refers to the group of all the degradation products with molecular weight less than that of triglycerides (Perkins, 1988).

During frying, oils degrade mainly via thermal oxidation and form volatile decomposition products (VDP) and nonvolatile decomposition products (NVDP). Hydroperoxides, the primary oxidation product, are very unstable and decompose via fission, dehydration, and formation of free radicals to form a variety of chemical products, both VDP and NVDP. NVDP are removed from the frying oil via absorption by the food being fried, by deposition on fry kettle parts, and possibly by filtration (Melton et al., 1994). Characterization of the VDPs is important because it may help in understanding the chemical reactions taking place during frying (White, 1991). However, measurement methods are time-consuming and tedious. Therefore, other methods have been researched.

Whereas VDP are responsible for deep-fried flavor, measurement of NVDP provides better methods for following degradation of a frying oil. NVDPs are absorbed by the food and eaten by the consumer. These are reliable indicators of oil abuse because they are nonvolatile and their accumulation is steady. They are the cause of physical (increased viscosity, color, and foaming) and chemical changes (increased FFA, carbonyl value, hydroxyl content, and saponification value) in the frying oil. Therefore, most methods for assessment of oil degradation are based on these physical and chemical changes.

It is believed that at a concentration of 25-27% of total polar materials (TPM), a frying oil has deteriorated and should be discarded. Although measurement of TPM is a popular method for following frying oil degradation, more research is needed to determine the TPM levels at which different frying oils should be discarded and to relate TPM levels to fried food quality for different types of frying oils (White, 1991). Other methods, such as measurement of FFA level, changes in oil color and polymer concentration, and various quick tests also have been used successfully for following degradation of frying oils. Measurements, such as the concentration of TPM (25–27%) or polymeric content (<10% or < 16%), or physical tests, such as smoke point (≤170°C), are often used in combination with the sensory evaluation. The most commonly used methods are (White, 1991):

1. Standard methods
   a. Polar components
   b. Conjugated dienoic acids
   c. Fatty acid analyses and 18:2/16:0 ratio
2. Quick tests
   a. Dielectric constant
   b. RAU test
   c. Fritest
   d. Spot test
   e. Alkaline contaminant materials, quick test
3. Complex procedures
   a. Gas-liquid chromatography
   b. Exclusion chromatography

One way to specify and evaluate the performance of a frying oil is to evaluate the relative sensitivity of a particular oil to degradation (oxidative, heat, time, etc.). An oil's resistance to oxidation is measured by the active oxygen method (AOM) or oil stability index (OSI). Both tests express the time (in hours) before oxidation products reach a specified value. They are also used in oil specifications to evaluate different oil choices. Because polyunsaturates are the most unstable fatty acids, oils with higher levels of these fatty acids are most likely to have lower AOM values. Monounsaturates are significantly more stable, and oils with high proportions of these fatty acids evidence lower AOM or OSI values (Stockwell, 1988).

Manufacturers of fried snack chips are well aware of the heavy demands put on oil by the fryer. For large-scale fryers, AOM values are simply one indicator of required functionality. Other requirements likely to be imposed on incoming oil include FFA maximums, peroxide values, anisidine value maximums, and trace mineral limits. Performance testing of the oil for finished food sensory analysis is also often required. Table 3–3 shows several quality control tests available to deep fat frying operations.

Indicators of Frying Oil Quality and Tests

Which test is the best indicator of oil quality is still unclear. The snack food industry tends to use FFA content as a chemical marker for predicting oil stability on the products as it passes through distribution to the consumer, but more and more individuals are becoming convinced that the TPM may be the best oil quality or end point index, as originally proposed by the Germans (Stier and Blumenthal, 1991b).

Table 3–3 Quality Control Tests Used in Deep-Fat Frying Operations

Physical Tests		Chemical Tests
Restaurants/ Food Services	Food Industries	Food Industries
Color	smoke point	FFA
Foaming	foam height	TPM
Odor	color	peroxide value
Length of oil use	refractation index	iodine number
Sensory: flavor, odor, texture of cooked food	sensory: flavor, odor, texture of cooked food	AOM
		dimers
		carbonyls
		anisidine value
		petroleum ether insolubles
		NUAF (nonurea-adduct-forming ester)

Note: NUAF, nonurea-adduct-forming ester.

Source: Adapted with permission from S. Stevenson, M. Vaisey-Genser, and N.A.H. Eskin, Quality Control in the Use of Deep Frying Oils, *Journal of the American Oil Chemists' Society*, Vol. 61, No. 6, p. 1106, © 1984, American Oil Chemists' Society.

Total polar materials range from 2% to 4% in an unused oil to more than 30% in an abused oil. Unused oils for frying should contain less than 0.1% of FFA. Typical limits for FFA in heated (used) oils are: 0.5% for potato chips, 1% for kettle-fried chips (double-fried, extra crisp snack), and 1.5–2.0% for pork rinds.

The U.S. Department of Agriculture's (USDA) guidelines for controlling the quality of processed foods (battered and breaded chicken, fish, or meat) require that a frying oil contain less than 2% FFA (USDA, 1997). Free fatty acids are readily oxidized during food storage to yield rancid by-products that can adversely affect the odor and flavor of many products (e.g., potato chips). Free fatty acids are primary initial breakdown products resulting from frying oil triglyceride degradation. Because the percentage of FFA is not linearly related to the degradation of the oil, it should not be the only index of oil quality during frying. This is because, due to heat, light, and oxidation, the FFA are in turn converted into a variety of other polar materials. Therefore, FFA is not a good end point test to determine when to discard an oil. The TPM test is superior for this purpose.

Table 3–4 shows the effect of oil degradation on the formation of FFA and TPM of soybean oil. The oil was degraded at 190°C by frying 300 g of tortilla chip masa (dough) with 55% (w.b.) for 10 minutes at hourly intervals for a total of 60 hours. The degradation levels were identified by number of hours the oil was heated during frying. Samples of degraded oils were collected every 10 hours (approximately 2 hours of exposure to product frying and 8 hours of heating) up to 60 hours (approximately 12 hours of exposure to product frying and 48 hours of heating), cooled down to room temperature, and stored in dark bottles for further analyses. The percentages of FFA and TPM during degradation were determined using standard methods.

FFA and TPM exhibited a linear relationship with degradation time (both indices increased with oil degradation time at 5% significance). After 60 hours of degradation, the percentage of FFA in the degraded oil was 0.61%, whereas TPM was 61.54%. According to the standard for oil quality control (1.0%), the oil quality was still good if it was judged by FFA content. However, if the end point criteria was based on the percentage of TPM (25–27%), the oil should have been discarded after about 25 hours of use. It was observed that between 20 and 30 hours of frying, foam began to form on the surface of the oil and a pungent off flavor was perceived. Based on these observations, the use of TPM as a standard for oil quality control seems to give a better indication for oil quality than does FFA (1.0%). The color readings changed significantly ($P<0.05$) with frying time. The correlation coefficients between color and degradation time were significantly high, i.e., 0.94 (Tseng et al., 1996).

Paul and Mittal (1996) measured the changes in chemical composition of canola oil during degradation in fast food restaurants. They measured concentrations of major degradation products, polymers, decomposition products, FFA, and total polar materials using standard methods. Some of their conclusions are:

Table 3–4 Effects of Degradation Time on the Formation of FFA and TPM of Partially Hydrogenated Soybean Oil Used To Fry Tortilla Dough

Degradation Time (h)	FFA (%)	TPM (%)	Red Lovibond Unit
0	0.03[a]	3.35[a]	0.2[a]
10	0.05[a,b]	15.88[b]	0.3[b]
20	0.15[b]	24.21[c]	0.7[c]
30	0.34[c]	33.94[d]	2.9[d]
40	0.43[d]	41.81	5.40[i]
50	0.51[e]	51.21[f]	10.0[e]
60	0.61[f]	61.54[g]	13.0[g]

[a–i]Averages in the same column within the same data set (category) that are not followed by the same superscript letter are significantly different ($P<0.05$). Values are average of three replications.

Note: FFA, free fatty acid; TPM, total polar material.

Source: Adapted with permission from Y.-C. Tseng, R.G. Moreira, and X. Sun, Total Frying—Use Time Effects on Soybean Oil Deterioration and on Tortilla Chip Quality, *International Journal of Food Science and Technology*, Vol. 31, pp. 287–294, © 1996, Blackwell Science Ltd.

1. The polymer concentration in the oil increases from 0.33–0.44% to 5–9.6% by mass in 9 days of frying. The polymer concentration decreases with high turnover ratio.
2. The concentration of decomposition products in the oil increases from 0% to 5.8–8.3% by mass. This increase is independent of high turnover ratio.
3. The concentration of FFA in the oil increases from 0% to 3.9–5.7% by mass. This increase is independent of high turnover ratio.
4. The concentration of TPM increases from 0.33–0.44% to 12.5–18.3% by mass. The TPM concentration decreases with high turnover ratio.
5. The decomposition products formed during the initial stages of degradation (up to 4–5 days) are almost entirely FFA. The FFA constitute the major portion (49–69%) of the decomposition.

Primarily, the U.S. is interested in quick tests. Quick tests are useful tools for monitoring frying processes. Because fats decompose when heated, monitoring frying oils and fats with quick tests and during frying is the best way to ensure that fried foods will consistently be of good quality (Litovsky et al., 1991).

The family of VERY-FRY test kits (Libra Laboratories) includes tests for TPM, titratable acidity (TA), and total alkaline materials (TAM). They provide rapid, inexpensive measurements of process and quality control parameters in refined, fresh, and used (heated) oils and fats. This technology is based on a gel of nontoxic reagents in a plastic tube. When the reagent is mixed with a hot oil sample, it develops a color in one step; the color can be visually compared against a color card to obtain a semiquantitative estimation. A quantitative measurement may also be obtained by putting the tube directly in a colorimeter and taking a reading. The tests are correlated with the Official methods of the American Organization of Analytical Chemists (AOAC) or the American Chemical Society (ACS) (Stier and Blumenthal, 1991b).

The FDA's *ideal frying oil quick test* should: (a) correlate with official and/or recognized methods; (b) be an objective index; (c) be simple and easy to use by unsophisticated operators anywhere; (d) be inexpensive for value received and save money by preventing waste; (e) be safe for use in a food service or food processing environment; (f) correlate to the quality of foods prepared and the stage of oil degradation; (g) work in any combination of foods, oils, and fryers in aggressive environments; (h) allow for remote reinspection of the results for a quality assurance program; and (i) be available from a vendor having no vested interest in food/oil supplies (Stier and Blumenthal, 1991b).

A novel approach is the development of inexpensive sensors for in situ measurement of frying oil quality. Research has been initiated on the commercial development of probes to assess oil quality based on viscosity measurements (Kress-Rogers et al., 1990).

How to Prevent Oil Degradation

The following recommendations are given to prevent oil degradation during frying (McGill, 1980):

1. The oil should not be overheated to above 191°C (in no case should it be heated above 204°C). The higher the temperature, the faster the breakdown reactions will go on, particularly oxidation and hydrolysis.
2. Food to be fried should be properly dried before frying. Wet foods, particularly potatoes, tend to increase the rate of hydrolysis.
3. Oil and frying equipment should be properly cleaned regularly.
4. Caution must be taken with metals. Iron, copper, and brass strongly enhance the rate of oxidation and should be avoided.
5. The food must be fried in the correct amount of oil. The general rule is to fry one part of food in six parts of oil.

Fats and oils also contain a number of other compounds, including phospholipids, waxes, sterols, and hydrocarbons. Most of these are reduced or eliminated during the refining process; however, certain trace elements remain, which give oils their distinctive odor and flavor or notes. As refined oils age, the background notes become more apparent. Many snack food producers are especially aware of the background notes of various oils and choose those that will enhance snack food's flavor, such as using corn oil to fry tortilla chips (Stockwell, 1988).

Methyl silicone retards oxidation and polymerization of frying oils. Methyl polysiloxane frequently is used for this purpose, and it is effective at the 0.5–2 ppm level. Frying stability increases 3–10 times (O'Brien, 1993). The mechanism of action is the formation of a protective film at the oil-air interface that limits access to oxygen. If the oil is stirred during the heating test, the protective effect disappears because the film is continuously disrupted (Sims, 1994). Except for its function as a defoamer, there would be little incentive to develop an alternate stabilizer for frying oils (Sims, 1994).

Several edible anionic surfactants can be functional in preventing oxidative polymerization. The term *protective index* is used to monitor the rate of destruction of polyunsaturates. If the drop in iodine value for the control oil containing no additive is greater than that of the oil containing the anionic surfactant, the protective index is greater than 1.0 (Sims, 1994). In comparing several anionic surfactants with methyl silicone, sodium salt of phosphated monoglycerides gave the best results in terms of protective index, as well as in preserving the light yellow color of the heated oil. Considerably higher levels of these compounds are required than with methyl silicone to obtain adequate protection. They function best when converted to their sodium salts. The mechanism of action appears to be the same as with the silicones (Sims, 1994).

In addition to the surfactants, the unsaponificable fractions isolated from olive, corn, and wheat germ oils can be effective polymerization inhibitors. The fraction responsible is largely sterol in nature. However, the mechanism for this behavior is still uncertain (Sims, 1994).

Modified Oils

Oil modification is required to further enhance the stability of natural oils for specific applications. A major application of high-stability oils is as frying media. A key factor is the nature of the antioxidant contained in the oil. Naturally occurring antioxidants (e.g., tocopherol) tend to be less volatile that the synthetic ones at normal frying temperatures, and most losses are due to absorption by the substrate (Miller, 1993). Fractionation techniques are used to maximize the concentration of tocopherol of canola oil.

Canola, high-oleic-acid sunflower oil, and high-oleic-acid safflower oil are successful developments of crop breeding technology. Oils with 50% or less saturated fat than commodity soybean, canola, and sunflower oils have also been developed commercially—especially a low-linoleic-acid soybean oil. This oil was tested for its performance under food service frying conditions. Fry life (average of 117.3 hours) was calculated as the sum of the number of elapsed hours to reach several end point criteria (Erickson and Frey, 1994).

A working oil (Appetize, a line of formulated fats produced by Bunge Foods, Bradley, Illinois) that is up to 50% more stable (64 hours vs. 20 for regular corn oil) in a deep-fat fryer was developed using a blend of corn oil with stripped tallow. This implies an extra day for frying because the breakdown is considerably reduced (Hayes, 1996). These formulated fats are modified natural saturated fats without the detrimental effect to lipoprotein metabolism (raising the low-density/high-density lipoprotein ratio), thus offering a solution for atherosclerosis (Hayes, 1996). They consist of blends of vegetable oil and animal fats stripped of cholesterol (and without *trans* fatty acids from hydrogenation) while maintaining the same texture, taste, and stability of the natural fats. These products may be an alternative to hydrogenated vegetable oils containing *trans* fatty acids (Hayes, 1996).

Biotechnological re-engineering affects the fatty acid profile of the base oil and, thus, customizes the functionality available before any processing changes. For example, Cargill Foods (Minneapolis) introduced a line of high-stability oils with oxidation resistance, and DuPont announced a high-oleic soybean oil, which could provide greater stability than traditional soybean oil (Stockwell, 1988). Biotechnology promises an entirely new generation of oils for food uses.

Health Issues

Fats and oils are the most concentrated form of food energy available to us (Stier and Blumenthal, 1991b). For some foods (potato chips, chicken, and fish), the process raises the fat levels considerably, but among others, the process makes no real difference in the level of fat.

Actual fat content of foods depends on a number of factors, especially the quality of oil—in particular, the surfactant concentration. The more poorly maintained the frying oil, the more fat the food will absorb. For example, cake doughnuts fried in badly abused fats have been found to contain as much as 60% fat. If a producer understands how oil degrades and maintains oil quality at a high level, it is theoretically possible to produce a reduced-fat fried food by simply paying closer attention to the condition of the frying oil (Stier and Blumenthal, 1991b).

The other concern that surfaces when frying is the toxicologic or mutagenic effect that heated oils have on individuals consuming them. Fresh frying oil, which is almost pure triglyceride, changes with repeated use. The type of compounds formed during deep-fat frying depends on a number of factors. Included are the type and volume of oil being used, the food or foods being fried, the amount of food being produced, the temperature at which the fryer is operated, the construction materials of the fryer baskets and fryer itself, the cleanliness of the fryer, the surface area of oil in contact with air, and even the lighting use over an open fryer (Stier and Blumenthal, 1991b).

Studies are currently being conducted to elucidate the mechanisms of volatile decomposition products that should aid in understanding those reactions occurring during frying of foods that may pose a potential health problem (Márquez-Ruiz et al., 1995; Takeoka et al., 1996).

There is some evidence that highly oxidized and heated oils may have carcinogenic properties because of potentially toxic substances. On the other hand, investigations of commercial frying have generally indicated that these oils have no deleterious effects on human health. Also, the nutritional value of frying oils is affected by loss of polyunsaturated fatty acids (PUFA), which supplement the essential fatty acids requirement in human metabolism. Soybean oil, because of its high content of PUFA, is considered to be superior to many vegetable oils and hydrogenated fats from a nutritional standpoint, but it is inferior in thermal stability at high temperatures (Tyagi and Vasishtha, 1996).

Nutritional studies have indicated that cholesterol, saturated fats, palmitic fatty acid and *trans* acids, among others, are health hazards. Many of the degradation products of frying oil are harmful to human health as they destroy vitamins, inhibit enzymes, potentially cause mutations, or cause gastrointestinal irritations (Clark and Serbia, 1991). Badly abused oils do contain compounds that may be harmful. These are found in the polar fraction of the oils. Oils used in restaurants are more abused than those used in commercial operations. In fact, if frying operations are ranked according to the average percentage of polar materials present when discarded, the degree of oil abuse changes according to business type: snack foods, 11–13% polar materials; par-fried and batter-breaded, 15–17%; food service/restaurant, 22–27% (Stier and Blumenthal, 1991b).

Studies, however, lead to the most prominent conclusion that deep-frying and consumption of deep-fried foods is not harmful unless the foods were fried in badly abused oils or if people consume too much fried food with their high caloric values (Stier and Blumenthal, 1991b).

PHYSICAL PROPERTIES OF FRYING OILS

In addition to the chemical alterations that occur during frying, changes in the oil's physical and thermal properties, such as surface tension, viscosity, density, specific heat, and convective heat transfer coefficient can affect food quality. The accumulation of alkaline materials was suggested by Stier and Blumenthal (1991a) to reduce food quality due to the decrease of interfacial tension between the product and the frying medium. Formo (1979) observed that the density of oil increased when the molecular weight of FFA decreased. Gutierrez and Dobarganes (1988) indicated that the increase in viscosity and density in frying oil was the consequence of a polymerization reaction.

Density of Frying Oils (ρ)

The density of a substance is equal to the mass of the substance divided by the volume it occupies. Density has a dimension of kg/m^3. Water has its maximum density of 1,000 kg/m^3 at 4°C. As the temperature increases above 4°C, the density will decrease. The addition of all solids (with the exception of fat) to water will increase its density. Density measurement can be used for pure substance as an indication of total solids (Lewis, 1996). However, it is often more convenient to measure the specific gravity (SG) of a liquid, as:

$$SG = \frac{\rho_L}{\rho_w} \qquad [1]$$

where ρ_L is the density of the liquid and ρ_w the density of water at a specified temperature. Specific gravity is a dimensionless quantity. The SG of liquids is measured using density bottles, pycnometers, or hydrometers. The pycnometer method consists of filling a 25–mL container with the unknown liquid, and SG is calculated as:

$$SG = \frac{W_o - W}{W_w - W} \qquad [2]$$

where W_o is the weight of the pycnometer filled with the unknown fluid, W the net weight of pycnometer, and W_w the weight of the pycnometer filled with water. The values of density and specific gravities for different frying oils are listed in Table 3–5. Note that the specific gravity of the oils changes less with temperature than do the density values.

The effect of oil degradation on the SG of soybean oil at 25°C was measured (Table 3–6). The SG of the soybean oil increased with oil degradation from 0.91 to 0.94, but the values were not significantly different ($P<0.05$) between 0–10 and 40–50 hours of degradation. The correlation between SG of soybean oil and TPM and FFA were 0.86 and 0.84, respectively (Tseng et al., 1996). Paul and Mittal (1996) found that, for degraded canola oil, specific gravity correlated well with the amount of polymers formed during frying.

Viscosity (μ)

In deep-fat frying processing operations, the viscosity of the oil changes considerably with frying time and oil temperature. This change must be taken into consideration when designing frying operations so that product quality can be controlled.

Viscosity can be defined as the internal friction acting within a fluid, i.e., its resistance to flow. Vegetable oils, in general, can be classified as Newtonian

Table 3-5 Density and Specific Gravity of Selected Frying Oils as a Function of Temperature

Temperature (°C)	Corn Oil (kg/m³)*	Sunflower Oil (kg/m³)	Sesame Oil (kg/m³)	Soybean Oil (kg/m³)	Cotton Oil (kg/m³)
-20	947 (1.030)	944 (1.027)	946 (1.029)	947 (1.030)	949 (1.032)
-10	940 (1.023)	937 (1.020)	939 (1.022)	941 (1.024)	942 (1.025)
0	933 (0.933)	930 (0.930)	932 (0.932)	934 (0.934)	935 (0.935)
10	927 (0.927)	923 (0.923)	925 (0.925)	927 (0.927)	928 (0.928)
20	920 (0.922)	916 (0.918)	918 (0.920)	920 (0.922)	921 (0.923)
40	906 (0.913)	903 (0.910)	905 (0.912)	907 (0.914)	908 (0.915)
60	893 (0.908)	899 (0.914)	891 (0.906)	893 (0.908)	894 (0.909)
80	879 (0.905)	876 (0.901)	878 (0.903)	879 (0.905)	881 (0.907)

*SG values are within parentheses.

Source: Adapted with permission from S. Rahman, *Food Properties Handbook*, p. 198, © 1995. Copyright CRC Press, Boca Raton, Florida.

Table 3–6 Effect of Degradation Time on Specific Gravity and on the Formation of FFA and TPM in Soybean Oil

Degradation Time (h)	Specific Gravity	FFA (%)	TPM (%)
0	0.91	0.03	3.35
10	0.91	0.05	15.88
20	0.92	0.15	24.21
30	0.92	0.34	33.94
40	0.93	0.43	41.81
50	0.94	0.51	51.21
60	0.94	0.61	61.54

Source: Adapted with permission from Y.-C. Tseng, R.G. Moreira, and X. Sun, Total Frying-Use Time Effects on Soybean Oil Deterioration and on Tortilla Chip Quality, *International Journal of Food Science and Technology*, Vol. 31, pp. 287–294, © 1996, Blackwell Science Ltd.

fluids, where the relationship between stress and strain is constant and equal to the Newtonian viscosity as:

$$\sigma = \mu \dot{\gamma} \qquad [3]$$

where σ is the shear stress, $\dot{\gamma}$ is the shear rate, and μ the proportionality constant appropriate for a Newtonian fluid (viscosity). The unit for viscosity is Pascal seconds (Pa.s), which is 1,000 centipoise (1 Pa.s = 1,000 cP = 1,000 mPa.s). *Dynamic viscosity* and *coefficient of viscosity* are other terms used for *Newtonian viscosity*. Newtonian fluids may also be described in terms of their kinematic viscosity (v) which is equal to the ratio of dynamic viscosity and density (μ/ρ).

The most common instruments used to measure fundamental rheological properties of liquids are classified into rotational type and tube type (Steffe, 1992). Tube types, such as the gravity-operated glass capillaries (e.g., the Cannon-Fenske viscometer), are suitable only for Newtonian fluids. Rotational viscometers, such as the cone-plate and concentric cylinder systems, can be used for Newtonian and non-Newtonian fluids and are also generally used to investigate time-dependent behavior. The reader is referred to Steffe (1992) for detailed discussion on rheological characterization of fluid food materials.

Table 3–7 shows the value of viscosity for selected frying oils as a function of temperature. For Newtonian fluids, viscosity decreases as temperature increases. This effect can be expressed in terms of an Arrhenius type equation involving the absolute temperature (T_a), the universal gas constant (R), and the energy of activation for viscosity (E_a):

Table 3–7 Viscosity of Selected Frying Oils as a Function of Temperature

Oil Type	Temperature (°C)	Viscosity (Pa.s)
Corn	25	0.0565
	38	0.0317
Cottonseed	20	0.0704
	38	0.0306
Olive	10	0.1380
	40	0.0363
	70	0.0124
Peanut	21	0.0647
	26	0.0656
	38	0.0251
	54	0.0268
Canola	0	0.1630
	20	0.0960
	30	0.0286
Safflower	25	0.0286
	38	0.0522
Soybean[a]	25	0.0532
	50	0.0222
	90	0.0081
	120	0.0049
	150	0.0031
	190	0.0020

Source: Data from J. Steffe, *Rheological Methods in Food Process Engineering*, © 1992, Freeman Press; and (a) Y.-C. Tseng, R.G. Moreira, and X. Sun, Total Frying-Use Time Effects on Soybean Oil Deterioration and on Tortilla Chip Quality, *International Journal of Food Science and Technology*, Vol. 31, pp. 287–294, © 1996, Blackwell Science Ltd.

$$\mu = f(T) = A \, \exp\!\left(\frac{E_a}{RT_a} \right) \qquad [4]$$

where E_a and A are determined from experimental data. The energy of activation for different frying oils is presented in Table 3–8. Energy of activation varied from

Table 3–8 Activation Energy for Selected Frying Oils

Oil Type*	Degradation Time (h)	Activation Energy (kJ/mol)
Canola	0	17.5 ± 0.2
	12	15.2 ± 0.3
	24	16.3 ± 0.1
	36	15.6 ± 0.4
Corn	0	15.7 ± 0.2
	12	16.7 ± 0.1
	24	16.9 ± 0.2
	36	17.1 ± 0.2
Palm	0	15.9 ± 0.3
	12	17.0 ± 0.5
	24	16.9 ± 0.5
	36	17.7 ± 0.2
Soybean	0	18.4 ± 0.2
	10	22.4 ± 0.3
	20	22.6 ± 0.2
	30	22.9 ± 0.4
	40	23.5 ± 0.3
	50	24.7 ± 0.2
	60	26.7 ± 0.1

*Temperature range from 170°C to 190°C for canola, corn, and palm; from 25°C to 190°C for soybean.

Source: Data from K.S. Miller, R.P. Singh, and B.E. Farkas, Viscosity and Heat Transfer Coefficients for Canola, Corn, Palm, and Soybean Oil, *Journal of Food Processing and Preservation*, Vol. 18, pp. 461–472, © 1994, Food and Nutrition Press, Inc.; and (a) Y.-C. Tseng, R.G. Moreira, and X. Sun, Total Frying-Use Time Effects on Soybean Oil Deterioration and on Tortilla Chip Quality, *International Journal of Food Science and Technology*, Vol. 31, pp. 287–294, © 1996, Blackwell Science Ltd.

15.7 to 18.4 kJ/mol for fresh oil and increased with degradation time, with the only exception of canola oil. A higher E_a indicates a more rapid change in oil viscosity with temperature (Table 3–9). This could be the result of polymerization during frying, which increases the weight of the molecules.

Viscosity of soybean oil increases as the degradation increases (see Table 3–9). The viscosity values (measured with a rotational viscometer–concentric cylinder configuration) between fresh and 60-hour degraded soybean oils at 25°C changed from 53.17×10^{-3} to 208.3×10^{-3} Pa.s, and from 2.04×10^{-3} to 4.39×10^{-3} Pa.s at 190°C.

Table 3–9 Effect of Degradation Time on the Viscosity of Soybean Oil

Degradation Time (h)	Viscosity (190°C) (Pa.s)	Viscosity (25°C) (Pa.s)
0	2.04 x 10[-3 a]	53.17 x 10[-3 a]
10	2.17 x 10[-3 b]	54.23 x 10[-3 b]
20	2.57 x 10[-3 c]	65.52 x 10[-3 c]
30	3.22 x 10[-3 d]	84.47 x 10[-3 d]
40	3.26 x 10[-3 d]	92.96 x 10[-3 e]
50	3.81 x 10[-3 e]	132.41 x 10[-3 f]
60	4.39 x 10[-3 f]	208.30 x 10[-3 g]

[a-g]Averages in the same column within the same data set (category) that are not followed by the same superscript letter are significantly different (P<0.05).

Source: Adapted with permission from Y.-C. Tseng, R.G. Moreira, and X. Sun, Total Frying-Use Time Effects on Soybean Oil Deterioration and on Tortilla Chip Quality, *International Journal of Food Science and Technology,* Vol. 31, pp. 287–294, © 1996, Blackwell Science Ltd.

In all cases, the oil, whether fresh or degraded, behaved as a Newtonian fluid. The change in viscosity occurred more rapidly after 30 hours of frying. Viscosity values were highly correlated to TPM (0.90). Paul and Mittal (1996) found that the kinematic viscosity of degraded canola oil was poorly correlated to TPM (0.65). Kinematic viscosity increased by 25% with an increase in polymer formation by 6.2% by mass.

Surface Properties

Deep-fat fried foods consist of two fluid phases (water and oil) in a porous solid medium. Therefore, there is a boundary between the two phases. This boundary is referred to either as a surface for gas-liquid systems or interface for liquid-liquid or liquid-solid systems. See Chapter 7 for detailed discussion on interaction of fluids with surfaces.

Surface tension (γ_{lg}) is defined as the force acting on the surface of a liquid, i.e., the force required to minimize the surface of the liquid. The units of surface tension are newtons per meter (N/m) or dynes per centimeter (dyn/cm). Water has a very high surface tension value, (72.8x10[-3] N/m) compared with cottonseed oil (35.4x10[-3] N/m), coconut oil (33.4x10[-3] N/m), and olive oil (33.0x10[-3] N/m), all measured at 20°C (Lewis, 1996).

There are many methods for determining surface tension, including: (a) direct measurement of capillary pull, or DuNouy's method; (b) capillary rise; (c) bubble pressure; (d) size of drops (volume or weight); (e) shape of drops or bubbles; and

(f) dynamic methods. The capillary rise method is the most accurate method, but accuracy to better than 0.5% can be obtained more conveniently with the drop-weight, ring detachment, or bubble pressure methods. Levitt (1973) described in detail the important practical aspects of these methods.

Capillary Rise

The phenomenon of capillarity described by the rise of the water in a capillary tube when it is immersed vertically in a pool of water (Figure 3–2) can be used to determine the surface tension of fluids. The water rises in the tube to a height h above the surface; the narrower the tube, the greater is the height. The angle between the tangent of the surface of a liquid at the point of contact with a surface, and the surface itself, is called the *contact angle* (Θ). The contact angle between clean glass and most aqueous fluids and alcohol is 0°. By applying force balances in the capillary tube at the fluid surface (see Chapter 7), the following equation results, allowing for determination of surface tension:

$$\gamma_{lg} = \frac{g\rho_L hr}{2\cos\Theta} \qquad [5]$$

where g is acceleration due to gravity, h the height to the bottom of the meniscus, and r the radius of the capillary tube. The contact angle can be estimated experimentally (Lewis, 1996). The heights of the meniscus and the tube radius are normally measured using a traveling microscope.

Bubble Method

The bubble method is based on the principle of forces acting on air bubbles in a liquid. Contraction or enlargement in the bubbles is prevented by the forces acting on the bubble, which are due to the liquid pressure p_1 plus the surface tension forces acting to reduce the size of the bubble, whereas the force due to the internal pressure p_2 acts to increase the size of the bubble. At equilibrium:

$$\gamma_{lg} = \frac{r}{2\cos\Theta}(p_2 - p_1) \qquad [6]$$

Thus, there is an excess of pressure within the air bubble. The magnitude of this pressure depends on the surface tension and, inversely, on the size of the bubble. If the excess of pressure inside a bubble can be measured and the radius is known,

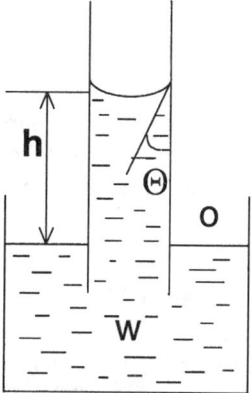

Figure 3–2 Capillary Pressure between Air and Water

the surface tension of the liquid can be calculated.

The Jaeger method is a simple, practical way of applying the bubble method principles. The method consists of displacing air of a container by slowly supplying water to the container. The air is directed to a piece of glass capillary tubing that is immersed below the surface of the test liquid. The water flow rate is regulated to give a slow but steady stream of air bubbles. A manometer is placed into the air line to measure the pressure. At the time the bubble breaks away from the tubing, the maximum pressure (p_2) is recorded. The pressure p_1 is given by the depth of immersion of the tubing. The method is simple and yields reliable results (Lewis, 1996).

Drop Weight Method

The method consists of measuring the mass of a droplet as it forms at the end of a capillary tube. The mass of the drop is directly related to the surface tension of the liquid (Levitt, 1973):

$$\gamma_{lg} = \frac{F_c mg}{\pi r}$$ [7]

where F_c is the correction factor and m the mass of one droplet. The correction factor is a function of the ratio V/r^3 where V is the volume of one drop and r the external radius of the tube. The assumption is that the drop has a cylindrical form, i.e., no spreading takes place when it is about to break away.

Direct Measurement Method

One commercial piece of equipment operating on the principle of the direct pull method is the DuNouy surface tension balance. The method accuracy is within $\pm 0.1 \times 10^{-3}$ N/m and the temperature is not easy to control. Its principle consists of measuring the force F required to detach a horizontal platinum wire ring from the surface of a liquid. The force required to tear the ring from the surface is determined by:

$$F = 4\pi R \gamma_{lg} \qquad [8]$$

where R is the radius of the ring and γ the surface tension. For accuracy, a correction factor F_c must be applied, ranging from 0.7 to 1.0. The magnitude of the correction factor depends on the R/r and R^3/V, where r is the radius of the wire and V the volume of the liquid raised (Shaw, 1970).

The surface tension of soybean oil decreases with increased degradation (frying time) (Table 3–10). At 25°C, the surface tension of the fresh oil is 30.1×10^{-3} N/m and of the 60-hour degraded oil is about 27.5×10^{-3} N/m. Surface tension correlates very well with TPM (-0.93) and FFA (-0.92). Degradation products seem to reduce the oil-air surface tension by 9% (Tseng et al., 1996). Silva and Singh (1995) observed the same phenomena when determining the surface tension of corn oil degraded up to 36 hours (Table 3–10).

The surface tension of most liquids decreases as temperature increases, and the relationship is almost linear (Lewis, 1996). Moreira and Barrufet (1998) measured the surface tension of fresh soybean oil from 25°C to 250°C using the ring method, and showed that, as expected, surface tension decreased linearly as oil temperature increased and varied from 32×10^{-3} N/m to 17×10^{-3} N/m for soybean oil and air (Figure 3–3).

In a deep-fat fried product consisting of water and oil in a solid porous media, the contact angle effects determine which pores will be filled by which liquid. At the boundary of these two emiscible liquids, there will be an imbalance of intermolecular forces, giving rise to the phenomena of *interfacial tension*. This governs whether the external oil would overcome the energy barrier to penetrate the porous structure, due to pressure gradients (Miller and Neogi, 1985). Young's equation describes the relationship between contact angle and interfacial tension for a system in equilibrium. Considering a system as shown in Figure 3–4:

$$\gamma_{lg} \cos \Theta = \gamma_{sg} - \gamma_{sl} \qquad [9]$$

where γ_{sg} is the interfacial tension between the solid and gas, γ_{sl} the interfacial tension between the solid and liquid, and γ_{lg} the surface tension between liquid

Table 3–10 Surface Tension of Soybean and Corn Oil as Function of Degradation Time

Product	Degradation Time (h)	Surface Tension (N/m)
Soybean oil[a]	0	30.14 x 10^{-3}
	10	30.18 x 10^{-3}
	20	29.79 x 10^{-3}
	30	29.52 x 10^{-3}
	40	29.31 x 10^{-3}
	50	28.15 x 10^{-3}
	60	27.49 x 10^{-3}
Corn oil[b]	0	34.60 x 10^{-3}
	3	34.60 x 10^{-3}
	9	34.40 x 10^{-3}
	15	34.20 x 10^{-3}
	21	34.10 x 10^{-3}
	27	34.00 x 10^{-3}
	33	33.90 x 10^{-3}
	36	33.90 x 10^{-3}

Source: Data from (a) Y.-C. Tseng, R.G. Moreira, and X. Sun, Total Frying-Use Time Effects on Soybean Oil Deterioration and on Tortilla Chip Quality, *International Journal of Food Science and Technology*, Vol. 31, pp. 287–294, © 1996, Blackwell Science Ltd.; and (b) M.G. Silva and R.P. Singh, Viscosity and Surface Tension of Corn Oil at Frying Temperatures, *Journal of Food Processing Preservation*, Vol. 19, pp. 259–270, © 1995.

and gas. In Eq.(9), $\cos\Theta$ is derived from interfacial tension at equilibrium conditions. For contact angle >90°, the liquid is said to not wet the solid, meaning that the liquid tends to ball up, runs off the surface easily, and does not enter the capillary pores; when the contact angle is 0°, the liquid wets the solid, i.e., it spreads over the solid easily.

The *drop weight method* can be modified to measure interfacial tension. A microsyringe is used to dispense the more dense fluid into the less dense fluid through a capillary tube of radius r. The interfacial tension γ is calculated from the following equation (Lewis, 1996):

$$\gamma = \frac{V(\rho_1 - \rho_2)}{r} F_c \qquad [10]$$

where ρ_1 and ρ_2 are the densities of the heavier and lighter fluids, respectively, V the volume of one drop, and F_c the correction factor (see information on drop weight method earlier).

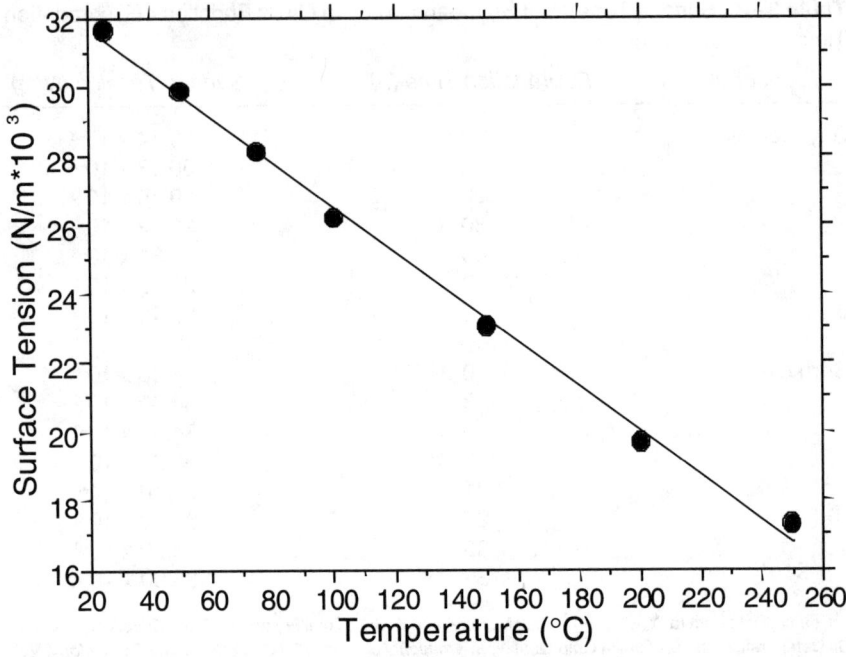

Note: Y = 0.03297–6.4684e–5*X; r^2= 0.996

Figure 3–3 Effect of Temperature on the Surface Tension of Fresh Soybean Oil

To calculate γ_{sl} using Eq. (9) the values of Θ, γ_{sg} and γ_{lg} must be known. The contact angle can be precisely measured using a goniometer (Pinthus and Saguy, 1994) and surface tension (γ_{lg}) by one of the methods described before; however, the values of γ_γ are difficult to determine.

Pinthus and Saguy (1994) used the Girifalco-Good-Fowkes-Young equation to determine the interfacial tension between potato products and different frying media (Table 3–11):

$$\gamma_{sl} = \frac{\gamma_{lg}}{4}(1-\cos\Theta)^2 \qquad [11]$$

They observed that the contact angle (Θ) was higher for the emulsifiers than for the soybean oil. Dipping the potato product in an emulsifier prior to frying in the soybean oil reduced the contact angle to practically zero (almost wetting), result-

Table 3–11 Interfacial Tension, Contact Angle, and Surface Tension Values for Different Frying Media for Potato Products

Frying Medium	Contact Angle	Surface Tension (N/m)	Interfacial Tension (N/m)
Soybean oil	38°	30.0×10^{-3}	0.34×10^{-3}
Polyoxyethylene sorbitan mono-oleate[a] (Tween 80)	77°	34.8×10^{-3}	5.23×10^{-3}
Sorbitan mono-oleate[a]	21°	30.9×10^{-3}	0.03×10^{-3}
(Span 80)	59°	31.7×10^{-3}	1.94×10^{-3}
Polyglycerol polyricinoleate[a] (PGPR)	0°	—	0.01×10^{-3}
Polyoxyethylene sorbitan mono-oleate/oil[b]	0°	—	0.01×10^{-3}
Sorbitan mono-oleate/oil[b]			

[a]Emulsifier; [b]Immersion in the emulsifier prior to frying in oil

Source: Adapted with permission from E.J. Pinthus and I.S. Saguy, Initial Interfacial Tension and Oil Uptake by Deep-Fat Fried Foods, *Journal of Food Science*, Vol. 59, No. 4, p. 805, © 1994, Institute of Food Technologists.

ing in a very low interfacial tension. Among the emulsifiers, *Tween 80*, a hydrophilic emulsifier, showed a larger interfacial tension, compared with oil. However, *Span 80*, a hydrophobic emulsifier, showed a significantly lower interfacial tension value.

Contact Angle Measurement

Several methods for measuring contact angle (Θ) have been suggested that are generally variations of the *tilting plate method* (Spelt and Vargha-Butler, 1996). Another method is a direct measurement of the angle made between a drop of liquid and the solid surface on which it rests using a telescope with a goniometer eyepiece. The contact angle can generally be obtained with high accuracy, even for shallow angles, using a Fabray-Perot interferometer. For direct measurement with a goniometer, even the trained eye gives reproducibilities of ±2°. An electronic micro balance is normally used as a force measurement sensor; the balance is the heart of a tensiometer and crucial to all measurements is its accurate and reproducible operation. Because the forces acting upon the surfaces and interfaces are extremely minute and dynamic, the balance must be capable of swift response to the lowest of force changes—as low as approximately 0.05 N.

Silva and Singh (1995) measured the stationary contact angle of oil droplets on a low-energy surface (1 mm thick polystyrene) by measuring the droplet dimensions with a dissecting microscope fitted with an ocular micrometer (height *h* and

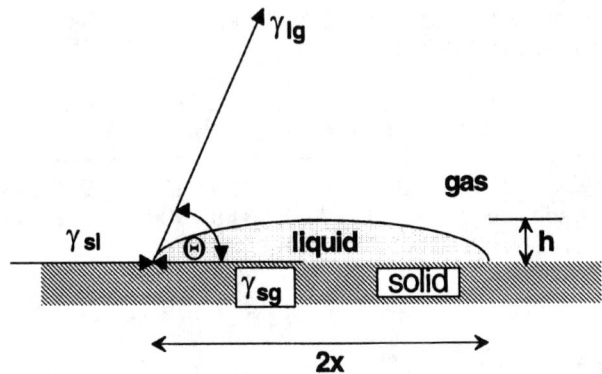

Figure 3–4 Forces Involved in Wettability of a Liquid Dropped on a Solid Surface

width *x*—see Fig. 3–4). The contact angle was calculated as $\Theta = 2arctg(h/x)$ for $(0<\Theta<90°)$.

Specific Heat (C_p)

Knowledge of the specific heat of frying oils is important for calculations involved in the processing and modeling of frying systems. This thermal property indicates how much heat is required to change the temperature of a material, and it has units of kJ/kg K. Classical calorimetric methods are traditionally used to determine the specific heat values; however, they are often tedious and time-consuming, and their precision is drastically reduced when operating at higher temperatures (Kowalski, 1988). The differential scanning calorimetry (DSC) method has been used for studying the thermal properties of fresh edible oils (Kowalski, 1988; Kasprzycka-Guttman and Odzeniak, 1991; Bhatnagar et al., 1994). The DSC method is very helpful in measuring C_p values, because it is more rapid and requires smaller sample sizes, providing greater accuracy.

Values of the specific heat of several frying oils and lard are shown as a function of temperature in Table 3–12. For all the oils, the specific heats do not vary substantially, and their values can be used for engineering design. Vijayan et al. (1997) showed that the specific heat of corn oil varied between 1.952 and 2.134 kJ/kg°C during frying. Paul and Mittal (1996) observed a significant decrease (14%) in specific heat of canola oil at higher levels of fat degradation (7 days of frying).

Convective Heat Transfer Coefficient (h')

In deep-fat frying, heat from the frying oil is transferred to the product surface by convection. In process engineering calculations, the most common method of

Table 3–12 Specific Heat of Vegetable Oils and Lard at Various Temperatures

Product	Temperature (°C)	Specific Heat (kJ/kg K)
Olive oil	70	2.072
	100	2.074
	120	2.139
	140	2.153
Sunflower oil	70	2.080
	100	2.173
	120	2.182
	130	2.179
Soybean oil	70	2.056
	100	2.047
	120	2.106
	140	2.142
Canola oil[a]	70	1.976
	100	1.980
	120	2.124
	140	1.061
Corn oil	70	2.033
	100	2.097
	120	2.115
	140	2.049
Lard	70	1.925
	100	1.885
	120	1.907
	140	1.904

Source: Adapted with permission from T. Kasprzycka-Guttman and D. Odzeniak, Specific Heats of Some Oils and Fat, *Thermochimica Acta*, Vol. 191, No. 1, pp. 44–45, © 1991, Elsevier Science Publisher; and (a) Kowalski (1988).

predicting the heat flux from the surrounding oil to the product is through establishing the heat transfer rate by means of a heat transfer coefficient obtained from dimensionless relation and actual temperature differences. However, in deep-fat frying, the heat transfer is accompanied by a mass (vapor) transfer, which must be taken into consideration. In addition, the vapor transport is associated with phase change, so the corresponding latent heat must be included. Therefore, the total convective heat flux in deep-fat frying is described by the heating of evaporated water from the surface, as well as the equation for heat transfer. The convective heat transfer can also be calculated through the use of the conservation equation (momentum plus mass and energy equations).

The convective heat transfer coefficient, h', relates the property of a boundary layer transport enthalpy from one side of a body/medium to the other. Mathematically, through the heat transfer equation, the heat transfer coefficient can be obtained from a dimensionless relationship:

$$q = h'(T_\infty - T_s) \qquad [12]$$

where T_s is temperature at the product surface and T_∞ the temperature of the thermal boundary layer (99% of the bulk temperature of the medium). Chapter 6 provides a detailed description of methods for measuring the convective heat transfer coefficients in deep-fat frying systems.

In the specific case of soybean oil, the convective heat transfer coefficient decreased nonlinearly from 274.37 to 225.03 W/m² °C at 190°C as the degradation time increased (Figure 3–5). The lumped capacity method (Chapter 6) with a copper spherical transducer was used to determine the h' values. The experimental data were well described by a first-order exponential equation as $h' = 276.5 - 0.525 \exp (0.076\ t)$; $(r^2 = 0.97)$. With up to 30 hours of frying, the changes in the convective heat transfer coefficient of the oils were, in almost all the cases, not significantly different; however, as the oil degradation increased, after 30 hours of frying, a more pronounced change in the values of h' occurred, and the values were significantly different. It was observed that h' was affected inversely by the viscosity of degraded oils, and the correlation coefficient between the two properties was high (-0.98).

Miller et al. (1994) used the lumped capacity method (with an aluminum spherical transducer) and observed that the convective heat transfer coefficient increased as the frying oil temperature increased. Within a degradation time of 36 hours, the convective heat transfer coefficients for corn, soybean, canola, and palm oils were not affected significantly by oil degradation time. Only corn oil heat transfer coefficient showed high correlation coefficients with FFA (-0.922) and percent polymers (-0.989).

FAT SUBSTITUTES—OLESTRA

The most extensively studied and publicized of the synthetic noncaloric fat substitutes are sucrose polyesters (assigned the generic name *Olestra*). Olean is Procter & Gamble's brand name for Olestra. Digestive enzymes do not release the fatty acid, so Olestra is noncaloric and not sweet. It tastes and behaves like fat

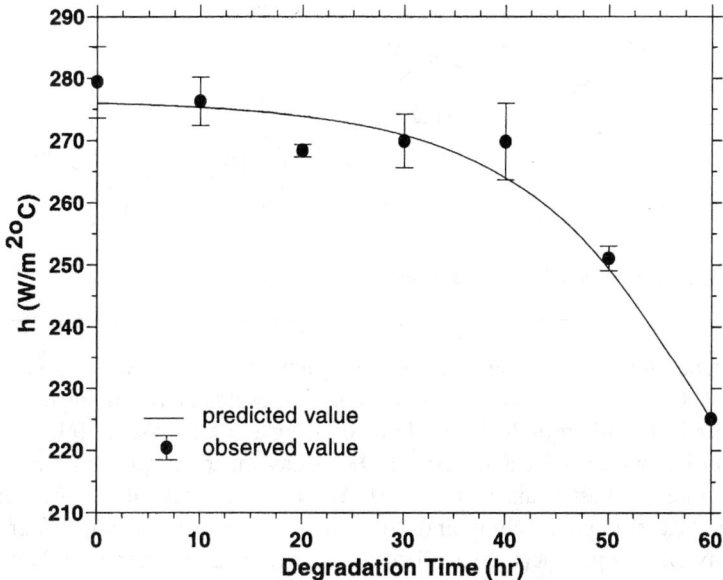

Figure 3–5 The Relationship between the Convective Heat Transfer Coefficient of Soybean Oil and Degradation Time. *Source:* Adapted with permission of Y.-C. Tseng, R.G. Moreira, and X. Sun, Total Frying-Use Time Effects on Soybean Oil Deterioration and on Tortilla Chip Quality, *International Journal of Food Science and Technology*, Vol. 31, pp. 287–294, © 1996, Blackwell Science Ltd.

when subjected to high temperatures (does not break down). That means it can be used to replace edible oils—partially or completely—in high-temperature frying to make chips and french fries, baked goods, and cookies (Henry et al., 1992).

What Is Olestra?

Olestra is a revolutionary molecule made from vegetable oil and sugar. Fatty acids (fat chains) in regular vegetable oil are attached three per fat molecule. To make Olestra, these vegetable oil fatty acids are removed from the vegetable oil molecule and simply reattached on a sugar molecule, so that there are about six or more fatty acid chains on each Olestra molecule. That makes a much bigger molecule out of the same fatty acids we eat every day. Figure 3–6 shows the generalized structure of Olestra.

Olestra (76.3% octaester, 23.3% heptaester, 0.2% hexaester, and 0.2% pentaester) is synthesized from sucrose and fatty acid methyl esters. Olestras have chemical and physical properties similar to those of triglycerides, and functionally

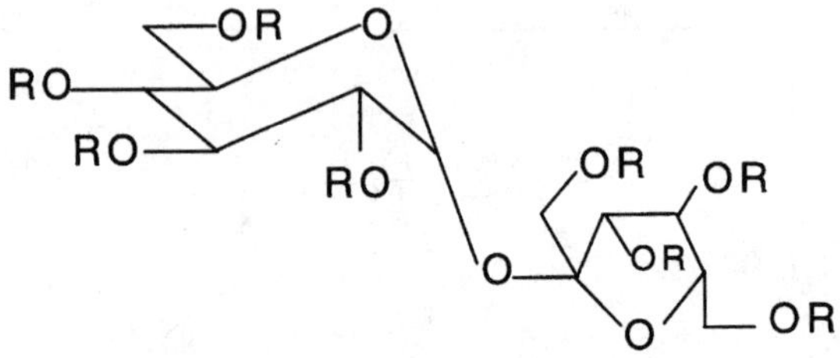

Figure 3–6 Generalized Structure of Olestra

they can be used for the same food processing applications. However, unlike trig-lycerides, Olestra is not metabolized and contributes no calories to the diet.

Olestra is the only fryable fat substitute offering all of the taste of full-fat snack products but with no fat and calories. Frito-Lay currently produces a clearly marked line of Olestra snacks called *MAX*, which includes Olestra versions of Lay's and Ruffle's potato chips, and Doritos and Tostitos tortilla chips. Currently, Frito-Lay Olestra products are available in only limited test markets (Gardner et al., 1992).

The U.S. Food and Drug Administration approved Olestra as safe for use in salty snacks, without restrictions (Henry et al., 1992). Olestra can cause gastrointestinal (GI) discomfort in some people at some times. This is because, after eating Olestra snacks, Olestra is mixed with other food products in the digestive system and may physically soften the stool—just like adding oil or water to bread dough. All available evidence indicates that the number of people who experience a real intestinal symptom from Olestra is low. For example, in a study of 3,357 people over 5 months, only 2% reported a GI effect. Importantly, symptoms are not an allergic reaction or a rejection of Olestra. Olestra has been shown not to cause any change in physiologic function of the digestive system (Henry et al., 1992).

The second side effect often discussed is the ability of pure Olestra to reduce the absorption of certain nutrients. Because Olestra is a fat, any fat-soluble nutrient eaten at about the same time could be in some amounts absorbed (or "sequestered") by any Olestra it comes in contact with in the digestive system. Also, because the Olestra itself is not absorbed, any sequestered nutrients would then be carried out and excreted with the Olestra. This sequestering effect happens to fat-soluble nutrients consumed within only about an hour or two of eating Olestra and to only those that are still in the digestive tract. Olestra has no effect on fat-soluble nutrients already absorbed by the body (Gardner et al., 1992).

Olestra snacks are a new option for consumers who are interested in eating savory snacks and in reducing fat, especially for health-oriented consumers who are not willing to sacrifice taste to reduce fat. Henry et al. (1992) determined that no unique components were created in Olestra when exposed to typical frying conditions, compared with refined, bleached, and deodorized soybean oil partially hydrogenated (Table 3–13). The product was tested for potato frying (oils heated to 185°C for 12 hours per day for 7 days and potatoes fried during 8 of these hours) (Gardner et al., 1992).

Olestra is just the latest development in edible fats. Research continues on the role that it and other fats play in the diet.

REGULATIONS

Frying is an area that regulatory agencies in the U.S. and other nations are looking at very closely. A driving force for regulatory activity is the mandate to protect the public from potential abuses of the food delivery system. In the U.S., regulating frying fats would fit into this later category (Stier and Blumenthal, 1991a). Frying oils are subject to control under the general provisions of the Federal Food, Drug and Cosmetic Act, which states that a food is considered to be adulterated if it "contains any poisonous or deleterious substance which may render it injurious to health" (Firestone, 1993).

Table 3–13 Fatty Acid Composition and Intact Analyses of Triglyceride and Olestra before and after Heating

Fatty acid (%)	Triglyceride (unheated)	Olestra (unheated)	Triglyceride (heated)	Olestra (heated)
C16	10.4	9.7	10.7	10.2
C16:1	0.1	0.1	0.1	0.1
C18	5.4	5.9	5.4	6.0
C18:1	61.7	62.3	62.0	62.4
C18:2	19.9	19.8	18.4	18.1
C18:3	0.4	0.5	0.4	0.4
C20	0.3	0.3	0.4	0.3
C20:1	0.3	0.3	0.3	0.3
C22	0.6	0.6	0.8	0.7
C22:1	—	—	0.1	0.2
C24	0.2	—	0.4	0.2
C24:1	—	—	0.2	0.2
FFA	0.0	0.0	0.5	0.3
Peroxide value	1.1	1.0	5.0	3.3
Polymer	<1	0.9	6.4	13.1

Source: Adapted with permission from D.E. Henry et al., Characterization of Used Frying Oils, Part 2: Comparison of Olestra and Triglyceride, *Journal of American Chemists' Society*, Vol. 69, No. 6, pp. 509–519, © 1992, American Oil Chemists' Society.

Concern for the effects that improperly fried foods may have on consumer health has motivated the governments of a number of countries to promulgate laws and regulations controlling frying conditions. The FDA has established no specific regulations to control frying fats in the U.S. However, the USDA has established guidelines for frying oil in fried meat operations (Firestone et al., 1991). The USDA Food Safety and Inspection Service's *Meat and Poultry Inspection Manual* contains some general guidelines for frying meat and poultry products. They allow antioxidants and antifoaming agents in frying fats. Large amounts of sediment and FFA content in excess of 2% are usually indications that frying fats are not wholesome and require reconditioning or replacement. Formal laws and regulations for control of frying fat quality have been adopted by only a few countries. However, several other countries employ practical guidelines and test procedures to control the quality of frying fats and fried foods. In addition, there is increasing awareness that good frying practice and proper control of frying fats improve the quality and acceptability of fried foods (Firestone et al., 1991). Lately, the tendency is primarily to forbid the use of fats and oils that have decomposed by more than about 25%, e.g., in Spain, Germany, and Belgium (Firestone, 1993).

The Germans proposed the first standards for frying oils, specifically for restaurant frying oils in 1973, due to a series of gastrointestinal problems encountered. Scientists in other countries also found that a significant percentage of restaurants and food service locations served food that had been cooked in abused oils. The standards stated that frying fats were considered unacceptable if either "the flavor or taste was unacceptable, the smoke point less than 170°C, and the oxidized fatty acid content was 0.75% or higher." They were revised, and a value of 27% polar materials correlated with the 0.7% oxidized fatty acid value proposed earlier. In 1990, the upper limit for polar content was reduced to 24–25% for restaurant oils (Stier and Blumenthal, 1991a). Countries using polar materials as a standard for regulation of restaurant frying oils are Germany, Belgium, Austria, France, Spain, Switzerland, and the Netherlands.

A worldwide set of rules and guidelines aimed at protecting the health of consumers, ensuring fair practices in food trade, and promoting harmonization of food standards is provided by the Codex. The CCFO (Codex Committee on Fats and Oils) met in 1993 and considered all standards: (a) palm olein and palm stearin; (b) edible oils and fats, both covered by individual standards; (c) products sold as an alternative to ghee; (d) named animal fats; (e) named vegetable oils; (f) fat spreads; (g) olive oils and olive-pomace oils; and (h) mayonnaise (Gillatt, 1994).

Another area of concern may be filter aids. "Active" filtering materials can be extremely beneficial in deep-fat frying because they can lengthen oil life, maintain it at a higher quality for longer periods, and reduce the time and materials needed for cleanup. However, they may also contain silicate or other inorganic materials that can break down during filtration or float in the oil, to be picked up on frying food. Some filter papers and plastic filters have been shown to leach phenolic binders and brand-identifying printing inks into fats and solvents, whereas various filter aids leached metals. Once in the oil, these could be construed as adulterants in the oil or as potentially toxic materials that could get into the food (Stier and Blumenthal, 1991a).

Changes incorporated from results of scientific research make the detection of adulteration much easier. The FDA may eventually move to establish regulatory limits based on food safety.

NOTATIONS

A	constant in Eq. (4) [Pa.s]
C_p	specific heat [kJ/kg K]
E_a	activation energy [kJ/mol]
F_c	correction factor
g	acceleration due to gravity [m/s^2]
h	height [m]
h´	convective heat transfer coefficient [W/m^2 °C]
m	mass [kg]
p	pressure [Pa]
q	heat transfer [W]
R	gas constant [kJ/mol K]
r	radius [m]
R	radius [m]
SG	specific gravity
T_∞	temperature of the medium [°C]
T_a	absolute temperature [K]
T_s	temperature at the surface [°C]
V	volume [m^3]
W	weight [kg]
$\dot{\gamma}$	shear rate [1/s]
γ_{lg}	surface tension [N/m]
γ_{sg}	interfacial tension between solid and gas [N/m]
γ_{sl}	interfacial tension between solid and liquid [N/m]
μ	viscosity [Pa.s]
Θ	contact angle [degree]
ρ_L	liquid density [kg/m^3]
ρ_w	water density [kg/m^3]
σ	shear stress [Pa]

REFERENCES

Bhatnagar, A.; Gennadios, M.; Hanna, A.; and Weller, C.L. 1994. Specific heats of selected lipids. Paper # 94-6035 presented at ASAE Summer Meeting at Kansas City, MO.

Blumenthal, M.M. 1991. A new look at the chemistry and physics of deep-fat frying. *Food Technol*, 45(2):68–94.

Brooks, D.D. 1991. Some perspectives on deep-fat frying. *INFORM*, 2(12):1091–1095.

Chang, S.S.; Peterson, R.J.; and Ho, C.-T. 1978. Chemical reactions involved in the deep-fat frying of foods. *JAOCS*, 55:718–727.

Clark, W.L., and Serbia, G.W. 1991. Safety aspects of frying fats and oils. *Food Technol*, 45(2):84–94.

Clarke, J.A., and Yermanos, D.M. 1980. *The use of jojoba oil in deep fat frying*. Proceedings of IV International Conference on Jojoba and Its Uses, Hermosillo, Mexico, 261–265.

Erickson, M.D., and Frey, N. 1994. Property-enhanced oils in food applications. *Food Technol*, 48(11):63–68.

Firestone, D. 1993. Worldwide regulation of frying fats and oils. *INFORM*, 4(12):1366–1371.

Firestone, D.; Stier, R.F.; and Blumenthal, M.M. 1991. Regulation of frying fats and oils. *Food Technol*, 45(2):90–94.

Formo, M.W. 1979. Physical properties of fats and fatty acids. In *Bailey's industrial oil and fat products*, ed. D. Swern. New York: John Wiley & Sons; 87–100.

Gardner, D.R.; Sanders, R.A.; Henry, D.E.; Tallmadge, D.H.; and Wharton, H.W. 1992. Characterization of used frying oils. Part 1: isolation and identification of compound classes. *JAOCS*, 69(6):499–508).

Gere, A. 1982. Studies of the changes in edible fats during heating and frying. *Die Nahrung*, 26(10):923–932.

Giese, J. 1996. Fats, oils, and fat replacers. *Food Technol*, 50(4):78–84.

Gillatt, P. 1994. Codex alimentarius' role in world trade. *INFORM*, 5(9):981–986.

Gutierrez, G.R., and Dobarganes, M.C. 1988. Analytical procedure for the evaluation of used frying fats. *In Frying of food, principles, changes, new approaches*, eds. Varela et al. Chichester, England: Ellis Horwood Ltd., 141–154.

Hayes, K.C. 1996. Designing a cholesterol-removed fat blend for frying and baking. *Food Technol*, 50(4):92–97.

Henry, D.E.; Tallmadge, D.H.; Sanders, R.A.; and Gardner, D.R. 1992. Characterization of used frying oils. Part 2: Comparison of olestra and triglyceride. *JAOCS*, 69(6):509–519.

Kasprzycka-Guttman, T., and Odzeniak, D. 1991. Specific heat of some oils and fat. *Thermochimica Acta*, 191:41–44.

Kowalski, B. 1988. Determination of specific heats of some edible oils and fats by differential scanning calorimetry. *J Thermal Anal*, 34:1321–1326.

Kress-Rogers, E.; Gillat, P.N.; and Rossell, J.B. 1990. Development and evaluation of a novel sensor for the *in situ* assessment of frying oil quality. *Food Control*, 1(3):163–178.

Levitt, B.P. 1973. *Findlay's practical physical chemistry*. London: Longman's.

Lewis, M.J. 1996. *Physical properties of food and food processing systems*. Cambridge, England: Woodhead Publishing Limited.

Litovsky, J.J.; Korbelak, T.; and Blumenthal, M.M. 1991. Quick tests for quality frying. A new tool to produce better fried foods in restaurants and industry. *ALIMENTARIA*, 28(225):97–104.

Márquez-Ruiz, G.; Tasioula-Margari, M.; and Dobarganes, M.C. 1995. Quantitation and distribution of altered fatty acids in frying fats. *JAOCS*, 72(10):1171–1176.

McGill, E.A. 1980. The chemistry of frying. *Bakers Digest*, 6(62):38–42.

Melton, S.L.; Jafar, S.; Sykes, D.; and Tigriano, M.K. 1994. Review of stability measurements for frying oils and fried food flavor. *J Am Oil Chem Soc*, 71:1301–1308.

Miller, C.A., and Neogi, P. 1985. *Interfacial phenomena equilibrium and dynamic effects*. New York: Marcel Dekker.

Miller, K.L. 1993. High-stability oils. *Cereal Foods World*, 38(7):478–482.

Miller, K.S.; Singh, R.P.; and Farkas, B.E. 1994. Viscosity and heat transfer coefficients for canola, corn, palm and soybean oil. *J Food Proc Preserv*, 18:461–472.

Moreira, R.G., and Barrufet, M.A. 1998. A new approach to describe oil absorption in fried foods: a simulation study. *J Food Eng*, 35:1–22.

O'Brien, R. 1993. Foodservice use of fat and oils. *INFORM*, 4(8):913–921.

Paradis, A. 1993. Nitrogen in total quality for snack food. *INFORM*, 4(12):1378–1386.

Paul, S., and Mittal, G.S. 1996. Dynamics of fat/oil degradation during frying based on physical properties. *J Food Process Eng*, 19:201–221.

Pavel, J. 1985. Introduction to food processing. Reston, VA: Reston Publishing Company, Inc.

Perkins, E.G. 1988. The analysis of frying fats and oil. *JAOCS*, 65(4):520–525.

Pinthus, E.J., and Saguy, I.S. 1994. Initial interfacial tension and oil uptake by deep-fat fried foods. *J Food Sci*, 59(4):804–807.

Rahman, S. 1995. *Food properties handbook*. New York: CRC Press.

Ranhotra, G.S. 1993. Nutritional and functional considerations of tropical oils. *Cereal Foods World*, 38(7):486–489.

Saguy, I.S.; Shani, A.; Weinberg, P.; and Garti, N. 1996. Utilization of jojoba oil for deep-fat frying of foods. Lebensmittel-Wissenschaftand U-Technologie. *Food Sci Technol*, 29(5/6): 573–577.

Shaw, D. 1970. Introduction to colloidal and surface chemistry, 2nd ed. Oxford, England: Butterworth-Heinemann.

Silva, M.G., and Singh, R.P. 1995. Viscosity and surface tension of corn oil at frying temperatures. *J Food Proc Preserv*, 19:259–270.

Sims, R. 1994. Oxidation of fats in food products. *INFORM*, 5(9):1020–1028.

Spelt, J.K., and Vargha-Butler E.I. 1996. Contact angle and liquid surface tension measurements: general procedures and techniques. In Surfactant series science. *Vol. 63, Applied surface thermodynamics*, ed. A.W. Neumann and J.K. Spelt. New York: Marcel Dekker; 379–440.

Steffe, J. 1992. *Rheological methods in food process engineering*. East Lansing, MI: Freeman Press.

Stevenson, S.; Vaisey-Genser, M.; and Eskin, N.A.M. 1984. Quality control in the use of deep frying oils. *JAOCS*, 61(6):1102–1108.

Stier, R.F., and Blumenthal, M.M. 1991a. Regulate fats and oils. *Baking and Snack*, 13(9):11–18.

Stier, R.F., and Blumenthal, M.M. 1991b. Frying and health. *Baking and Snack*, 13(9):27–30.

Stockwell, A.C. 1988. Oil choices. *Baking and Snack*, 20(2):74–80.

Swern, D. 1964. *Bailey's industrial oil and fat products*. 4th ed, Vol. I. New York: Interscience Publisher.

Takeoka, G.; Perrino Jr., C.; and Buttery, R. 1996. Volatile constituents of used frying oils. *J Agr Food Chem*, 44:654–660.

Tseng, Y.-C.; Moreira, R.G.; and Sun, X. 1996. Total frying-use time effects on soybean oil deterioration and on tortilla chip quality. *Int J Food Sci Technol*, 31:287–294.

Tyagi, V.K., and Vasishtha, A.K. 1996. Changes in the characteristics and composition of oils during deep-fat frying. *JAOCS*, 73(4):499–506.

U.S. Department of Agriculture. 1997. *Agricultural statistics*. National Agricultural Statistics Service. Washington, DC: Government Printing Office.

Vijayan, J.; Sebben, E.; and Singh, R.P. 1997. Specific heat of corn oil during frying. IFT Annual Meeting. Orlando, FL.

White, P.J. 1991. Methods for measuring changes in deep-fat frying oils. *Food Technol*, 42:75–80.

CHAPTER 4

Fried Product Quality

The quality of fried products depends both on the quality of the frying oil and the type of product being fried. A high-quality frying oil provides the best flavor interaction with the food being fried; however, some products are more affected by the oil flavor—fried potatoes, for example—than are other food products, such as breaded fish and chicken, where the product flavor tends to overwhelm the flavor of the oil. The desirable quality characteristics of potato and tortilla chips, french fries, doughnuts, and fried chicken are not the same; also, the quality of raw material requirements differs from product to product. The quality characteristics of raw materials such as cereal grains and potatoes affect the quality of the fried products. Factors such as (1) environmental conditions during the growing period, (2) time and system of harvesting, (3) postharvest treatment (e.g., drying), (4) storage practices, and (5) transportation procedures can affect the quality of raw materials. Fried product quality is affected by the process steps and operation conditions involved in manufacturing of the product, from cutting to blanching, baking, frying, cooling, and packaging. The effects of processing conditions on the various quality characteristics of fried products are the primary topics of this chapter. See Chapter 3 for discussion on oil quality properties and measurement. Raw material quality factors will not be considered in detail in this book. The reader is referred to Lesińska and Leszczyński (1989), Brooker et al. (1992), and Varmam and Sutherland (1995).

QUALITY PROPERTIES OF FRIED PRODUCT

According to Bourne (1982), the four principal quality factors in foods are: (1) appearance, including color, shape, gloss, etc.; (2) flavor, including taste and odor; (3) texture; and (4) nutrition. Appearance, flavor, and texture refer to *sensory acceptability factors* because they are directly perceived by the senses. Nutrition is not described by sensing.

75

In general, the frying industry controls product quality by product appearance and flavor. These quality characteristics can be determined by measuring the related product properties. The properties that determine the overall quality of a fried food product include:

1. moisture content
2. color
3. oil content
4. flavor
5. texture
6. yield
7. nutrition
8. shelf-life stability

Not all consumers or manufacturers of fried products are interested in the same quality characteristics. Consumers of potato chips like a tasty product with a crispy golden (yellow) crust (not oily), so items 2, 3, 4, and 5 are important. The food processor is interested in the product yield and the flavor stability during storage, thus items 6 and 8 are also important.

In this chapter, fried product quality attributes are defined, and the techniques for their measurement are discussed. Factors included in the grade standards of french fries and onion rings in the U.S. are also presented.

FRIED PRODUCT GRADE STANDARDS

According to the USDA standards (USDA, 1997) for grades of processed fruits and vegetables, the purpose of the grade standards is twofold: (1) to facilitate marketing and (2) to serve as a basis for inspection and grading of this commodity by federal inspection. These standards are for the voluntary use of producers, suppliers, buyers, and consumers.

The first federal grades and standards for frozen french fries and onion rings in the U.S. were adopted in 1966 (amended in 1967) and 1969, respectively. Since then, they have not changed.

These U.S. grade standards describe frozen french fries as those products that are prepared from "mature, sound, white, or Irish potatoes." They are classified into two types: (a) retail type, intended for household consumption and normally packed in small packages and (b) institutional type, intended for the hotel, restaurant, or other large establishment; primary containers weigh usually 2.27 kg or more. The frozen onion ring is the product prepared from "clean and sound, fresh onion bulbs." There are two types of onion rings: (a) french fried—onion rings that have been deep fried prior to freezing and (b) raw breaded—onion rings that have not been oil blanched (USDA, 1997).

The U.S. grade standards are based on sensory evaluation to determine the overall quality characteristics of the product. The grade of a sample unit (450 g for retail and 907 g for institutional types) of frozen french-fried potatoes is determined by considering the ratings for: color, uniformity of size and symmetry, defects, and texture; flavor is used but not scored. The U.S. frozen french-fried potatoes standards are given in Table 4–1. For frozen breaded onion rings, the factors for quality in the U.S. grade standards (Table 4–2) are: color, defects, character, and flavor, which is not scored. See Color Plate 3 and Appendix 4–A for color standards for frozen french-fried potatoes and score sheets for frozen french-fried potatoes and breaded onion rings.

The U.S. standards contain only a limited number of factors. There is a significant number of fried product properties in addition to color, shape, defects, flavor, and texture that are of interest to end users.

More recently, with the growing healthy consciousness of the consumer, demands for nutritious and lower-fat-content fried foods have challenged the food industry to produce the so-called better-for-you products. Therefore, in-depth evaluation of the quality characteristics of food products becomes very important, and a number of product properties have to be taken into consideration. The use of sensory panels is still the most reliable measure of food quality today in the food industry. However, this method is time-consuming, expensive, and requires well-trained personnel. The subjectivity of human judgment indicates the need for a more scientific, objective way of specifying product quality. Therefore, new methods of measuring accurately food properties objectively and on-line are required.

FOOD QUALITY PROPERTIES

Factors affecting food quality will be defined, and measurement techniques (including on-line sensors) will be discussed in this section. These factors are: moisture content, color, oil content, flavor, texture, yield, stability, and nutritional value.

Moisture Content

Moisture content is an important property in fried food product quality maintenance. The moisture content of fried foods denotes the quantity of water per unit mass of either wet or dry product; it is usually expressed on a percentage basis. The moisture content wet basis is defined as:

$$MC_{wb} = \frac{\text{mass of water}}{\text{mass of wet product}} \times 100 \qquad \textbf{[1]}$$

Table 4–1 Grades and Grade Requirements for Frozen French-Fried Potatoes

Grade	Color	Uniformity of Size and Symmetry	Defects	Texture	Flavor	Score
U.S. Grade A	27–30	18–20	18–20	27–30	good flavor	≥ 90
U.S. Grade A short	27–30	18–20	18–20	27–30	good flavor	≥ 90
U.S. Grade B	24–26	16–17	16–17	24–26	reasonably good flavor	≥80 but < 90
Standard	0–23	0–15	0–15	0–15	not good flavor	<80

Color: The color of the product is evaluated before complete defrosting but after any exterior frost has evaporated. The color numbers refer to a color that is visually similar to the designated color in the *USDA Color Standards for Frozen French Fried Potatoes.*

Good color (27–30 points): bright, characteristic color of properly prepared potatoes; the fry color may be extra light (lighter than USDA No. 1), light (similar to USDA No. 1 color), medium light (mostly similar to USDA No. 2, predominately lighter than No. 3 but may include No. 1), or medium (mostly similar to USDA No. 3, may include units of No. 4 and/or No. 2 fry color).

Reasonably good color (24–26 points): dull but not off color; the fry color may be variable, exceeding the uniformity criterion of extra light, light, medium light, or medium.

Sstd (0–23 points): frozen french-fried potatoes that fail to meet the requirements above.

Uniformity of size and symmetry: the length designations described to strip styles only ($\frac{1}{4}$ x $\frac{1}{4}$ in, $\frac{3}{8}$ x $\frac{3}{8}$ in, $\frac{1}{2}$ x $\frac{1}{2}$ in, or $\frac{3}{8}$ x $\frac{3}{8}$ in).

Extra long: 80% or more are 2″ in length or longer and 30% or more are 3″ in length or longer; **Long:** 70% or more are 2″ in length or longer and 15% or more are 3″ in length or longer; **Medium:** 50% or more are 2″ in length or longer; **Short:** less than 50% are 2″ in length or longer.

Uniform size and symmetry (18–20 points): not more than 15% may consist of small pieces, slivers, and/or irregular pieces.

Reasonably uniform size and symmetry (16–17 points): not more than 30% may consist of small pieces, slivers, and/or irregular pieces.

Sstd (0–15 points): frozen french-fried potatoes that fail to meet the requirements above.

Defects: This factor is concerned with imperfections in the product, such as, necrosis, crushed units, discolored eyes, callous areas, and discolorations that affect appearance or debility.

Free from defects (18–20 points): in a 1-lb sample unit retail type (5 minor and major defectives, and 1 limit for major defectives); in a 2-lb sample unit food service unit (18 minor and major defectives, and 4 limit for major defectives).

Reasonably free from defects (16–17 points): in a 1-lb sample unit retail type (9 minor and major defectives, and 2 limit for major defectives); in a 2-lb sample food service unit (28 minor and major defectives, and 8 limit for major defectives).

Sstd (0–15 points): frozen french-fried potatoes that fail to meet the requirements above.

Texture: the factor of texture is evaluated within 3 min after the product has been heated or fried and while it is well above room temperature.

Good texture (27–30 points): the external surfaces of the units are moderately crisp, show no noticeable separation from the inner portion, and are not excessively oily; the interior portions are well cooked, tender, and practically free from sogginess; except that shoestring style and dices may be moderately crisp throughout.

Reasonably good texture (24–26 points): the external surfaces of the units may be slightly hard or slightly tough, show no more than a moderate separation from the interior portion, and are not excessively oily; the interior portions are well cooked, reasonably tender, and reasonably free from sogginess.

Sstd (0–15 points): frozen french-fried potatoes that fail to meet the requirements above.

Flavor: this factor is not scored.

Good flavor: good characteristic flavor and odor of properly prepared french-fried potatoes; such flavor is free from rancidity and bitterness, from pronounced scorched or caramelized flavors, and from off flavors and off odors of any kind.

Reasonably good flavor: a flavor that may be somewhat lacking in good flavor and odor but is free from objectionable flavors and objectionable odors of any kind.

Source: Adapted from the United States Department of Agriculture, 1997.

Table 4-2 Grades and Grade Requirements for Frozen Breaded Onion Rings

Grade	Color	Defects	Character	Flavor	Score
U.S. Grade A	25–30	34–40	26–30	good flavor	≥85
U.S. Grade B	21–24	28–33	21–25	reasonably good flavor	≥70 but <85
Standard	0–20	0–27	0–20	not good flavor	<70

Color: The color of the product is evaluated before complete defrosting but after any exterior frost has evaporated. The color numbers refer to a color that is visually similar to the designated color in the *USDA Color Standards for Frozen French Fried Potatoes.*

Good color (25–30 points): the units possess a characteristic cream to golden color typical of properly prepared frozen onion rings; the product is bright, practically uniform in color; and, after heating in a suitable manner, is practically free from units that vary markedly from the predominating color.

Reasonably good color (21–24 points): the units may possess a light cream to brown color typical of frozen onion rings and may be variable in such typical color; the product may be dull but not off color; and, after heating in a suitable manner, the variation in color of the units does not seriously affect the appearance of the product.

Sstd (0–20 points): frozen breaded onion rings that fail to meet the requirements above.

Defects: This factor of defects refers to the degree of freedom from harmless extraneous vegetable material, dark carbon specks, imperfect rings, and blemished units.

Free from defects (34–40 points): the surfaces of the units are practically free from carbon specks; no more than 25%, by weight, of the units may be imperfect rings; and for each 16 oz, on an average, of the frozen product, there may be present no more than 1 piece of harmless extraneous vegetable material and 1 blemished unit.

Reasonably free from defects (28–33 points): the surfaces of the units are practically free from carbon specks; no more than 40%, by weight, of the units may be imperfect rings; and for each 16 oz, on an average, of the frozen product, there may be present no more than 2 pieces of harmless extraneous vegetable material and 2 blemished units.

Sstd (0–27 points): frozen breaded onion rings that fail to meet the requirements above.

Character: the factor of texture is evaluated within 3 min after the product has been heated or fried and while it is well above room temperature.

Good texture (26–30 points): after heating in a suitable manner, the external surfaces of the units are at least moderately crisp; the appearance and eating quality is not materially affected by cracking or unbreaded areas; the units are not oily, soggy, or dry; and the onion ingredient is succulent and tender.

Reasonably good texture (21–25 points): after heating in a suitable manner, the external surfaces of the units are fairly crisp; the appearance and eating quality is not seriously affected by cracking or unbreaded areas; the units are not oily, soggy, or dry; and the onion ingredient is reasonably tender.

Sstd (0–20 points): frozen breaded onion rings that fail to meet the requirements above.

Flavor: this factor is not scored.

Good flavor: good characteristic flavor and odor of properly prepared frozen onion rings; such flavor is free from rancidity and bitterness, from pronounced scorched or caramelized flavors, and from objectionable flavors and objectionable odors of any kind.

Reasonably good flavor: a flavor that may be somewhat lacking in good flavor and odor but is free from objectionable odors of any kind.

Source: Adapted from United States Department of Agriculture, 1997.

The moisture content dry basis is defined as:

$$MC_{db} = \frac{\text{mass of water}}{\text{mass of dry matter}} \times 100 \qquad [2]$$

Conversion from a dry basis to a wet basis moisture content, and vice versa, is done as follows:

$$MC_{wb} = \frac{100\ MC_{db}}{100 + MC_{db}} \qquad [3]$$

$$MC_{db} = \frac{100\ MC_{wb}}{100 - MC_{wb}} \qquad [4]$$

The methods for determining the moisture content of different products are classified as either direct and indirect. Direct methods determine the amount of water in the product by removing the moisture. These include: (a) chemical reaction, (b) heating—oven, (c) distillation, (d) microwave radiation, and (e) infrared radiation. Indirect methods require the measurement of an electrical property of the product, either resistance or capacitance. These methods are described in detail in Brooker et al. (1992) for cereal grains and oil seeds.

The *chemical reaction* method consists of extracting the water from the sample chemically and calculating the amount of water stoichiometrically; this is time-consuming and complicated.

The *air-oven* method is the standard procedure for moisture content determination. The time and temperature during and at which a product sample should remain in an air-oven have to be determined to make sure that the product has lost all but the chemically bound moisture and a small amount of volatile matter. Table 4–3 presents the recommended temperature and heating time for moisture content determination of various food products.

Distillation methods require toluene or oil for moisture determination. These chemicals boil under atmospheric conditions at temperatures well above 100°C. The water in the sample evaporates during heating, and the vapor is collected by condensation. Moisture content is calculated by measuring the water volumetrically. This method requires a laboratory-type setting.

Infrared and *microwave meters* are faster than the oven method, requiring only minutes for moisture determination. Infrared meters must be calibrated for ground sample size and heating time against the air-oven to obtain accurate measure-

Table 4–3 Oven Temperature and Heating Time for Moisture Content Determination

Product	Oven Temperature (°C)	Heating Time (hr)
Potato chips[a]	103–105	12
Tortilla chips[b]	103–105	24
French fries[c]	70 and 105	12 and 1
	110	2
Meats[d]		
Raw	100–102	16–18
Fried	100–102	16–18
Doughnuts[e]	105	12
Falafel balls[e]	105	12
Akara (fried cowpea balls)[f]	70 (vacuum)	24
Plantain[g]	102	24

Source: Adapted from (a) Gamble and Rice (1988) and Baumann and Escher (1995); (b) Moreira et al. (1995); (c) Lamberg et al. (1990) and Mazza and Qi (1992); (d) ASAE (1986); (e) Pinthus et al. (1993); (f) Tan et al. (1995); and (g) Diaz et al. (1996).

ments. Microwave meters are easier to calibrate and take less than 1 minute for the sample to be analyzed; this technique is also nondestructive.

Among the *indirect methods* for measuring moisture content, the capacitance type is the most accurate. Precise weighing of the sample is essential; if the product temperature factor differs from the standard value (usually 25°C), a temperature correction factor has to be applied to the moisture content value.

On-line moisture sensors are essential in the food industry. They are necessary not just for continuous product quality monitoring but also for process control by providing continuous and reliable signals to a feedback controller system. These sensors use near infrared reflectance (NIR) techniques (see description below) and do not touch the food during measurement, being particularly suitable to continuous processing lines (Brescia and Moreira, 1997). The sensor measures the product surface moisture and must be carefully calibrated (e.g., selection of the correct wave lengths) for correct reading.

Color

Color is among the major factors influencing consumer acceptability of a fried product. It can indicate high-quality product (such as the golden yellow of a potato chip) and can also influence flavor recognition. Panel evaluation and comparison to standards is the most common approach for determining color consistencies or differences in fried foods in the food industry.

Color is a three-dimensional (3-D) characteristic of appearance. The human vision responds to a tristimulus in the sensory perception of color. The eye possesses three types of light-sensing devices, each corresponding to a different band of wave lengths, i.e., red, green, and blue. The International Committee on Illumination (CIE) has adopted the X, Y, Z standard values as numerical representatives of the three sensory color responses.

There are many systems for specification of colors in 3-D. The spectrometric methods, for example, describe color in terms of three attributes: reflectance (lightness or *value*), dominant wave length (*hue*), and purity (intensity or *chroma*). The *value* attribute refers to the difference between *light* colors and *dark* colors; *hue* is the attribute of color by which an object is judged to be red, yellow, green, blue, etc.; and *chroma* is used to specify the position of the color between gray and the pure *hue*. Colors can be distinguished from one another by specifying these three visual attributes (Francis and Clydesdale, 1975).

The first system for color was developed in 1905 by Munsell. A cylindrical representation of the Munsell system consists of 10 basic hues, which are spaced around the vertical lightness axis called *value*; *chroma* increases from the central gray axis to outside of the color solid. Human judgment is required to express how close a particular color is to a certain Munsell color sample. The Munsell system is used by the standards color chart to compare frozen french fries (see Color Plate 3).

Today, benchtop and on-line colorimeters exist that can be used objectively to determine the color of a great variety of products. These color sensors use the L*a*b color scale to express color differences among samples. The L dimension defines the lightness, the a refers to the red-green hues, and the b dimension refers to the blue-yellow hues.

On-line colorimeters have been used to measure color of snack food products in processing lines. The signal from the sensor is used to automatically control a continuous process. An example is the Colorex sensor (Infrared Engineering Inc., Waltham, Massachusetts), which uses the L*a*b color scale. Generally, this sensor is positioned after the cooling section of the process and the b value, which is positive when the product is yellow and negative when the product is blue, is preferred for use over a and L for modeling and control because it is affected most when process variables are varied (Schonauer and Moreira, 1995).

Oil Content

Oil content is one of the most important quality attributes of fried products. A high oil content is costly to the processor and results in an oily and tasteless product. However, the texture of a low-oil-content product can be hard and unpleasant. The oil content of fried foods denotes the quantity of oil per unit mass of either wet or dry product; it is usually expressed on a percentage basis (see Eqs. 1–4).

The methods used for oil content determination can be classified as: (1) extraction (2) refractometric method, (3) hydraulic press, and (4) NIR spectroscopy.

The *extraction method* uses solvents such as diethyl ether, petroleum ether, hexane, acetone, ethanol, and chloroform/methanol (2/1) to extract the oil from the product (AACC, 1986). Petroleum ether, diethyl ether, and hexane give similar oil extraction results, ethanol dissolves carbohydrates and inorganic salts besides oil, and chloroform/methanol mixtures have been used to also extract some chemically bound fat (Tecator manual, 1983). The particle size distribution of the sample material also affects extraction yield. A finely ground sample gives a higher oil value compared with a coarser one, due to the larger available surface area.

The extraction is done with equipment of Soxhlet type, requiring at least 4–16 hours of extraction time. New instruments as the Soxtec System (Perstorp, Inc., Silver Spring, Maryland) can produce accurate results in a shorter time (0.5–2 hours). Lulai and Orr (1979) showed that only 2-hour Soxhlet extraction using petroleum ether gave accurate and reproducible results for potato chips.

Oil content in a Soxhlet system is determined as follows. Ground samples (2–5 g) are placed in an extraction thimble and extracted with petroleum ether (50–150 mL) for 0.5–16 hours. After cooling and evaporation, the flask is dried for 15 minutes (at 105°C). The oil content is then gravimetrically determined by:

$$OC_{wb} = \frac{\text{mass of oil}}{\text{mass of product}} \times 100 \qquad [5]$$

Supercritical fluid extraction (SFE), another technique to measure oil content, uses a relatively nonpolar solvent to dissolve oils. SFE uses CO_2, a nontoxic solvent, and it is faster than the Soxhlet method. It is an official American Oil Chemists' Society (AOCS) method for oil extraction from oilseeds (Food Technology, 1997).

Baumann and Escher (1995) used the *refractometric method* to measure oil content in fried potato slices. Dried ground samples (2 g) and 3 mL of l-chloronaphthalene were placed into centrifuge tubes and the mixture homogenized for 30 seconds. The solution was then allowed to settle for 8 minutes, then was centrifuged again for 5 minutes. The clear supernatant was transferred to a refractometer and the refractive index (RI) read at 25°C. The fat content of the sample, in dry basis, was calculated as:

$$OC_{db} = \frac{V_{sol} d_f (n_s - n_{25})}{W_s (n_s - n_f)} \times 100 \qquad [6]$$

where V_{sol} is the volume of solvent, n_s the refractive index of solvent, n_{25} the refractive index of solution at 25°C, W_s the weight of the dried sample, and n_f the refractive index of oil. This method is rapid, and the results showed less than 1% deviation with the Soxhlet method.

Later, Ufheil and Escher (1996) modified the refractometric method by replacing the solvent with n-heptane. Extraction was carried out in a water-bath shaker at 70°C for 20 hours, and the tubes were cooled to 20°C. Refractive index readings from extract and pure solvent were taken in a refractometer at 20°C. Instead of calculating oil content from RI of oil and solvent, the following calibration curve, in wet basis, was developed for peanut oil and n-heptane:

$$OC_{wb} (kg/kg) = (-19.31768 + 13.92377 * n_{20}) * \frac{W_{sol}}{W_s} \qquad [7]$$

where n_{20} heptane = 1.38784 and n_{20} peanut oil = 1.47154. The method was calibrated against the Soxhlet using petroleum ether.

Extraction by pressing the sample using a *hydraulic system*, known as cold pressing, can also be used to determine oil content faster than the standard method; however, the values must be calibrated against the Soxhlet method. A calibrated carver laboratory press (Model C, Fred S. Carver Inc., Menomonee Falls, Wisconsin) was used to measure oil content of tortilla chips (Lee, 1991). The carver press method tends to underestimate the oil content of the product.

Near infrared reflectance is a nondestructive technique; it is quick and requires no reagents for oil content determination. The principle of NIR involves energy absorbance by chemical bonds. Specific bonds vibrate at particular frequencies and absorb photons of coincident energy. This energy leads to elevation of the energy level of the molecule. Some of this energy is transmitted, and the remainder is reflected. The amount of energy reflected at different wave lengths can be correlated with the number and types of functional groups present. Fundamental absorption bands for organic molecules occur primarily in the infrared region.

The NIR technique has been successfully used for fat determination in eggs (Wehling et al., 1988), cheddar cheese (Wehling and Pierce, 1989), bread (Osborne et al., 1984), and tortilla chips (Dudley, 1993). On-line NIR sensors have been used to measure oil content in snack food production lines (Brescia and Moreira, 1997).

Flavor

Flavor is an important quality factor to the food processor interested in flavor stability of fried products during storage. It comprises taste (perceived in the tongue) and odor (perceived in the olfactory center in the nose); it is defined as the response of receptors in the oral cavity to chemical stimuli (Bourne, 1982). Odor plays the dominant role in the flavor sensation. Under normal conditions, only volatile chemicals can reach the olfactory epithelium, and the sense of taste is used to detect nonvolatile chemicals (Dodd et al., 1992).

The human sense of smell is still the primary instrument used by various industries for evaluating quality of a wide range of food products. At present, the sensory qualities of food products are evaluated by organoleptic tests, i.e., sensory panels. This is an expensive process requiring intensive training as human sense of smell varies with age, health, and diet. Gas-liquid chromatography (GLC) and mass spectroscopy (MS) technologies have been used to correlate taste panel scores to volatile profiles. Both GLC and MS techniques are expensive and of limited value (i.e., not sensitive enough for routine quality control) (Bett and Dionigi, 1997).

The GLC technique uses an inert carrier gas to move an evaporated sample through a long coated column; components are separated along the column and emerge individually; a detector at the end of the column measures the amount of each compound. In MS, the sample is ionized, and the ions formed are measured by the mass spectrometer to produce a mass spectrum; this can then be used to determine the identity and quantity of the original compound by its mass.

There is a great need in the food industry for an instrument that is capable of mimicking the human sense of smell. Such a sensor, called *electronic nose*, can be used to detect odors that are sensed by the human nose (Dodd et al., 1992). The sensor consists of: (1) hardware—gas sensors and electronics; and (2) software—for processing the responses of the sensors, comparing them with stored data and identifying and interpreting them. It is an integration of multiple gas sensors and artificial intelligence.

When a chemical input is presented to the sensor array, the sensors convert the chemical signal into an electronic signal (e.g., conductance) that is then analyzed and identified using, for example, supervised patter recognition techniques (e.g., artificial neural networks). The calibration process is tedious, and the process is complex. The reader is referred to Gardner and Bartlett (1992) for detailed discussion on the electronic nose.

Texture

The term *texture* is still not well defined in food processing technology; however, it is a very important quality characteristic of a fried product. It consists of a number of different physical sensations or group of physical characteristics that: (1) arise from the structural elements of the food; (2) are sensed by touching; (3) are related to the deformation, disintegration, and flow of food under force; and

(4) are measured objectively by functions of mass, time, and distance (Bourne, 1982).

Instrumental methods to measure texture in food can be classified as fundamental, empirical, or imitative (Szczesniak, 1963). *Fundamental* tests measure simple physical parameters such as strain, stress, viscosity, and elasticity, but usually correlate poorly with sensory evaluation of texture properties. *Empirical* tests measure parameters that are poorly defined but correlate well with textural quality. *Imitative* methods mimic the conditions to which the food is subjected in practice. One example is the General Foods Texturometer, which duplicates the chewing action of the teeth.

An important texture characteristic for fried foods is crispness. Crispness denotes freshness and high quality (Szczesniak, 1988). A crisp food should be firm and should snap easily when deformed, emitting a crunchy sound (Christensen and Vickers, 1981).

Moisture content affects the texture of dry snack foods by plasticizing and softening the starch/protein matrix, which alters the mechanical strength of the product (Katz and Labuza, 1981). Nelson and Labuza (1993) point out that dry cereals have a crisp texture in the glassy state, but plasticization by an increase in water content or temperature may transfer the material to the rubbery state, causing sogginess. Kärki et al. (1994) showed that crispness for an extruded snack was lost when the T_g (glass transition temperature) was depressed to below ambient temperature as a result of water plasticization.

Many scientists have worked to define and measure the sensation of crispness. Crispness for a long time had been regarded as a force-deformation sensation until the work of Vickers and Bourne (1976) proposed that crispness was an auditory sensation. Since then, research indicates clearly that crispness is a combination of both force-deformation and auditory sensations.

The Universal testing machines such as the Instron and the Texture Analyzer (Texture Technologies Corp., New York) are typically used for force-deformation studies, whereas a variety of equipment has been used for acoustical measurement. Vickers (1988) used an oscilloscope to record the crispness of potato chips; Povey and Harden (1981) used an ultrasonic pulse echo technique to measure ultrasonic velocity in foods; and Tesch et al. (1995) and Richardson et al. (1997) used the sound program *Edit* to record the acoustics of cheese balls and croutons, and potato chips, respectively.

Force-Deformation Measurements

The Instron and Texture Analyzer operate under constant strain rate to measure the force necessary to keep that deformation constant. Instruments that operate under constant stress rate have the advantage of being more similar to the application of force in the mouth (Vickers, 1988); for details of this instrumentation, see

Jowitt and Mohamed (1980). The measurements usually used to determine food texture can be classified as simple compression, texture profile analysis (TPA), or stress relaxation.

Simple Compression. Biological materials may be evaluated in terms of a bioyield point and a rupture point (Mohsenin, 1984). Figure 4–1 illustrates a typical curve for fruits and vegetables when a cylindrical sample is tested in *simple compression.* The initial portion of the curve (I–II) is a straight line up to the linear limit (II). The modulus of elasticity, Young's modulus, of that curve is calculated from the slope of stress (σ) and strain (ε),

$$E = \frac{\sigma_b}{\varepsilon_b}$$

[8]

where

$$\sigma_b = \frac{F}{A} \text{ and } \varepsilon_b = \frac{dh}{h_o}$$

[9]

and E is the Young's modulus, σ_b the stress, ε_b the strain, F the force, A the cross-sectional area, h the height, and h_o the initial height. This portion of the curve is often taken as an index of firmness. When stress and strain cannot be calculated, data may be plotted in terms of force and deformation.

The point (III), observed at σ_{III} and ε_{III}, is called *bioyield* and is related to a failure in the microstructure of the material associated with an initial disruption of cellular structure. The rupture point (IV), defined by σ_{IV} and ε_{IV}, is related to the microscopic failure in the sample. For a brittle material, the rupture point may be close to the yield point, whereas for tough samples, these points may be widely separated. A standard method for the compression testing of food materials of convex shape has been established by the American Society of Food, Biological and Agricultural Engineers (ASAE). The method is similar to the one described here.

Texture Profile Analysis. The TPA was developed by Szczesniak (1963) and extended by Bourne (1974). The test consists of compressing a food sample two times (two bites cycle test), usually 80% of its original height. Compression is achieved using parallel plates where one plate is fixed and the other plate moves with reciprocating linear cycling motion. A generalized texture profile curve is illustrated in Figure 4–2. Several parameters can be obtained from the TPA curve; Table 4–4 presents the meaning of each of those parameters (Bourne, 1982).

Texture profile analysis has demonstrated to be a very useful technique for test-

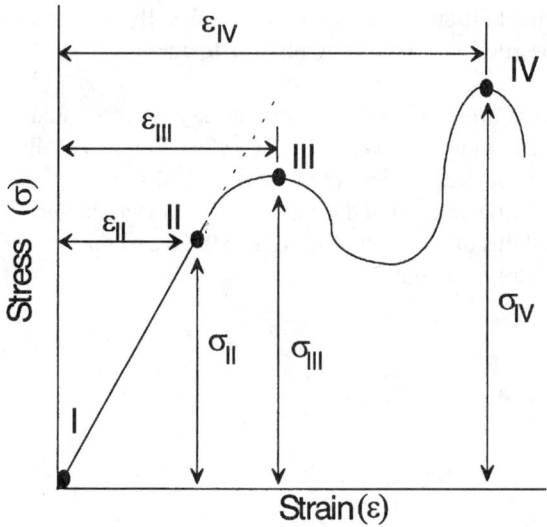

Figure 4–1 Generalized Compression Curve for a Biological Solid

ing products that fracture; however, interpretation of the experimental data is very difficult because of the variability of the texture profile curves (Steffe, 1992).

Stress Relaxation. Foods, in general, show time-dependent mechanical behavior, which is outside the scope of purely elastic or viscous mechanical theories. The elastic theory accounts for materials that have the capacity to store mechanical energy without dissipating it. A Newtonian viscous fluid dissipates energy and has no capacity to store it. Materials that have the capacity to store some of the energy, but not all, fall between the elastic and viscous categories and are said to exhibit viscoelastic behavior.

In characterizing viscoelastic materials such as foods, three types of tests are commonly used: (a) step strain input and measurement of stress relaxation output; (b) step stress input and measurement of creep strain output, and (c) sinusoidal stress, or strain, input and measurement of strain, or stress, output.

In a *stress relaxation* test, the sample is given an *instantaneous strain*, and the stress required to maintain the deformation is observed as a function of time. Figure 4–3 illustrates the wide range of behavior that one may observe in a stress relaxation test. No relaxation would be observed in ideal elastic materials (solid), whereas ideal viscous materials (liquid) would relax instantaneously. Viscoelastic materials would relax gradually with the end point, depending on the molecular structure of the tested material; stress would decay to an equilibrium values for viscoelastic solids and would decay to zero in the case of viscoelastic liquids (Steffe, 1992).

Table 4–4 Definition of Textural Parameters Obtained from the TPA Curve

Parameter	Definition	Popular Term
Fracturability	force at the first major drop in force curve	crunchy, crumbly, brittle
Hardness 1	force at maximum compression during first bite	soft, firm, hard
Area 1	work done in the sample during the first bite	—
Area 3— adhesiveness	work caused from a tensile force, needed to pull food apart and separate it from compression plates	sticky, tacky, gooey
Adhesive force	maximum negative force	—
Stringiness	distance food extends before it breaks away from the compression plates	—
Hardness 2	force at maximum compression during second bite	—
Area 2	work done on the sample during the second bite	—
Springiness	distance or length of compression cycle during the second bite	plastic, elastic
Cohesiveness	ratio of area 2 divided by area 1	—
Gumminess	the product of hardness times cohesiveness	short, mealy, pasty, gummy
Chewiness	the product of gumminess times cohesiveness times springiness	tender, chewy, tough

Source: Data from M.C. Bourne, *Food Texture and Viscosity,* © 1982, Academic Press.

In the stress relaxation test, when an input of constant strain is induced, it is assumed that $\varepsilon(t) = \varepsilon_0 H(t)$, where $H(t)$ is the unit step function:

$$H(t-\tau) = \begin{cases} 0, t < \tau \\ 1, t > \tau \end{cases} \qquad \text{[10]}$$

The stress relaxation modulus $E(t)$ is defined as:

$$E(t) = \frac{\sigma}{\varepsilon_o} \qquad \text{[11]}$$

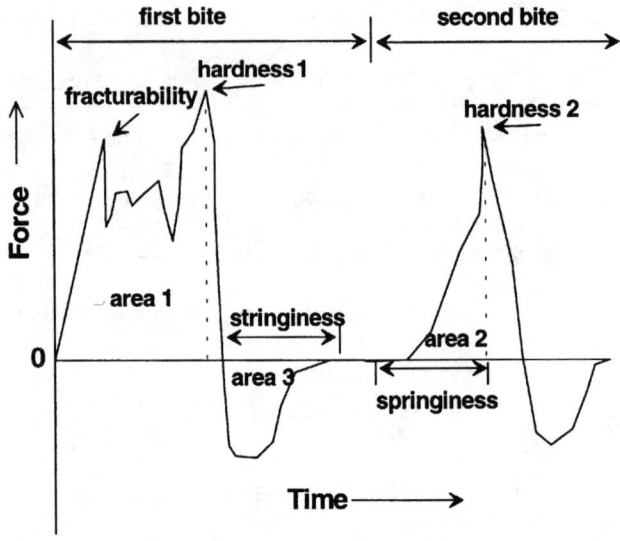

Figure 4–2 Generalized Texture Profile Curve Obtained from the Instrom Universal Machine

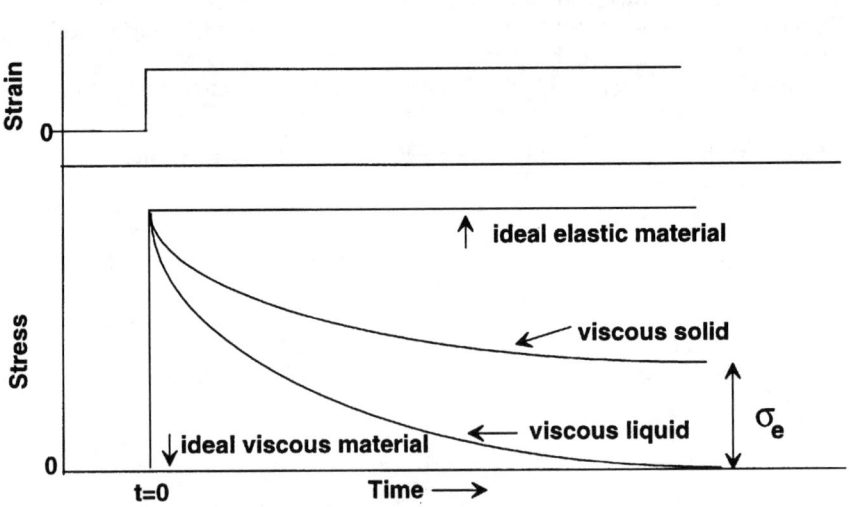

Figure 4–3 Stress Relaxation Curves

In the *creep test*, when an input of constant stress is induced, it is assumed that $\sigma(t) = \sigma_0 H(t)$. The creep compliance modulus $D(t)$ is then:

$$D(t) = \frac{\varepsilon}{\sigma_o}$$ [12]

The *dynamic* response is an experimental technique that consists of applying an oscillating stress or strain input at constant frequency. The strain or stress could be specified as $\varepsilon_o e^{iwt}$ or $\sigma_o e^{iwt}$ respectively. ε_o is real and represents the maximum amplitude of the sine wave, and the response is $\sigma^* e^{iw}$, with σ^* a complex function of frequency. The opposite case applies when $\sigma_o e^{iw}$ is specified, with σ_o being real and representing the maximum load of the sine wave. In this chapter, we will discuss the stress relaxation test in detail. The reader is referred to Steffe (1992) for explicit explanation on creep and sinusoidal tests.

The difference between viscoelastic and elastic media lies essentially in the relationship between stress and strain. Normal elastic analysis is based on a constant, called the *Young modulus*, that relates stress and strain, and is described by an ideal spring element. Conceptually, in the case of viscoelastic materials, the incorporation of time effects into the mathematical model is attained by providing dashpot elements in addition to the elastic spring elements. The two elements that form the building blocks of viscoelastic mechanical models are shown in Figure 4–4. They are assumed to have unit cross-sectional area and initial length. Thus, stress and strain can be used instead of force and displacement.

When the spring and the dashpot are placed in series, a Maxwell model (Figure 4–5) is obtained, which is characterized by Eq. (13). The Maxwell model corresponds to an unlimited deformation under an applied load.

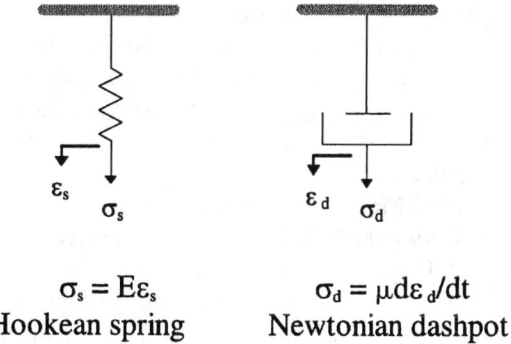

$$\sigma_s = E\varepsilon_s \qquad\qquad \sigma_d = \mu d\varepsilon_d/dt$$
Hookean spring   Newtonian dashpot

Figure 4–4 The Building Blocks of Visoelastic Mechanical Models

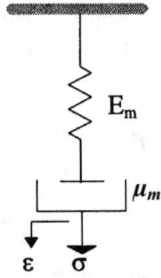

Figure 4–5 Maxwell Model

$$\varepsilon = \frac{\sigma}{E_m} + \int_0^t \frac{\sigma}{\mu_m} dt \qquad [13]$$

where E_m is the elastic modulus, μ_m the viscosity, and t the time.

The Maxwell model is not general enough to represent most real materials. The simplest model that possesses most of the general features of viscoelastic deformation is the three-element model shown in Figure 4–6 (Schapery, 1974). The governing equation for this model is:

$$\sigma = \left[\frac{E_m}{\mu_m} t + \left(\frac{E_m + E_e}{E_m} - 1 \right) \left(1 - e^{\frac{tE_m}{\mu_m}} \right) \right] E_e \frac{\mu_m}{E_m} \frac{d\varepsilon}{dt} \qquad [14]$$

This model (Figure 4–6) exhibits an instantaneous response, as well as delayed elasticity and recovery. The model has three material constants—two springs and one dashpot—with which to curve fit the experimental response of an actual material, which many times are not sufficient. To improve the accuracy, it is necessary to increase the number of elements so the model can imitate the mechanical behavior of the material under study.

If an infinite number of Maxwell elements were placed in parallel with an equilibrium spring representing the long time modulus, a Wiechert model would be obtained (Figure 4–7). For practical reasons, it is more convenient to use the approach given by Schapery (1962), who took advantage of the exponential character of viscoelastic deformations and proposed an effective collocation scheme using prony series expansion of the stress relaxation, assuming that stress relaxation modulus $E(t)$ can be represented by:

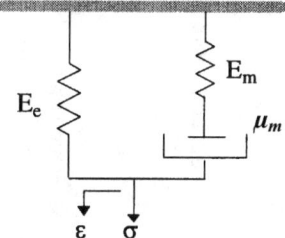

Figure 4–6 Maxwell Elements in Parallel with a Spring—One Element Maxwell and a Free Spring

$$E(t) = \sum E_i e^{-\frac{\mu_d}{E_s}t} + E_\infty \qquad \text{[15]}$$

where E_i are the elastic moduli for individual spring elements; μ_i are viscous parameters for individual dashpot elements; and E_∞ is the long-time equilibrium modulus. Schapery (1962) then proceeded to select n decades of time over the transition range of the experimental data (until equilibrium is reached) and observed that, for small times, the glassy modulus, where the writhing thermal motions essentially cease, must be recovered. This led him to collocate experimental data at n-1 points. In this manner, a material property could be characterized by replacing n-1 time values, with a decade interval, into the appropriate Eq. (15) and equating to respective experimental points. Another way of obtaining the coefficients for Eq. (15) is by nonlinear regression curve fittings of experimental data.

Force-Deformation Tests

Table 4–5 shows a summary of force-deformation tests used to measure crispness of food products. The type of tests commonly used to determine crispness of foods are puncture and snapping.

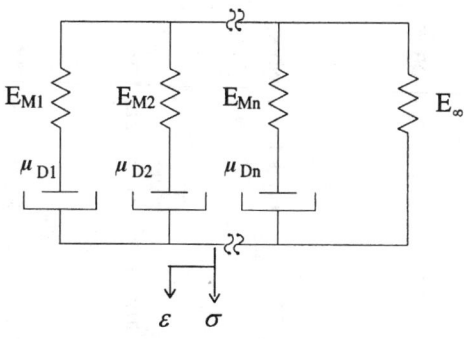

Figure 4–7 Generalized Maxwell Model

Table 4–5 A Summary of Force-Deformation Tests Used to Measure Crispness

Test	Parameter Measured	Food Product	Method	Best Correlation with Crispness	Reference
Snapping	peak force, slope, deformation to fracture	cookies and crackers	• triple bean snap • cross head speed: 5,10,20,25 cm/min	slope	Vickers and Christensen (1980)
Puncture	initial slope, peak force	potato chips	• 0.47 cm diam flat-ended cylindrical probe • cross head speed: 2 cm/min	inconclusive	Katz and Labuza (1981)
TPA	peak force, work, cohesiveness	popcorn	• shear compression plate with four metal blades • cross head speed: 2 cm/min	cohesiveness	Katz and Labuza (1981)
Crushing	work, peak force	biscuits	• constant loading rate	ratio of work done during fracture to the total work	Povey and Harden (1981)
Crushing	peak force, work	dry crisp foods	• Krammer shear	all had large negative correlation with crispness	Seymour (1985)

Test	Parameters	Product	Method	Results	Reference
Snapping	peak force, slope, work	potato chips	• two horizontal bars, 100 mm long, connected to aluminum bars 25 mm high • cross head speed: 50 cm/min	all had small negative correlation with crispness	Vickers (1987)
Puncture	peak force, work	tortilla chips		peak force	Moreira et al. (1995)
Puncture	peak force, slope, work	tortilla chips	• 0.203 cm diam flat-ended cylindrical probe • cross head speed: 10 mm/s	peak force	Lujan and Moreira (1997)
Snapping	peak force, work, no. peak forces	baked tortilla chips	• 0.635 diam ball-ended cylindrical probe • cross head speed: 2 cm/min	no. of peak forces	Quintero-Fuentes (1997)
SR	relaxation time modulus of elasticity	tortilla chips	• cantilever bending • cross head speed: 7 mm/s	inconclusive	Chen (1995)
Snapping	flexural	french fries	• 3% deformation • three-point bending	stress	Lima and Singh (1995)

Note: SR, Stress relaxation.

The *puncture test* measures the force required to push a probe into a food. Using the Universal Testing Machine, the test consists of penetration of the probe into the food, causing crushing. The parameters generally evaluated during the tests are shown in Figure 4–8. *Fracturability*, or peak force, is the maximum force required to fracture the product; *modulus of deformation* is defined by the slope of the force-deformation curve from the beginning of the compression to the fracturability point; and *work* is defined by the area under the curve from the beginning of the compression to the fracturability point.

The *snapping and bending tests* are usually applied to foods that have a rectangular or circular cross-sectional area (Steffe, 1992). Figure 4–9 shows the most common test configurations. The three-point beam test consists of placing a food on the two supports and compressing it at the midway point with a third bar. The food is bent until it snaps. With the cantilever beam, the product is held at one end and is allowed to bend freely throughout its length. Snappy foods such as snack products have a rigid, unbending texture that breaks suddenly once the rupture force has been reached. For a uniform product with a rectangular or a circular

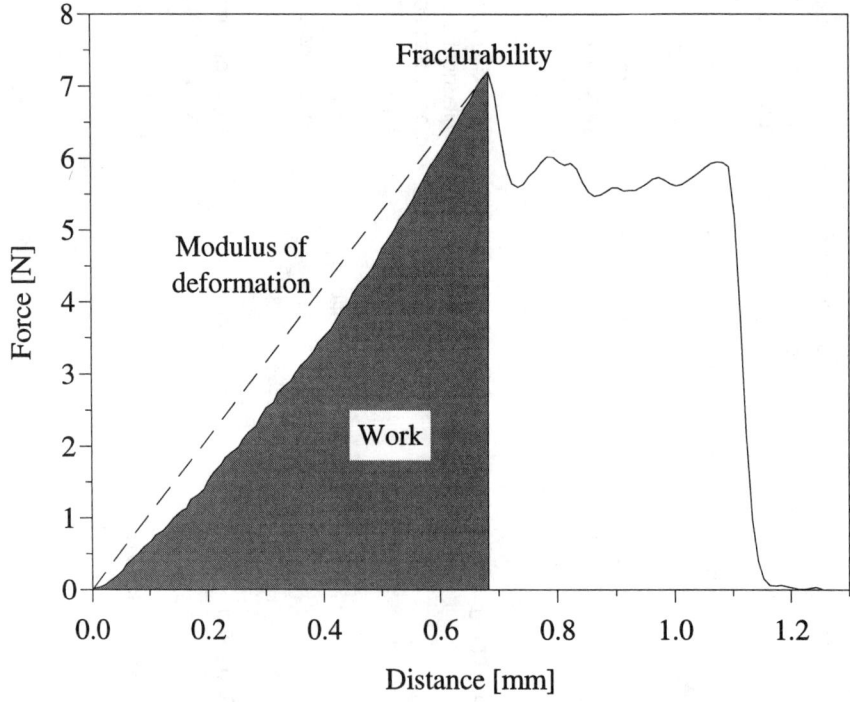

Figure 4–8 Force-Deformation Curve

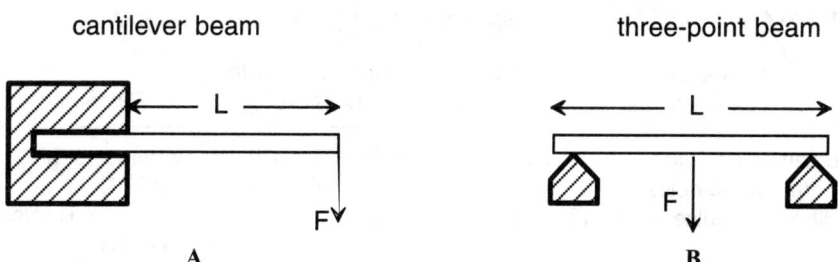

Figure 4–9 Snapping and Bending Tests. **A**, Cantilever Beam Test; **B**, Three-point Beam Bending Test

cross-sectional area, the snapping force can be calculated as (Bruns and Bourne, 1975):

$$\text{rectangular} \quad F = \frac{2}{3}\sigma_c b \frac{d^2}{L} \qquad \text{[16]}$$

$$\text{circular} \quad F = \sigma_c \pi \frac{R^3}{L} \qquad \text{[17]}$$

where d is the height of the beam, L the length of the beam, R the radius of the beam, b the width of the beam, *and* σ_c the failure stress.

As has been shown, mechanical tests have been used to correlate crispness to a physical parameter in a force-distance deformation curve. These objective methods, including force deformation and similar tests using various measurement systems, are quick and easy to use but have not typically produced correlation with sensory evaluations. Another problem is that many foods are too small or have irregular sizes and shapes (Vickers, 1988).

Acoustic Measurements

Researchers have studied the instrumental acoustic measures of crispness. Some of their results are summarized in Table 4–6. These data show that measures of loudness or total amount of sound tend to correlate positively with crispness.

Although sensory research also indicates that acoustic measurements are very important for crispness (Vickers and Wasserman, 1980), sensory study also shows that force-deformation sensations contribute to crispness (Edminister and Vickers, 1985). Some combinations of the two types show better results for predicting crispness (Vickers, 1988).

Table 4–6 A Summary of Acoustic Tests Used To Measure Crispness

Parameter Measured	Food Product	Correlation with Crispness	Reference
Extent of deviation from white noise	crisp bread	0.74 (correlated to brittleness)	Anderson et al. (1973)
Ultrasonic pulse echo	biscuits	0.52	Povey and Harden (1981)
Equivalent sound level (average sound energy)	different foods	0.7	Mohamed et al. (1982)
Log (no. of sound occurrences) X (sound amplitude)	different foods	0.47 (correlated with the log of crispness)	Edminister and Vickers (1985)
Log (no. of sound occurrences) X (sound amplitude)	dry crispy foods	0.66 (correlated with the log of crispness)	Edminister and Vickers (1985)
Log of duration of sound	wet crisp foods	0.51 (correlated with the log of crispness)	Edminister and Vickers (1985)
Mean sound pressure in 0.5–3.3 kHz	potato chips	0.91 and 0.80	Seymour (1985)
Mean sound pressure level 1.2–1.98 kHz	crunch twist	0.89	Seymour (1985)
Mean sound pressure level 1.91–2.59 kHz	potato chips	0.85	Seymour (1985)
Acoustic intensity in 0.5–3.3 kHz	potato chips	0.92	Seymour (1985)
Number of sound occurrences	potato chips	0.92	Vickers (1987)

Source: Adapted with permission from Z.M. Vickers, Instrumental Acoustical Measurements of Crispness in Foods, *Journal of Texture Studies,* Vol. 16, pp. 153–167, © 1988, Food and Nutrition Press, Inc.

Vickers (1987) used a combination of sensory, acoustic, and force-deformation methods to measure the crispness of potato chips and found that crispness is inversely related to force-deformation parameters and positively related to acoustic parameters, as:

$$Crisp = -15.6 + 5.35NP + 133MHP - 6.21Peak \qquad [18]$$

where NP is the number of sound occurrences, MHP the mean height of peaks, and *Peak* the peak of the force-deformation curve. In a follow-up study, Vickers (1988) suggested that toughness or hardness could be correlated to auditory sensations such as crispness and crunchiness. Seymour (1985) found that the sensory measure of hardness was inversely related for potato chips, indicating that other textural sensations, such as hardness and/or toughness, seem to counteract crispness.

Yield

The mass of final product per mass of 100 kg of raw product constitutes the total yield of a fried product. The processing operations that affect the yield of french fries, for example, are peeling, trimming, cutting, and frying. Peeling and trimming losses usually fall within the range of 15–40%; removal of slivers and nubbins may lower the yield by another 10%. Further losses result from frying, due to water loss, which more than offsets the weight of oil absorbed. The final yield of finished french fries obtained from 100 kg of unpeeled raw potato is in the range of 35–45 kg (Lesińska and Leszczyński, 1989).

The yield of fried products depends on the quality of the raw materials. For example, the size and shape of raw potatoes affect trimming and peeling losses; high specific gravity of potato tubers (dry matter content) result in an increased yield of chips (Lulai and Orr, 1979).

Processing, frying, packaging, handling, and transporting frequently lead to physical damages in the product. These damages are caused by excessive stresses (compression, tension, shear) occurring during processing and storage and resulting in yield reduction. Quintero-Fuentes (1997) used a tumbler to measure the breakage susceptibility of baked tortilla chips. Chips that broke in small pieces after the test were considered more susceptible to breakage than those that broke into larger pieces.

Breakage of products during storage can be reduced by using stiff packaging materials, making the packaging plump by flushing with gases, and avoiding crunching during transportation (Matz, 1993).

Shelf-Life Flavor Stability

In the food industry, the most important test of product quality is shelf stability, i.e., how the product tastes and smells after few months of storage.

Rancidity resulting from oxidation of lipids is a primary concern during storage of fried products. The amount of fat a food must have to develop oxidation off flavor depends not only on the type of frying oil but also on the type of food and storage conditions. Snack products should be stored at temperature around 21°C (Paradis, 1993).

Initial studies (Warner et al., 1974) on flavor of potato chips show that pentane correlated with flavor scores. Samples of potato chips having at least 0.1 ppm of pentane was an indication that the oil was rancid. Pentane formation was lower in samples with 3.8% w.b. moisture content in comparison with the chips with 0.95% moisture.

It is well known that lipid oxidation is affected by unsaturation, surface area, pro-oxidants, antioxidants, oxygen, light, and temperature (Chapter 3). These apply not to lipids alone but also to lipids in foods. Fritscher (1994) discussed that in addition to those variables, lipid distribution in foods also affects the rate of lipid oxidation. In food with a continuous lipid phase (most high-fat foods), the rate of oxidation is similar to that of the same lipids; however, lipid type and surface area affect the rate of oxidation. In low-fat foods, where the lipids are generally dispersed in the polar components, the rate of oxidation is affected by water activity, the ratio of free to bound lipids, and pH.

At both very high and very low water activities, lipid oxidation rates are high, compared with the rate at intermediate water activities; reaction rates within a polymer system should be higher when the system is above the glass transition temperature (T_g) than when below this temperature, due to an increase in free volume associated with the system (Nelson and Labuza; 1992). Diffusion of oxygen (which causes lipid oxidation) in glassy and in amorphous matrices above T_g is affected by the food physical state; i.e., if the polymers are in the rubbery state the lipids are free to move and react with oxygen; however, when they are in the glassy state, the lipids are encapsulated and will not oxidize (Roos and Graf, 1995).

In many noncontinuous lipid phase foods, a portion of the lipids oxidize rapidly and the other portion either slowly or not at all. Fritscher (1994) suggested that the ratio between the oxidable and stable lipids fractions can be manipulated by varying formulation and processing conditions thus reducing the labile lipids. Generally, the lowest rate of lipid oxidation in foods with noncontinuous lipid phase occurs near pH 7. As the pH decreases, more transition metals are solubilized; the lipid encapsulated may be weakened; the rate of nonenzymatic browning reaction (which produces antioxidant) is low at lower pH; and less oxygen may be depleted by other reactions.

Flavor stability can be controlled by correct choice of packaging materials and design. In the case of fried snack foods, opaque and gas-flushable package materials are important to minimize oxidation. Generally, these packages are flushed with nitrogen (having less than 2% oxygen headspace) and have low permeability.

In the industry, shelf-life stability monitoring requires analytical tests and sensory taste panels. The analytical tests assess: (1) oil quality (see Chapter 3); (2) package performance—including moisture vapor transmission rate, oxygen transfer rates, and seal quality; and (3) product quality attributes—including moisture content, texture, and rancidity development.

A number of analytical methods exists for measuring lipid oxidation in food. These methods can be classified according to what they measure: (1) oxygen, (2) unsaturated fatty acids, and (3) peroxides. Methods (1) and (2) are unreliable for measuring the early stage of lipid oxidation in low-fat foods. Method (3) yields reliable results if the food has at least 10% oil content, is stored under controlled conditions, and the extraction is carried out with purified solvents under conditions that will not affect the peroxide values (Fritscher, 1994).

Nutrition

Manufacturing of fried products, such as snack foods, requires complex mechanical and thermal operations that can result in enormous losses of nutritional components of the material. On the other hand, when foods are fried, they become fat-enriched, thus increasing their energy content. This fat can also help the transport of liposoluble components such as unsaturated fatty acids and liposoluble vitamins (Moreiras-Varela et al., 1988).

In addition, lipids influence flavor through their effect on flavor perception (mouthfeel, taste, and aroma), flavor stability, and flavor generation. A serious flavor defect of low-fat food is its quick disappearance in the mouth (Roos, 1997). A reduction in fat content will result in higher flavor loss during processing and storage, due to the increase in flavor volatility (Roos and Graf, 1995); it can also result in a decrease of the chemical stability of the flavor.

The changes that occur in the composition (increasing oil content) of a food during frying are affected by frying oil composition, texture, size and shape of the food, and frying operating conditions (temperature, residence time, etc.). As most of the water is evaporated from the food, the product surface temperature rises to that of the frying oil, causing the food to turn golden brown (Maillard-type reactions) and to form a porous and crispy surface (a crust). These changes make the product more palatable to the consumer.

Moreiras-Varela et al. (1988) looked to factors such as palatability, digestibility, and metabolic utilization to evaluate the effects of frying on the nutritive values of foods. No palatability changes were detected by a trained panel in potato and fish fried in either fresh and used frying oils (Varela et al., 1983). The variations in digestibility and metabolic utilization of protein in fried foods (fish, pork, beef) were found to be minimal (Table 4–7).

Potato chips contain 50–70% of the nutrients present in raw potatoes. In french fries, losses of total nitrogen are on the order of 29–43%, nonprotein nitrogen 20–35%, and amino acids 45% (Lesińska and Leszczyński, 1989).

As long as the oil used to frying the foods is not abused, fried foods pose no danger to health. Frying makes food more palatable, it is fast, and fried products keep better. The negative aspects of frying are due to overheating. If frying is done

Table 4–7 Protein Metabolic Utilization of Raw and Fried Foods

Product	Biologic Value	Net Protein Utilization
Fish		
raw	0.67	0.63
fried	0.66	0.64
Pork		
raw	0.78	0.72
fried	0.80	0.73
Meatballs		
raw	0.72	0.65
fried	0.68	0.69

Source: Adapted with permission from O. Moreiras-Varela, B. Ruiz-Roso, and G. Varela, Effects of Frying on the Nutritive Value of Food, in *Frying of Foods: Principles, Changes, New Approaches*, G. Varela, A.E. Bender, and I.D. Morton, eds., p. 100, © 1988, VCH Publishers.

gently, frying does not produce toxic agents that depress intake and growth (Cuesta el al., 1988).

The trend today is to produce low-fat fried products with the same taste and texture of the regular fried product. Health-conscious consumers are looking for more nutritious products. Government regulations and labeling laws will improve the nutritious aspects of most fried products, having more impact on the snack food industry.

NOTATIONS

A   area $[m^2]$ - Eq.(9)
b   width of beam [m] - Eq.(16)
d   height of beam [m]- Eq.(16)
d_f specific gravity of oil $[kg/m^3]$ - Eq.(6)
dh   decrease in height [m] - Eq.(9)
E   Young's modulus [Pa] - Eq.(8)
E_e equilibrium modulus [Pa] - Eq.(14)
E_m elastic modulus [Pa] - Eq.(13)
F   force, or snapping force [N] - Eq.(9), (16), (17)
H(t)   unit step input - Eq.(10)
h_o initial height [m] - Eq.(9)
L   length of beam [m] - Eq.(16) and (17)
MC_{db} moisture content [% d.b.] - Eq.(2)
MC_{wb} moisture content [% w.b.] - Eq.(1)

MHP mean height of peaks - Eq.(18)

n_{20} refractive index of solution at 20°C - Eq.(7)

n_f refractive index of oil - Eq.(6)

NP number of peaks or sound occurrences - Eq.(18)

n_s refractive index of solvent - Eq.(6)

OC_{db} oil content [% d.b.] - Eq.(6)

OC_{wb} oil content [% w.b.] - Eq.(5)

Peak peak of the force-deformation curve - Eq.(18)

R radius of beam [m] - Eq.(16)

V_{sol} volume of solvent [mL] - Eq.(6)

W_s weight of dried sample [kg] - Eq.(6) and (7)

W_{sol} weight of solvent [kg] - Eq.(7)

ε strain [ ] - Eq.(8)

μ_m viscosity [Pa.s] - Eq.(13)

σ stress [Pa] - Eq.(8)

σ_c failure stress [Pa] - Eq.(16) and (17)

REFERENCES

AACC, 1986. *Approved methods of the American Association of Cereal Chemists*. St. Paul, MN: AACC.

Anderson, Y.; Drake, B.; Granquist, A.; Halldin, L.; Johansson, B.; Pangborn, R.M.; and Akesson, C. 1973. Fracture force, hardness and brittleness in crisp bread with generalized regression analysis approach to instrumental-sensory comparison. *J. Texture Studies*, 4:119–123.

ASAE. 1986. *Standards for moisture measurement—meat and meat products*, ed. Hahn and Rosentreter. St. Joseph, MI: The Society for Engineering in Agriculture.

Baumann, B., and Escher, F. 1995. Mass and heat transfer during deep-fat frying of potato slices. I. Rate of drying and oil uptake. *Lebensm-Wiss U-Technol*, 28:395–403.

Bett, K.L., and Dionigi, C.P. 1997. Detecting seafood off-flavors: limitations of sensory evaluation. *Food Technol*, 51(8):70–76.

Bourne, M.C. 1974. Texture changes in ripening peaches. *J Can Inst Food Sci Technol*. 7:11–15.

Bourne, M.C. 1982. *Food texture and viscosity*. New York: Academic Press.

Brescia, L., and Moreira, R.G. 1997. Modeling and control of a continuous frying process: I. Dynamic analysis and system identification. *Trans Chem Eng*, 75(c):3–11.

Brooker, D.B; Bakker-Arkema, F.W.; and Hall, C.W. 1992. *Drying and storage of grains and oilseeds*. New York: Van Nostrand Reinhold.

Bruns, A.J., and Bourne, M.C. 1975. Effects of sample dimensions on the snapping force of crisp foods. Experimental verification of mathematical model. *J Texture Studies*, 6:445–458.

Chen, Y. 1995. *Texture analysis of tortilla chips*. Unpublished paper. Food Engineering Laboratory. College Station, TX: Texas A&M University.

Christensen, C.M., and Vickers, Z.M. 1981. Relationships of chewing sounds to judgment of food crispness. *J Food Sci*, 46:574.

Cuesta, G.; Sánchez-Muniz, J.; and Varela, G. 1988. Nutritive value of frying fats. In *Frying of foods: principles, changes, new approaches*, ed. Varela, Bender, and Morton. New York: VCH Publishers.

Diaz, A.; Totte, A.; Giroux, F.; Reynes, M.; and Raoult-Wack, A.L. 1996. Deep-fat frying of plantain. I. Characterization of control parameters. *Lebensm-Wiss U-Technol*, 29:489–497.

Dodd, G.H.; Bartlett, P.N.; and Gardner, J.W. 1992. Odours—the stimulus for an electronic nose. In *Sensory and sensory systems for an electronic nose*, ed. Gardner and Bertlett. NATO ASI Series. Kluwer Academic Publishers.

Dudley, P.R. 1993. *The effect of processing parameters on oil content of corn tortilla chips*. M.S. Thesis. College Station, TX: Texas A&M University.

Edminister, J.A., and Vickers, Z.M. 1985. Instrumental acoustical measurement of crispness in foods. *J Food Texture Studies*, 16:153–160.

Food Technology, 51(6):42–45.1997. Laboratory instrument preview.

Francis, F.J., and Clydesdale, F.M. 1975. Food colorimetry: theory and applications. New York: Van Nostrand Reinhold.

Fritscher, C.W. 1994. Lipid oxidation—the other dimensions. *INFORM*, 5(4):423–436.

Gardner, J.W., and Bartlett, P.N. 1992. Sensory and sensory systems for an electronic nose, ed. Gardner and Bertlett. NATO ASI Series. Kluwer Academic Publishers.

Gamble, M.H., and Rice, P. 1988. The effect of slice thickness on potato crisp yield and composition. *J Food Eng*, 8:31–46.

Jowitt, R., and Mohamed, A.A.A. 1980. An improved instrument for studying crispness in foods. In *Food process engineering*, Vol. 1, ed. Linko, Malkki, Olkku, and Larinkari. New York: Applied Science Publishers.

Kärki, M.K.; Roos, Y.H.; and Tuorila, H. 1994. *Water plasticization of crispy snack foods*. Paper presented at the Annual Meeting of Institute of Food Technologists, Atlanta, GA.

Katz, E.E., and Labuza, T.P. 1981. Effect of water activity on the sensory crispness and mechanical deformation of snack food products. *J. Food Sci*, 46:403–409.

Lamberg, I.; Haltrom, B.; and Olsson, H. 1990. Fat uptake in a potato drying/frying process. *Lebensm-Wiss U-Technol*, 29:295–300.

Lee, J.K. 1991. *The effects of processing conditions and maize varieties on physicochemical characteristics of tortilla chips*. PhD diss. Department of Crop and Soil Science. College Station. TX: Texas A&M University.

Lima, M.I., and Singh, R.P. 1995. *Measurement of textural properties of french fries*. IFT Annual Meeting. Anaheim, CA.

Lesińska, G., and Leszczyński, W. 1989. *Potato science and technology*. New York: Elsevier Science.

Lujan, F.J., and Moreira, R.G. 1997. Effects of different drying processes on oil absorption and microstructure of tortilla chips. *Cereal Chemistry*, 74(3):216–223.

Lulai, C.E., and Orr, P.H. 1979. Influence of potato specific gravity on yield and oil content of chips. *Am Potato J*, 56:379–390.

Matz, S.A. 1993. *Snack food technology*. New York: Van Nostrand Reinhold/AVI.

Mazza, G., and Qi, H. 1992. Effect of after-cooking darkening inhibitors on stability of frying oil and quality of french fries. *JAOCS*, 69(9):847–853.

Mohamed, A.A.A.; Jowitt, R.; and Brennan, J.G. 1982. Instrumental and sensory evaluation of crispness in friable foods. *J Food Eng*, 1:55–63.

Mohsenin, N.N. 1984. *Physical properties of plant and animal materials*. New York: Gordon and Breach Science Publishers.

Moreira, R.G.; Palau, J.; Sweat, V.; and Sun, X. 1995. Thermal and physical properties of tortilla chips as a function of frying time. *J Food Process Preserv*, 19:175–189.

Moreiras-Varela, O.; Ruiz-Roso, B.; and Varela, G. 1988. Effects of frying on the nutritive value of food. In *Frying of foods: principles, changes, new approaches*, ed. Varela, Bender, and Morton. New York: VCH Publishers.

Nelson, K.A., and Labuza, T.P. 1992. Relationship between water and lipid oxidation rates: water activity and glass transition theory. *Symp Ser Am Chem Soc*. 93–103.

Nelson, K.A., and Labuza, T.P. 1993. Glass transition theory and the texture of cereal foods. In *The Glassy State of Foods*, ed. Blanshard and Lillford. Lougborough, England: Nottingham University Press.

Osborne, B.G.; Barret, G.M.; Cauvain, S.P.; and Fearn, T. 1984. The determination of protein, fat and moisture in bread by near infrared reflectance spectroscopy. *J Sci Food Agr*, 35:940–945.

Paradis, A. 1993. Nitrogen in total quality for snack food. *INFORM*, 4(12):1378–1382.

Pinthus, E.J.; Pnina, W.; and Saguy, I.S. 1993. Criterion for oil uptake during deep-fat frying. *J Food Sci*, 58(1):204–205, 222.

Povey, M.J.W., and Harden, C.A. 1981. An application of the ultrasonic pulse echo technique to the measurement of crispness of biscuits. *Food Technol*, 16:167–175.

Quintero-Fuentes, X. 1997. *A method to determine ingredient functionality in baked tortilla chips*. Masters thesis. College Station, TX: Texas A&M University.

Richardson, J.; Aradteh, N.; Sultanbawa, Y.; and Durance, T. 1997. *Sensory and instrumental measurement of potato chip crispness, crunchiness, and strength*. Paper presented at the Annual Meeting of Institute of Food Technologists, Orlando, FL.

Roos, K.B. 1997. How lipids influence food flavor. *Food Technol*, 51(1):60–62.

Roos, K.B., and Graf, E. 1995. Nonequilibrium partition model for predicting flavor retention in microwave and convection heated foods. *J Agric Food Chem*, 43:2204–2211.

Schapery, R.A. 1962. *Approximate methods of transform inversion for viscoelastic stress analysis*. Proc US Natl Cong Appl Mech, 4(2):1075–1085.

Schapery, R.A. 1974. Viscoelastic behavior and analysis of composite materials. In *Mechanics of composite materials*, ed. G.P. Sendeckyj, 85–168. New York: Academic Press.

Schonauer, S., and Moreira, R.G. 1995. *Development of a fixed-GPC controller for a food extruder based on PQA. I: System identification*. Transactions of the Institute of Chemical Engineers, 73(c):189–199.

Seymour, S.K. 1985. *Studies of the relationship between the mechanical, acoustical and sensory properties in low moisture food products*. PhD diss. North Carolina State University.

Steffe, J.F. 1992. *Rheological methods in food process engineering*. East Lansing, MI: Freeman Press.

Szczesniak, A.S. 1963. Classification of textural properties. *J Food Sci*, 28(1):385–392.

Szczesniak, A.S. 1988. The meaning of crispness as a textural characteristic. *J Texture Studies*, 19:51–59.

Tan, P.; Hung, Y.C.; and McWatters, K.H. 1995. Akara quality as affected by frying/reheating conditions. *J Food Sci*, 60(6):1301–1306.

Tecator Manual. 1983. Fat extraction on feeds with the Soxtec system HT—The influence of sample preparation and extraction media. Application Note 67/83. Silver Spring, MD: Perstorp, Inc.

Tesch, R.; Normad, M.; and Peleg, M. 1995. On the apparent fractal dimension of soundburst in acoustic signatures of two crunchy foods. *J Texture Studies*, 26:685–694.

Ufheil, G., and Escher, E. 1996. Dynamics of oil uptake during deep-fat frying of potato slices. *Lebensm-Wiss U-Technol*, 29:640–644.

USDA. 1997. *United States standards for grades of processing fruits and vegetables*. Washington, DC: U.S. Department of Agriculture.

Varela, G.; Moreiras-Varela, O. and Ruiz-Roso, B. 1983. Utilización de algunos aceites en frituras repetidas, Cambios en las grasa y análisis sensorial de los alimentos fritos. *Grasa y Aceites*, 34:101–106.

Varmam, A.H., and Sutherland, J.P. 1995. *Meat and meat products—technology, chemistry and microbiology*. New York: Chapman and Hall.

Vickers, Z.M. 1987. Sensory, acoustic and force-deformation measurements of crispness and crunchiness of food sounds. *J Texture Studies*, 16:85–95.

Vickers, Z.M. 1988. Instrumental acoustical measurements of crispness in foods. *J Texture Studies*, 16:153–167.

Vickers, Z.M., and Bourne, M.C. 1976. A psychoacoustic theory of crispness. *J Food Sci*, 41(5):1158–1164.

Vickers, Z.M., and Christensen, C.M. 1980. Relationship between sensory crispness and other sensory instrumental parameters. *J Texture Studies*, 11:291–307.

Vickers, Z.M., and Wasserman, S.S. 1980. Sensory qualities of food sounds based on individual perceptions. *J Texture Studies*, 10:319–332.

Warner, K.; Evans, C.D.; List, G.R.; Boundy, B.K.; and Kwolek, W.F. 1974. Pentane formation and rancidity in vegetable oils and potato chips. *J Food Sci*, 39:761–765.

Wehling, R.L., and Pierce, M.M. 1989. Application of near infrared reflectance spectroscopy to determination of fat in cheddar cheese. *J Assoc Off Anal Chem*, 72(1):56–58.

Wehling, R.L.; Pierce, M.M.; and Froning, G.W. 1988. Determination of moisture, fat, and protein in spray-dried whole egg by near infrared reflectance spectroscopy. *J Food Sci*, 53(5):1356–1359.

Score Sheets for
Frozen Fried Potatoes and
Frozen Breaded Onion Rings

Score Sheet for Frozen Fried Potatoes

| Size and Kind of Container . . . . |
| Container Mark of Identification . . . . |
| Label . . . . |
| Net Weight (ounces) . . . . |
| Style and Cross-Sectional Dimension . . . . |
| Type . . . . |
| Length . . . . |

Factors			Score Points
Color	30	A	27–30
		A short	27–30
		B	24–26*
		SStd	0–23*
Uniformity of size and symmetry	20	A	18–20
		A short	18–20
		B	16–17*
		SStd	0–15*
Defects	20	A	18–20
		A short	18–20
		B	16–17*
		SStd	0–15*
Texture	30	A	27–30
		A short	27–30
		B	24–26*
		SStd	0–23*
Total Score	100		

| Flavor . . . . |
| Grade . . . . |

*Indicates limiting rule.

Source: Reprinted from *United States Standards for Grades of Frozen French Fried Potatoes and Frozen Onion Rings*, United States Department of Agriculture.

Score Sheet for Frozen Breaded Onion Rings

Size and Kind of Container			
Container Marks			
Sample			
Cases			
Label			
Net Weight (ounces)			
Type: ( ) French Fried: ( ) Raw Breaded			
Factors		**Score Points**	
Color	30	A B SStd	25–30 21–24 0–20*
Defects	40	A B SStd	34–40 28–33* 0–27*
Character	30	A B SStd	26–30 21–25 0–20*
Total score	100		
Similar varieties			
Flavor			
( ) good; ( ) reasonably good; ( ) off flavor			
Grade			

*Indicates limiting rule.

CHAPTER 5

Introductory Analysis of Frying Systems

In the previous chapters, the principles of oil degradation and product quality characteristics have been discussed. This chapter presents an introduction to the frying process of foods with relation to water loss and temperature changes during frying. The effect of oil temperature, residence time, fryer type, and product characteristics on moisture loss and oil absorption are also presented.

DEEP-FAT FRYING

The speed and efficiency of the frying process depend on the temperature and quality of the frying oil. The temperature of the oil is usually between 150°C and 190°C. Oil turnover time (mass of used oil/oil usage rate) is about 10 hours. It is important to understand what happens to the temperature, moisture, and oil content of the product during the frying process to determine safe temperatures and turnover times of the frying oil for a given fryer type.

Frying is a complex process involving numerous factors that depend on: (1) the process, (2) the frying oil, and (3) the food. For the *process*, these factors include temperature of the heated oil, the product residence time in the fryer, and the fryer used (batch or continuous). For the *oil*, they include chemical composition, the physical and physicochemical constants, and the presence of additives and contaminants that affect the frying process. Additives or contaminants can have a marked effect on the palatability, digestibility, and metabolic utilization of a fried food.

Finally, there are factors that are dependent on the *food*. Each food must be fried in a specific way with regard to the process itself and the frying oil. The food weight/frying oil volume and surface area/volume ratios are also important as they will determine the extent to which the oil penetrates the food. In addition, frying acts directly on the food's thermolabile components or indirectly on the interac-

111

tions that occur between the different nutrients in the food, and between those nutrients and the frying oil (Varela et al., 1988).

THE PROCESS

The frying process can be defined as cooking and drying of a product by contact with a hot medium—the frying oil. During the process, starch is gelatinized, tissues are softened (in the case of raw potatoes), and enzymes are partially inactivated. As moisture is lost during the process, a crust is formed at the product surface that is characterized by having very low moisture content, temperature above 100°C, porous structure, and crispy texture. This crust is about 1–2 mm thick for french fries but comprises the whole product in the case of potato and tortilla chips. The effect of frying conditions, fryer types, oil quality, and product characteristics on the moisture loss and oil content of fried products will be presented.

Water Loss During Frying

The drying behavior of a food material in a fryer depends on the physical characteristics of the product. Food materials are hygroscopic capillary-porous products in which the pores are partially filled with water and partially filled with an air/water-vapor mixture. During the frying process, the moisture evaporates at the product surface and leaves the product, due to the partial vapor pressure difference between the product and the frying oil. The rates at which a food product or a batch of food products loses moisture under different frying conditions and in different fryer types are important parameters in fryer design.

Effect of Oil Temperature and Residence Time on Water Loss

Examples of the drying behavior of potato chips and tortilla chips under different frying oil temperatures are shown in Figures 5–1 and 5–2, respectively.

The drying rate for potato chips is accelerated with oil temperature (Figure 5–1), being higher at a temperature rise from 150°C to 160°C than from 170°C to 180°C. It takes about 140 seconds for the product to reach a final moisture content of about 2 % (w.b.) when fried at 150°C in comparison with only 80 seconds when potato chips are fried at 180°C, i.e., a reduction in drying time of 60 seconds.

The frying of tortilla chips follows the same trend, with drying rate increasing at higher frying oil temperature (Figure 5–2). It requires only 40 seconds for tortilla chips to dry to 2.9 % (w.b.) when fried at 190°C, but it would require longer frying time when tortillas are fried at 130°C. This phenomenon will be much more pro-

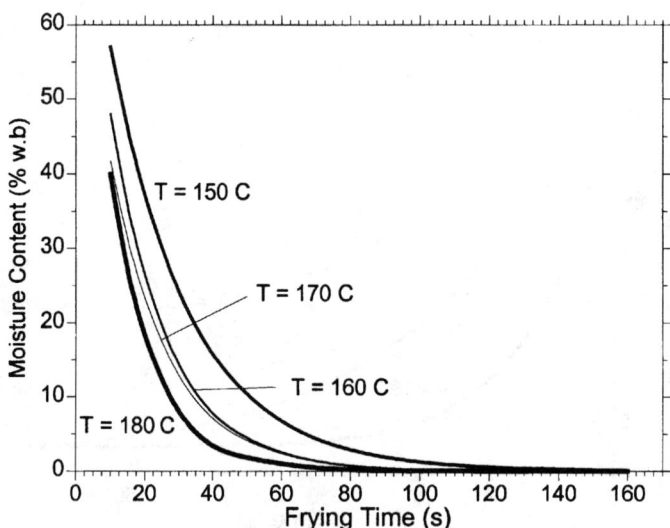

Figure 5–1 Drying Behavior of Potato Chips (1.2 mm thickness) with Oil Temperature at 160°C–180°C under Isothermal Conditions. *Source:* Adapted from *Lebensmittel-Wissenschaft und Technologie*, Vol. 28, B. Baumann and F. Escher, Mass and Heat Transfer During Deep-Fat Frying of Potato Slices, Rate of Drying and Oil Uptake, pp. 395–403, © 1995, by permission of the publisher Academic Press.

nounced in industrial operations, where the mass of product per volume of oil is larger. Large temperature variations within continuous fryers are very common in industrial operations. Therefore, at a given temperature, the drying time is longer in industrial fryers than at isothermal conditions.

Effect of Oil Temperature and Frying Time on Oil Uptake

During frying, the rapid drying is critical for ensuring the desirable structure and texture characteristics of the final product. However, the loss of water results in a substantial absorption of oil by the product. Table 5–1 presents oil content of several food products as a function of oil temperature and frying time. A lower oil temperature results in a lower oil content in the early stages of frying (for all products). Higher differences were observed between 145°C and 165°C than between 165°C and 185°C (for potato chips) and between 130°C and 160°C than between 160°C and 190°C (for tortilla chips). Higher oil temperatures lead to a faster crust formation, thus favoring the conditions for oil absorption.

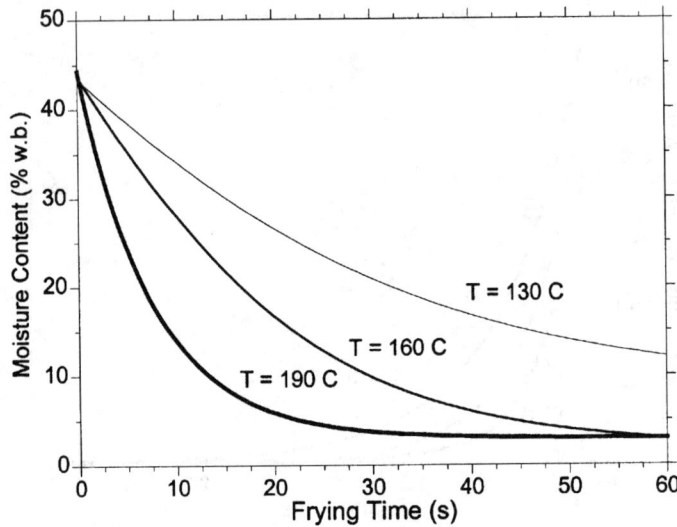

Figure 5–2 Drying Behavior of Tortilla Chips with Oil Temperature at 130°–190°C under Isothermal Conditions

Oil content is temperature-independent when potato chips are fried at a specific temperature range (Figure 5–3). As the moisture is reduced with frying time, the ratio of oil content to the amount of water or moisture removed (OC/MR) becomes independent of the oil temperature (Table 5–1), indicating that the oil content is not directly related to the oil temperature but to the remaining moisture present in the product (Gamble et al., 1987).

The effect of frying time on the quality attributes of fried frozen french fries was observed by Du Pont et al. (1992). Oil content increased by almost 50% with frying time up to 4 minutes, remained constant between 4 and 6 minutes, then increased after 9 minutes of frying (Table 5–2). The crust thickness increased by about 12% with frying time up to 4 minutes, remained constant up to 6 minutes of frying, then was unidentifiable at 9 minutes. Moisture content also showed to be slightly constant between 4 and 6 minutes of frying, indicating that the process reached a steady state under that condition. After 6 minutes of frying, disruption of crust-core structure caused an increase in oil uptake (20.8–23.8%). The product became harder and pliable as frying time increased.

Modulus of elasticity increased with frying time. Sensory attributes showed an increase in crispness of crust, mealiness of core, softer inside texture, and darker color, but also an increase in oiliness with frying time (Table 5–2).

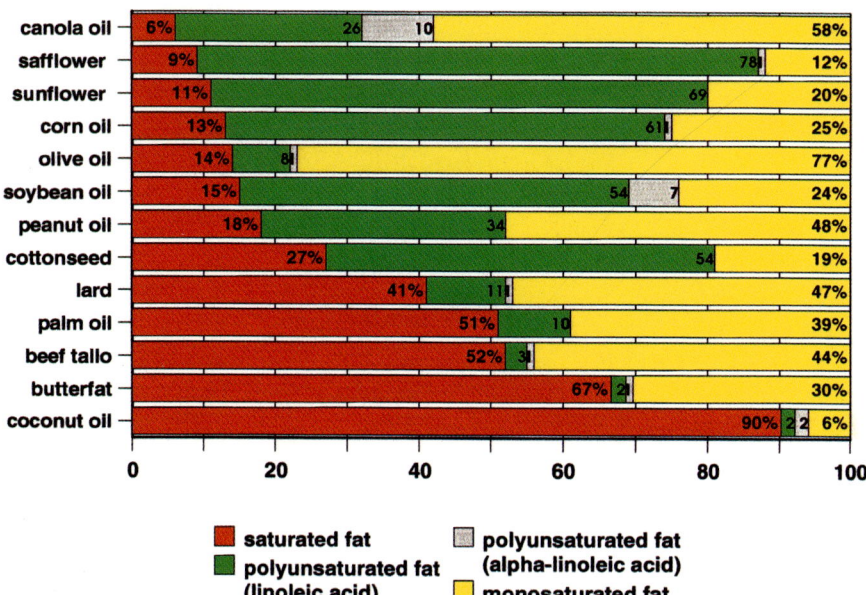

Color Plate 1 Comparison of Frying Oils. *Source:* Adapted from *Agricultural Handbook* No. 8-4 and Human Nutrition Information Service, 1979, United States Department of Agriculture.

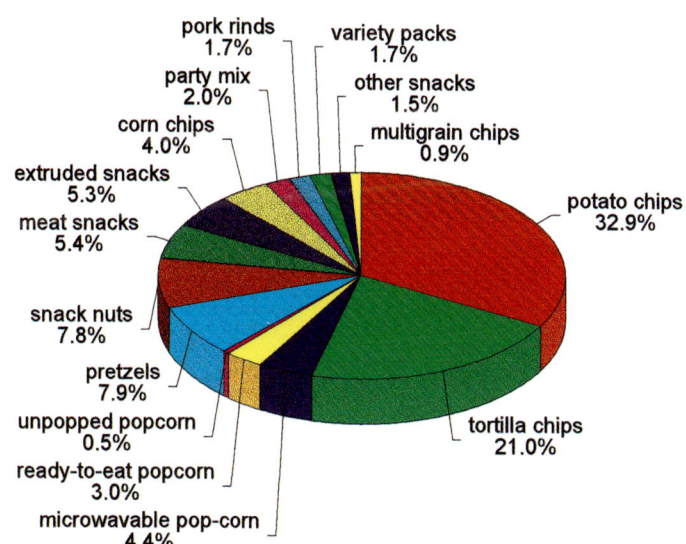

Color Plate 2 Snack Segment Share of Dollar Sales. *Source:* Adapted with permission from *State of the Snack Food Industry Report*, p. 3, © 1997, Snack Food Association.

Color Plate 3 Color Standards for Frozen French Fried Potatoes

Making the color evaluation:
(1) Classify the product where it best fits in the series of fry colors
   - No. 000 in the series has no shading of brown due to the frying process. A strip matching the No. 000 standard would be designated as being USDA No. 000 color.
   - Colors 00, 1, 2, 3, and 4 show color of increasing intensity.
(2) Designate the colors of a unit in terms of the color standard
   - A unit exactly matching the color of a standard would be designated as the color. For example, "USDA No. 3 Color."
   - A sample that does not exactly match might be designated as "Darker than No. 3, but not as dark as No. 4," or any appropriate manner.

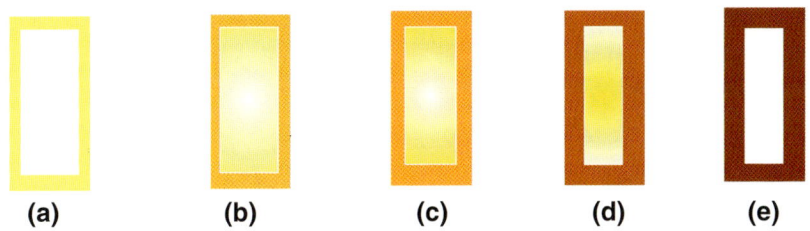

(a)   (b)   (c)   (d)   (e)

Color Plate 4 Quality of French Fries When Fried in Degraded Oil. **A**, break-in, **B**, fresh, **C**, optimum, **D**, degraded, **E**, runaway.

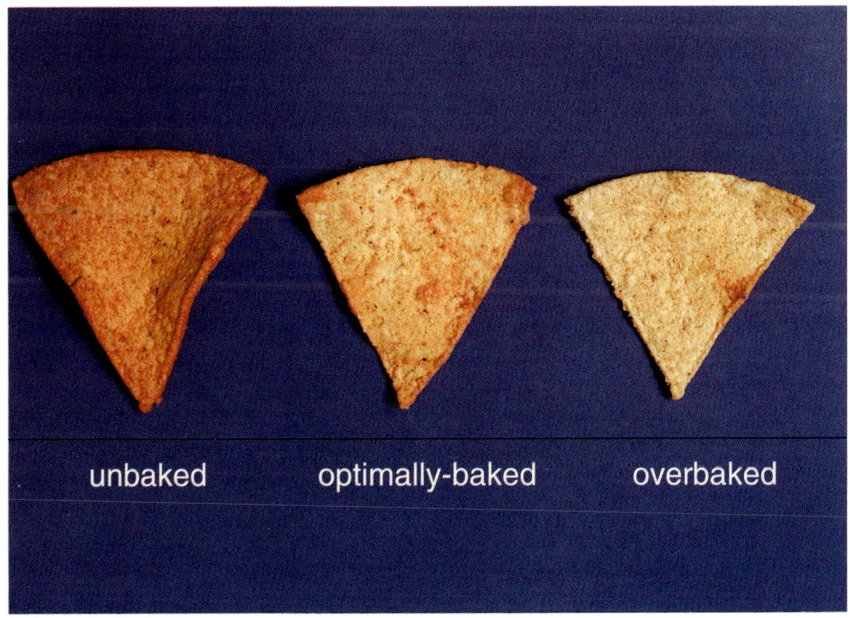

Color Plate 5 Oil Distribution in Tortilla Chips during Frying

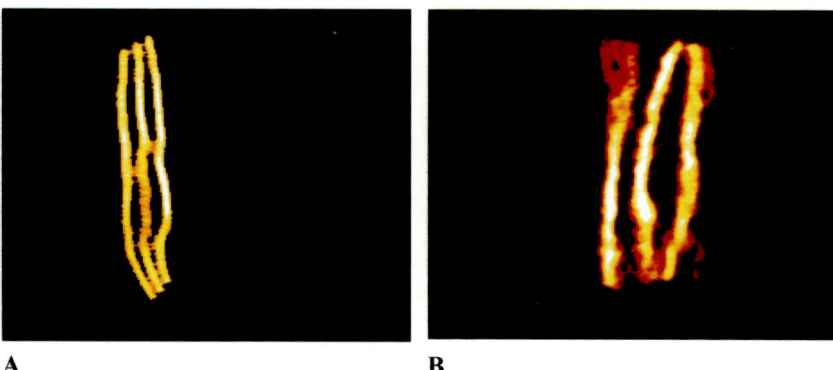

A   **B**

Color Plate 6 Magnetic Resonance Image of the Cross-Sectional Area of Three Tortilla Chips. **A**, Raw Tortilla Chips; **B**, Tortilla Chips Fried for 60 s. *Source:* Reprinted with permission from R.G. Moreira et al., Deep-Fat Frying of Tortilla Chips: An Engineering Approach, *Food Technology*, p. 147, © 1995, Institute of Food Technologists.

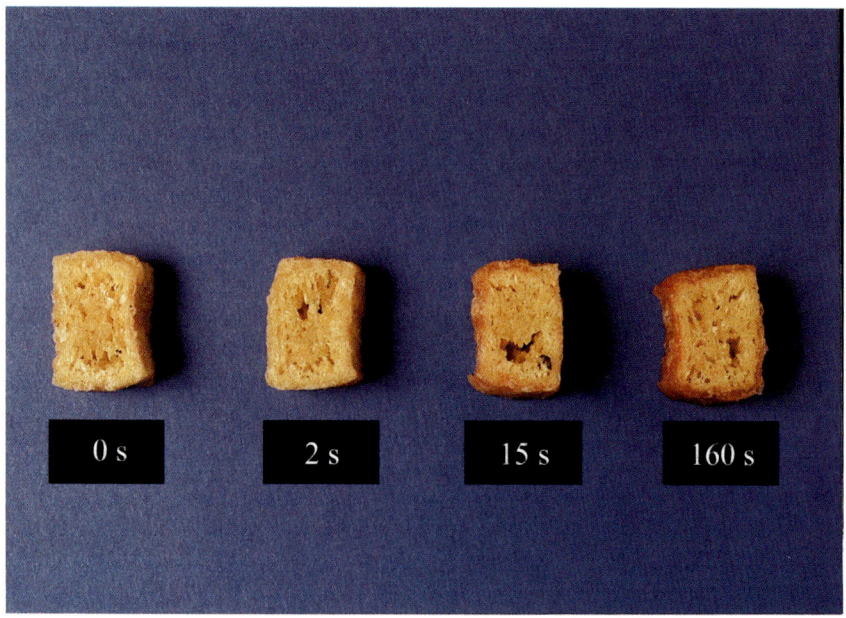

Color Plate 7 Oil Distribution in Fried Tortilla Cubes during Cooling. Fried for 5 min at 190°C.

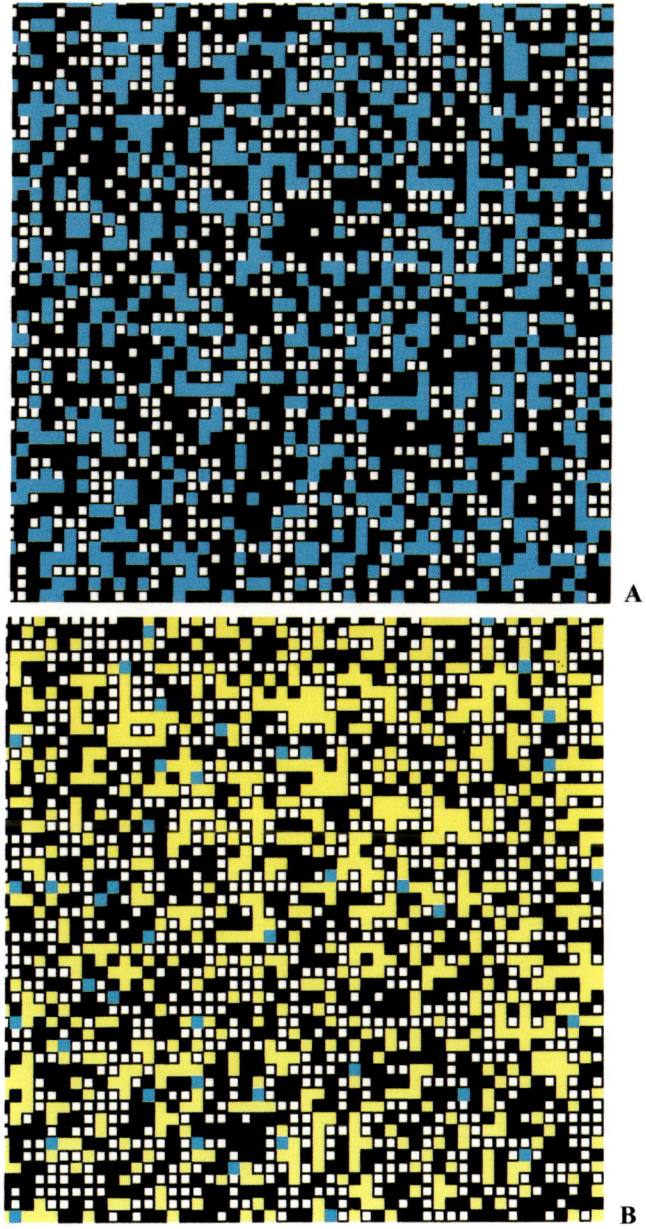

Color Plate 8 Midplane Perpendicular to Flow Direction Showing Composition Distribution in Tortilla Chips. **A,** before frying; **B,** after frying. *Source:* Reprinted with permission from R.G. Moreira et al., Deep-Fat Frying of Tortilla Chips: An Engineering Approach, *Food Technology*, p. 149, © 1995, Institute of Food Technologists.

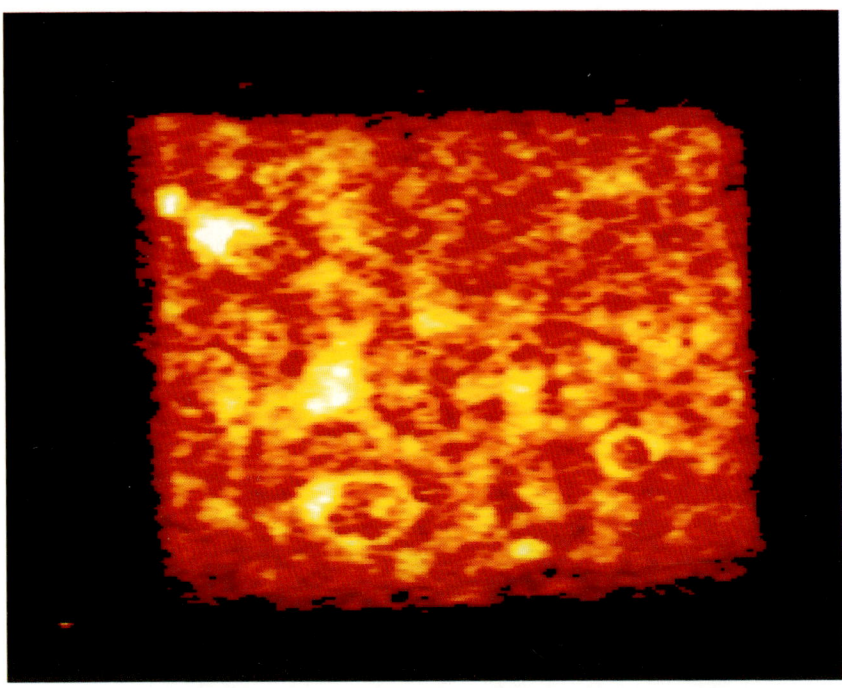

Color Plate 9 Magnetic Resonance Imaging of Tortilla Chips. *Source:* Reprinted with permission from R.G. Moreira et al., Deep-Fat Frying of Tortilla Chips: An Engineering Approach, *Food Technology*, p. 149, © 1995, Institute of Food Technologists.

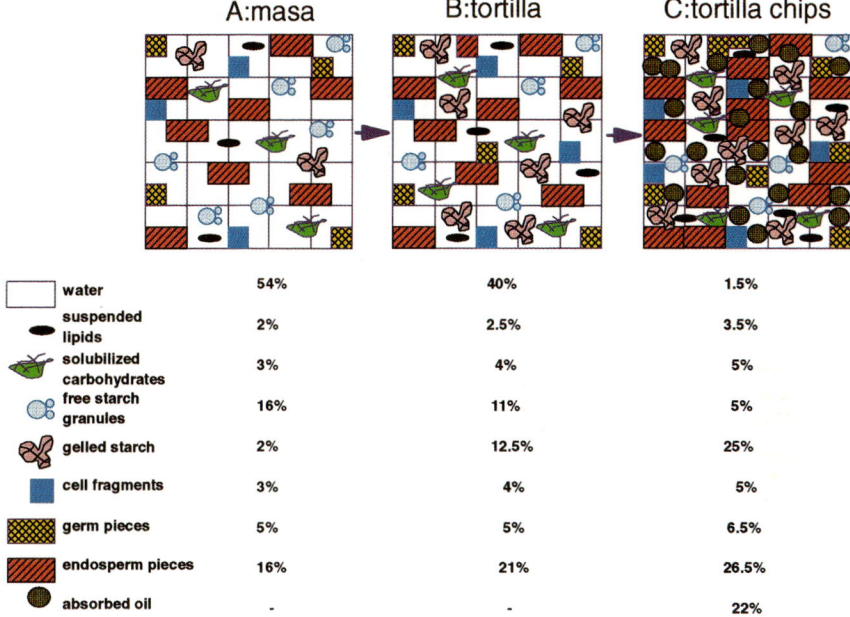

		A:masa	B:tortilla	C:tortilla chips
	water	54%	40%	1.5%
	suspended lipids	2%	2.5%	3.5%
	solubilized carbohydrates	3%	4%	5%
	free starch granules	16%	11%	5%
	gelled starch	2%	12.5%	25%
	cell fragments	3%	4%	5%
	germ pieces	5%	5%	6.5%
	endosperm pieces	16%	21%	26.5%
	absorbed oil	-	-	22%

Color Plate 10 Composition of **A**, Masa, **B**, Tortilla, and **C**, Tortilla Chips. *Source:* Adapted from American Association of Cereal Chemists Inc., 1992.

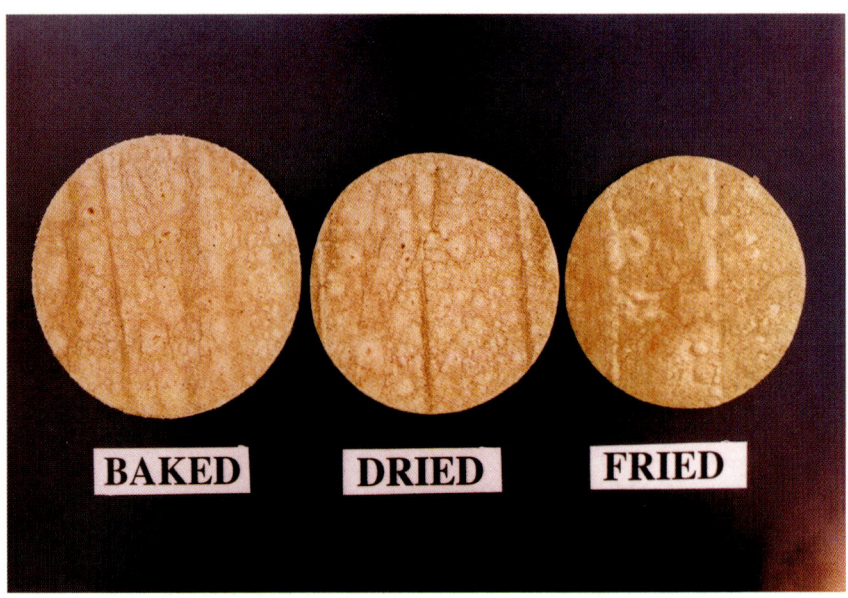

Color Plate 11 Convection-Oven-Baked, Impingement Air-Dried, and Fried Tortilla Chips. *Source:* Reprinted from *Lebensmittel-Wissenschaft und Technologie*, Vol. 30, No. 8, F.J. Lujan-Acosta and R.G. Moreira, Reduction of Oil in Tortilla Chips Using Impingement Drying, pp. 834–840, © 1997, by permisson of the publisher Academic Press.

Table 5–1 Effect of Frying Conditions on Oil Uptake of Several Products

Product	Frying Temperature (°C)	Frying Time (s)	Moisture Content (% w.b.)	Oil Content (% w.b.)	OC/MR
Potato chips[a]	145	60	66.5	3.0	0.05
(IMC = 80%	145	300	14.7	31.8	0.16
w.b.)	165	60	58.0	11.8	0.15
	165	300	4.9	35.9	0.15
	185	60	37.7	14.1	0.09
	185	300	0.7	36.2	0.14
Tortilla chips[b]	130	10	34.1	10.2	0.68
(IMC = 44 %	130	60	10.9	19.5	0.42
w.b.)	160	10	28.7	17.5	0.85
	160	60	4.5	25.1	0.48
	190	10	15.3	17.2	0.42
	190	60	2.9	26.4	0.49
French fries[c]	182	120	58.7	9.6	0.33
(IMC = 70%	182	135	50.0	12.1	0.24
w.b. par-	182	150	46.2	13.9	0.24
fried)	182	165	44.5	15.7	0.25

Note: OC/MR, oil content/moisture removed.

Source: (a) Gamble et al. (1987); (b) Moreira et al. (1997); (c) Zak and Holt (1973).

Effect of Isothermal and Nonisothermal Frying on Oil Uptake

Isothermal frying conditions are obtained by maintaining a very low ratio of mass of product to the frying oil. Generally, this situation is found when frying a single piece of product. Nonisothermal conditions are encountered in batch frying systems, typical of commercial operations. The temperature gradient within the frying oil during the process can affect the residence time of the product in the fryer and the quality of the final product.

Totte et al. (1996) fried plantain chips into three different frying systems, i.e, free-frying, forced-immersion, and stirring systems. In the free frying system, the chips introduced into the fryer floated freely in the oil bath. The forced-immersion system is characterized by placing the chips in single/multiple layers in the fryer basket and covering with a lid to keep the chips immersed in the oil during frying. The stirring system has a semi-helicoidal stirrer that moves the oil continuously during frying. Two different batches of plantain chips (300 g/3 kg oil, Ratio =

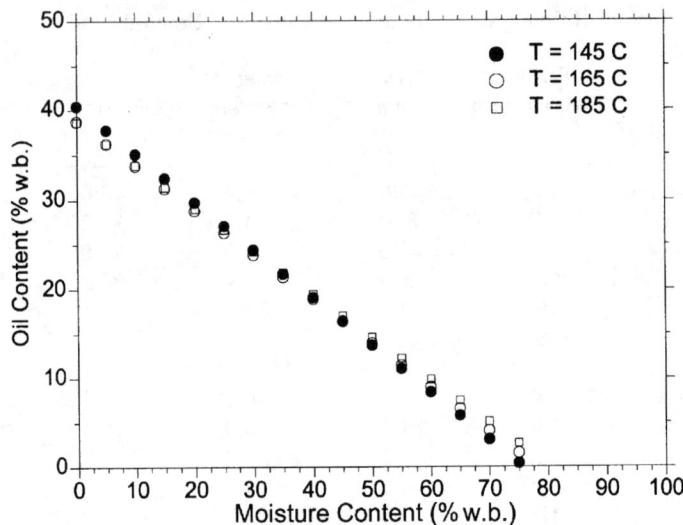

Figure 5–3 Oil Content Against Moisture Content for Potato Chips Fried at Different Temperatures. *Source:* Adapted with permission from M.H. Gamble, P. Rice, and J.D. Selman, Relationship Between Oil Uptake and Moisture Loss During Frying of Potato Slices from c.v. Record U.K. Tuber, *International Journal of Food Science and Technology*, Vol. 22, p. 239, © 1987, Blackwell Science Ltd.

1/10; and 75 g/3 kg oil, Ratio = 1/40) were used to evaluate temperature changes on the bottom and surface of the oil bath.

Figures 5–4 to 5–6 show how the temperature changes during frying of plantain chips using the three deep-fat frying systems. The stirring system seemed to be the more stable, with a maximum oil temperature decrease (DT) in the order of 7°C (R = 1/40) and 31°C (R = 1/10) at the bottom of the fryer (Figure 5–6). The stirring system provided better oil circulation, reduced temperature gradient within the fryer, and decreased the frying time. With a stirring system, the contact between the medium and the product is enhanced—mainly for large batches of products—and the water vapor in between the chips is released easily, thus reducing the problem of the chips sticking to each other during the process.

Without stirring the oil, large variations in the free-frying system occur (Figure 5–4), with temperatures ranging from 11°C to 51°C for the small (R = 1/40) and large (R = 1/10) batch systems, respectively. Changes were more pronounced at the surface than at the bottom of the fryer, due to water vapor and ambient air temperature difference. The immersion system, on the other hand, reduced the temperature gradient between the surface and the bottom of the fryer during the process (Figure 5–5).

Table 5–2 Effect of Drying Time on Quality Attributes of Frozen French Fries Fried in Vegetable Oil at 187°C

Measurement	Frying Time (min)				
	2	*3*	*4*	*6*	*9*
Physical properties					
Oil content (% w.b.)	10.8	17.1	20.7	20.8	23.8
Moisture content (% w.b.)	54.0	40.3	36.4	35.2	21.9
Sample width (mm)	9.3	9.3	9.3	9.8	10.2
Bulk density (kg/m^3)	520.0	420.0	300.0	350.0	280.0
Crust thickness (mm)	0.4	0.7	0.9	0.9	—
Modulus of elasticity (N/m^2 x 10^2)[a]	7.4	7.7	23.9	43.4	47.3
Flexural strength (N/m^2)[a]	38.8	44.2	87.3	199.4	283.1
Deflection to fracture (mm)[a]	5.0	5.0	4.8	4.2	2.1
Sensory attributes[b]					
Crispness crust (soggy to crispy)	12.9	24.1	39.1	58.5	81.3
Mealiness core (firm to powdery)	43.2	46.6	51.1	58.2	65.3
Inside texture (hard to soft)	55.6	56.8	60.7	61.4	63.8
Oiliness (nongreasy to greasy)	59.6	47.3	48.6	38.5	41.3
Color (pale to dark)	12.1	18.8	35.2	60.1	75.2

[a]Using Instron 3-point bending test; [b]Values are means on a 0–100 scale.

Source: Adapted with permission from M.S. DuPont, A.R. Kirby, and A.C. Smith, Instrumental and Sensory Tests of Texture of Cooked Frozen French Fries, *International Journal of Food Science and Technology*, Vol. 27, p. 291, © 1992, Blackwell Science Ltd.

Drying was better with the stirring system (Table 5–3) than with the immersion and free-frying systems. Increasing the stirring velocity resulted in higher moisture loss. For the high mass ratio (R = 1/10), product moisture content was higher for the forced-immersion system than for the free system. The forced-immersion system had a cover in the baskets to force the chips to be submerged during the process; this may have blocked the escape of water vapor, thus resulting in the product having higher water content. Oil content was related to the moisture content remaining in the product (Table 5–3). Therefore, the stirring system yielded chips with higher oil content (7–4% more oil content).

Diaz et al. (1996) studied the effect of product/oil mass ratio on the frying kinetics of plantain chips fried in a 3-kg batch fryer equipped with a stirring system. Table 5–4 shows changes in oil content, moisture content, and oil bath temperature during the frying process at the center point of the fryer. For the 1/10 product/oil mass ratio, a temperature drop of 15°C was observed at the beginning of the

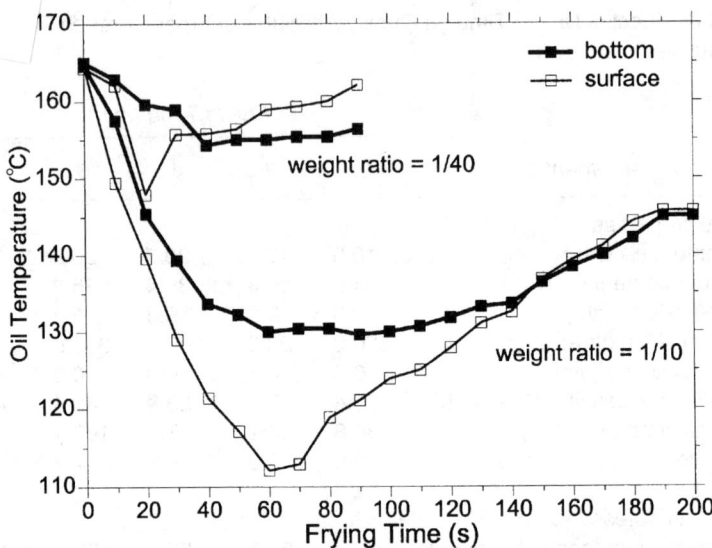

Figure 5–4 Oil Temperature History on the Bottom and Surface in the Free-Frying System. Product, plantain chips—300 g/3 kg oil or 75 g/3 kg oil; oil temperature set point, 165°C. *Source:* Adapted from *Lebensmittel-Wissenschaft und Technologie*, Vol. 29, A. Totte et al., Deep-Fat Frying of Plantain, II. Experimental Study of Solid/Liquid Phase Contacting Systems, p. 603, © 1996, by permission of the publisher Academic Press.

process, and recovery to the setpoint took 60 seconds, which is not negligible as compared with the 240 seconds required to evaporate 95% of the water from the product.

In the case of low mass ratio (1/40), homogeneity could be obtained in the process (Table 5–4). There was a decrease of only about 6°C on the oil bath temperature in the first 60 seconds of frying, when most of the moisture dropped to 14% w.b. Oil content reached an equilibrium after 90 seconds, when moisture loss rate was greatly reduced.

In conclusion, stirring uniformizes the oil temperature in the fryer, reduces adhesion between slices, and improves heat and mass transfer. A combination of stirring and forced-immersion systems, typical of continuous processes, would make the frying process more efficient.

THE FRYING OIL

Chapter 3 explains that changes in frying oil characteristics are due to the presence of oil degradation products from food components or oxidation and hydrolysis of chemicals in the oil itself. These substances alter the heat transfer by acting as surfactants at the oil/product interface, thus affecting the product quality.

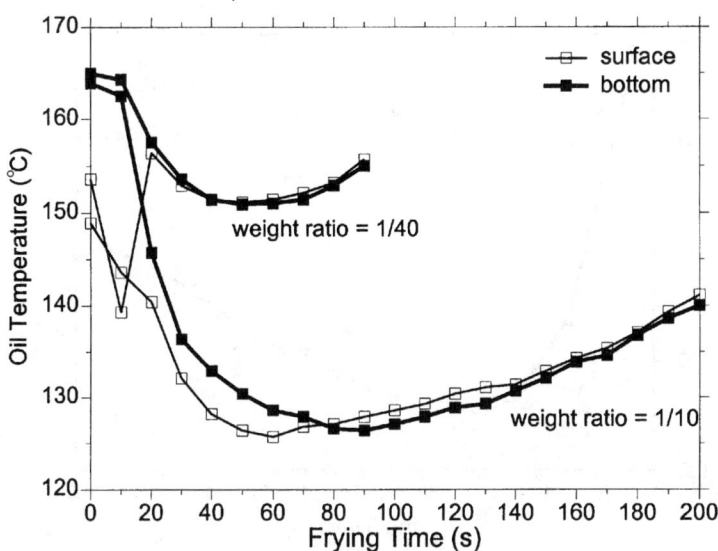

Figure 5–5 Oil Temperature History on the Bottom and Surface in the Forced Immersion System. Product, plantain chips—300 g/3 kg oil or 75 g/3 kg oil; oil temperature set point, 165°C. *Source:* Adapted from *Lebensmittel-Wissenschaft und Technologie*, Vol. 29, A. Totte et al., Deep-Fat Frying of Plantain, II. Experimental Study of Solid/Liquid Phase Contacting Systems, p. 603, © 1996, by permission of the publisher Academic Press.

Surfactant Theory

Blumenthal (1991) suggested that surfactants are responsible for the surface and interior differences in fried foods that are induced by aging oils. The Blumenthal's surfactant theory of frying consists of seven basic assumptions:

1. Frying is basically a dehydration process. When food is fried, water and materials suspended or dissolved in the water are heated and "pumped" from the food to the frying oil.
2. The heat-transfer medium, the frying oil, is a nonaqueous material, whereas the food can be assumed to be almost water. Water and oil are immiscible.
3. For frying to occur, heat must be transferred from the nonaqueous medium—the oil—to the mostly aqueous medium—the food.
4. Any changes in the frying or heat-transfer characteristics of the oil must result from degradation products formed from the oil.
5. Food materials leaching into the oil, thermal and hydrolytic breakdown of the oil, and oxygen absorption at the oil–air interface all contribute to altering the oil from a medium that is almost pure triglyceride to a mixture of hundreds of compounds.

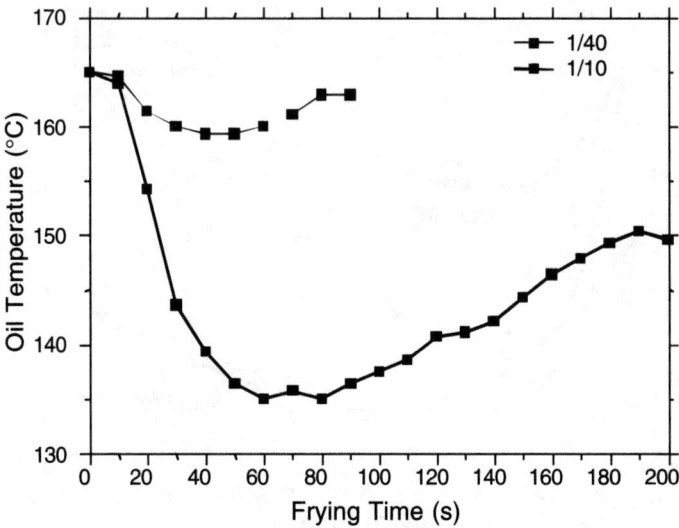

Figure 5–6 Oil Temperature History at the Bottom in the Stirring System. Product, plantain chips—300 g/3 kg oil or 75 g/3 kg oil; oil temperature set point, 165°C, 5.24 rad/8. *Source:* Adapted from *Lebensmittel-Wissenschaft und Technologie,* Vol. 29, A. Totte et al., Deep-Fat Frying of Plantain, II. Experimental Study of Solid/Liquid Phase Contacting Systems, p. 603, © 1996, by permission of the publisher Academic Press.

Table 5–3 Effect of Frying System on Oil and Moisture Content of Plantain Chips

System	Weight Ratio*	Stirring Speed (rad/s)	Moisture Content (% w.b.)	Oil Content (% w.b.)
Stirring	1/40	5.24	6.7 ± 1.2	28.9 ± 1.5
	1/10	5.24	6.6 ± 0.9	26.4 ± 1.5
	1/40	7.33	5.0 ± 0.5	—
	1/10	7.33	4.4 ± 0.7	—
Forced immersion	1/40	5.24	11.8 ± 1.0	23.9 ± 2.1
	1/10	5.24	13.5 ± 2.3	22.5 ± 1.1
Free frying	1/40	—	13.6 ± 1.7	21.1 ± 1.5
	1/10	—	9.8 ± 3.3	23.3 ± 1.4

*300 g/3 kg oil, or ratio 1/10; and 75 g/3 kg, or ratio 1/40.

Source: Adapted from *Lebensmittel-Wissenschaft und Technologie,* Vol. 29, A. Totte et al., Deep-Fat Frying of Plantain, II. Experimental Study of Solid/Liquid Phase Contacting Systems, p. 606, © 1996, by permission of the publisher Academic Press.

Table 5–4 Effect of Batch Size on Moisture and Oil Content of Fried Plantain Chips Fried in a Stirring System

Batch Size[a]	Frying Time (s)	Oil Temperature (°C)	Moisture Content (% w.b.)	Oil Content (% w.b.)
1/10[b]	0	150.0	64.1	0.2
	120	135.0	17.0	22.5
	240	149.1	4.6	25.4
	360	150.0	1.6	35.4
	600	150.0	1.6	40.5
1/40[c]	0	165.0	64.1	0.2
	30	160.0	17.1	21.2
	60	159.0	14.0	23.1
	90	163.3	3.1	37.0
	120	165.0	1.7	36.1

[a]Product/oil mass ratio; [b]300 g/3 kg oil; set point temperature of oil, 150°C; stirring speed, 5.2 rad/s; [c]75 g/3 kg; set point temperature of oil, 165°C; stirring speed, 5.2 rad/s.

Source: Data from A. Diaz et al., Deep-Fat Frying of Plantain. I. Characterization of Control Parameters, *Lebensmittel-Wissenschaft und Technologie*, Vol. 29, pp. 489–497, © 1996, Academic Press.

6. Substances that affect heat transfer at the oil-food surface reduce the surface tension between the two immiscible materials (water and oil). These substances act as wetting agents and are regarded as *surfactants*.

7. As the oil degrades, more surfactants are formed, causing increased contact between the food and oil. This causes excessive oil uptake by the food and an increased rate of heat transfer to the surface of the food. Eventually, excessive darkening and drying of the surface occur before the food is cooked, as the rate of conduction of heat to the interior of the food is constant and cannot be sped up by changes in the oil.

Surfactants are simply any materials that are soluble in both oil and water—that is, they will have hydrophilic and lipophilic components in their makeup. Some examples of surfactant materials commonly found in degrading frying oils are soap (magnesium oleate), phospholipids (lecithin), inorganic salts, mono- and diglycerides, slightly polar thermal polymers, and highly polar oxidative polymers (Blumenthal and Stier, 1991). Each of these compounds reduces the interfacial tension between the food and the frying oil and between the oil and the surface of the fryer heater. The most surfactants are probably the alkaline soaps, which are formed from interactions between free fatty acids and metal ions in the presence of traces of water.

The surfactant compounds not only affect heat transfer and, thus, food quality, but also directly affect the rate of oil degradation. As surfactant concentrations increase, the degree of foaming increases. As the foam bubbles increase in size, a large oil surface is exposed, at which oxidative reactions can readily occur. Furthermore, as the foam bubbles burst, they release peroxidation products into the oil, resulting in further reactions. The stability, interfacial surface area and stacking of the foam bubbles ("steam domes") all increase with increasing surfactant concentrations in degrading frying oil.

Oil Quality

The data shown in Table 5–5 describe how the frying oil in a fast-food restaurant changes over time. The frying quality of an oil changes with age. Oil quality increases after the oil is "broken in" and subsequently decreases as degradation reactions take place. An oil that does not contain a significant level of additives contains 96–98% triglycerides. During frying, the triglycerides, which make up the bulk of the oil, are exposed to food particles and components, water, heat, light, and oxygen, and react to form many compounds, including triglycerides that contain oxidized fatty acids. These degradation products may be defined as the "polar fractions." The key components that strongly influence the frying process are the alkaline soaps, which are present at low levels, and polymers.

As the concentration of polar materials increases in a frying oil, the physicochemical properties of the oil change. Viscosity, thermal conductivity, and specific gravity increase (see Chapter 3), while heat capacity, surface tension, and interfacial tension decrease. Aged oil behaves more like a gelatinous fluid than a homogenous liquid. The increase in the level of polar substances allows some water to remain in emulsion in heated frying oil (Blumenthal and Stier, 1991).

Change in Food Quality

The quality of the oil as a frying medium and the quality of the food processed in it are intimately bound. Figure 5–7 shows the five stages of oil degradation and relates them to food quality (Blumenthal and Stier, 1991). A statistical percentage of oil's contact time with food during frying are:

1. new oil, < 5%
2. break-in oil, 10%
3. fresh oil, 20%

Table 5–5 Oil Degradation in a Mixed-Use Fast-Food Fryer

Oil Quality	Triglycerides (%)	Polar Comp (%)	Polymers (%)	Free Fatty Acids (%)	Oxidized fatty acids (%)	Soaps (ppm)
New Oil	>96	<4	0.5	0.02	0.01	0.7
"Break-in"	90	10	2.0	0.50	0.08	10.0
Fresh	85	15	5.0	1.00	0.20	35.0
Optimum	80	20	12.0	3.00	0.70	65.0
Degraded	75	25	17.0	5.00	1.00	>150.0
"Runaway"	65	35	25.0	8.00	2.00	>200.0

Source: Adapted from *Trends in Food Science and Technology*, M.M. Blumenthal and R.F. Stier, Optimization of Deep-Fat Frying Operations, Pages No. 144–148, Copyright 1991, with permission from Elsevier Science.

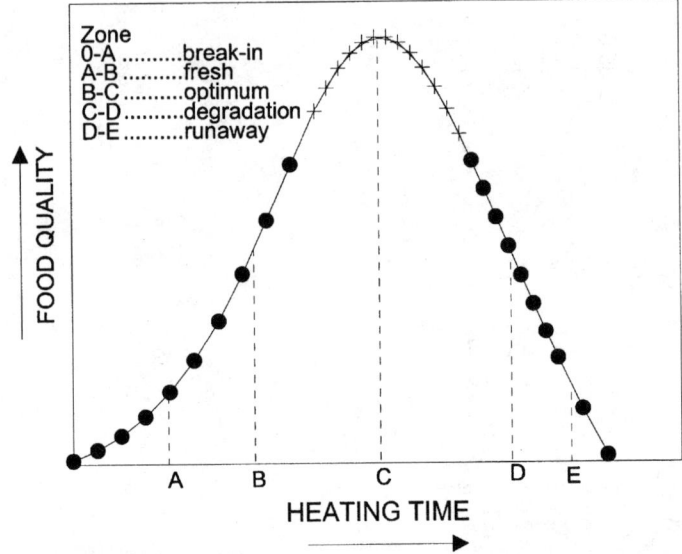

Figure 5–7 Frying Oil Quality Curve. *Source:* Adapted from *Trends in Food Science and Technology*, M.M. Blumenthal and R.F. Stier, Optimization of Deep-Fat Frying Operations, Pages No. 144–148, Copyright 1991, with permission from Elsevier Science.

4. optimum oil, 50%
5. breakdown oil, 80%
6. runaway oil, 90%

A qualitative analysis on quality of french fries fried in different levels of degraded oil was made by Blumenthal and Stier (1991) who observed (Color Plate 4):

a. *New and break-in oils*: white product; raw, ungelatinized starch at the center of the french fry; no cooked odors; no crisping of the surface; little oil pickup by the food.
b. *Fresh oil*: slight browning at the edges of the product; partially cooked (gelatinized) centers; crisping of the surface; slightly more oil absorption.
c. *Optimum oil*: golden-brown color; crisp, rigid surfaces; delicious potato and oil odors; fully cooked centers; optimal oil absorption.
d. *Degrading oil*: darkened and/ or spotty surfaces; excess oil pickup; product moving toward limpness; hard surfaces.
e. *Runaway oil*: dark, case-hardened surfaces; excessive oily product; surface collapsing inward; centers not fully cooked; off odor and flavors (burned).

THE PRODUCT

The effect of raw material properties, composition, and pre-frying treatment affect the quality of the final fried product.

Specific Gravity

High specific gravity of raw potatoes and dry matter and starch contents produces potato chips with lower oil content. The relationship between chip's oil content and raw potato SG is shown in Figure 5–8. Specific gravity of raw potatoes shows a high correlation with the fat content of potato chips. French fries made from potatoes with 24% dry matter contained 9% less oil than did french fries made from varieties containing about 19.5% dry matter (Lesińska and Leszczyński, 1989).

The external qualities of raw potato, such as uniform size, smooth shoulders, shallow eyes, and lack of defect, are the main factors affecting the yield of fried potatoes because they reduce the losses due to peeling and slicing/cutting. The yield of chips and french fries is also connected to the SG of raw potatoes. In-

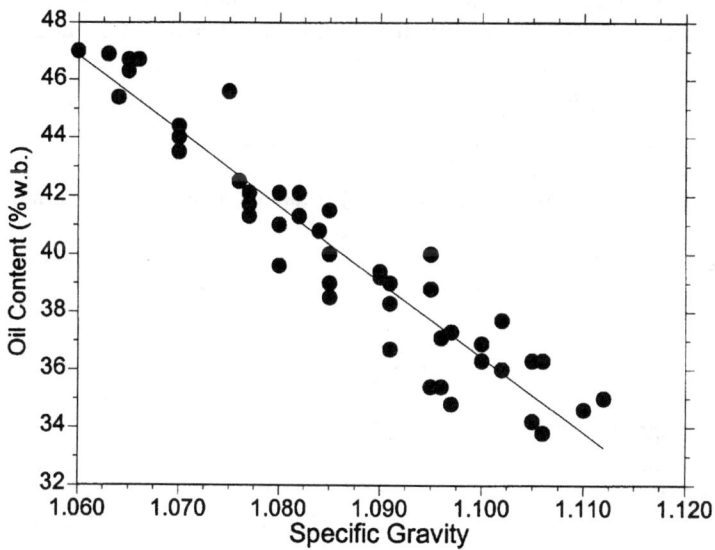

Figure 5–8 Effect of Specific Gravity on Potato Chip Oil Content. *Source:* Adapted with permission from E.C. Lulai and P.H. Orr, Influence of Potato Specific Gravity on Yield and Oil Content of Chips, *American Potato Journal*, Vol. 56, p. 384, © 1979, Potato Association of America.

creases in dry matter content result in increased yield of chips and french fries (Table 5–6).

High dry matter content (above 25%) can produce potato chips with good texture. French fries made from high SG potatoes are also of desirable textural quality. The texture of french fries can be affected not only by starch content of the tuber but also by the ratio of amylose to amylopectin and the size of starch granules (Lesińska and Leszczyński, 1989). Variations in amylose content can significantly affect the viscosity of the gelatinized starches, thus influencing the textural characteristics of fried products.

Slice Thickness

Gamble and Rice (1988) observed that increasing the thickness of a potato slice will reduce the volume of oil in the chip (Table 5–7). The volume of oil as a percentage of the total volume of slice decreased from 25.2% at 0.88 mm to 14.8% at 1.58 mm thickness. The total volume of oil taken up was related to the available thickness of the slices, i.e., as the surface area increased, the volume of oil absorbed increased.

This decrease in oil content with increasing thickness of potato slices has been found to be nonlinear (Baumann and Escher, 1995). There was a larger increase in oil content between 0.8 and 1.1 mm than between thicknesses of 1.3 and 1.6 mm (Table 5–8). The increased drying time (to reach a final moisture content of 2% w.b.) with increase in thickness is evident in Figure 5–9.

Table 5–6 Effect of Specific Gravity of Raw Potatoes on Potato Chip Yield

Specific Gravity	Yield of Chips (%)[a]	Yield of French Fries (%)[b]
1.060	22.44	46.80
1.065	29.22	-
1.070	30.00	47.90
1.075	30.78	50.80
1.080	31.56	-
1.085	32.33	50.70
1.090	33.11	50.30
1.095	33.89	52.20
1.100	34.67	55.70
1.105	35.45	-
1.110	36.23	56.20

Source: Data from (a) E.C. Lulai and P.H. Orr, Influence of Potato Specific Gravity on Yield and Oil Content of Chips, *American Potato Journal*, Vol. 56, pp. 379–390, © 1979, Potato Association of America; and (b) M.E. Kirkpatrick et al., *Tech. Bull.* 1142, pp. 1–46, © 1956, United States Department of Agriculture.

Table 5–7 Effect of Slice Thickness and Surface Area on Oil Content of Potato Chips

Thickness/ Slice (mm)	No. Slices/ 100 g	SA/100 g Slices (m²)	Volume/ Slice (cm³)	Oil/100 g Slices (g)	Oil Content* (%)
0.88	82	2,076	1.11	23.2	25.2
0.95	77	1,933	1.19	20.9	22.7
1.12	65	1,627	1.39	18.9	20.9
1.18	61	1,544	1.48	17.9	19.7
1.38	53	1,326	1.73	16.6	18.0
1.48	49	1,236	1.86	14.9	16.2
1.58	46	1,153	1.99	13.5	14.8

*Percentage of total slice volume occupied by oil, SA = surface area.

Source: Adapted from *Journal of Food Engineering*, Vol. 8, M.H. Gamble and P. Rice, The Effect of Slice Thickness on Potato Crisp Yield and Composition, Pages No. 31–46, Copyright 1988, with permission of Elsevier Science.

Prefrying Treatment

The oil uptake varies with pretreatment of raw product before frying.

Blanching/Drying

Lamberg et al. (1990) quantified the effect of prefrying treatment on the oil content of french fries fried for 5 minutes in vegetable oil at 180°C. The highest oil content was obtained for potatoes that were blanched without drying and the lowest oil content when the drying conditions were 80°C, 2% relative humidity (RH) and 15 minutes drying time (Table 5–9).

Table 5-8 Effect of Thickness on Oil Content of Potato Chips

Thickness of Slice (mm)	Frying Time (s)	Oil Content (% w.b.)
0.8	90	57.0
1.1	165	51.0
1.3	160	42.7
1.6	180	46.7

Note: DM = 0.18 kg/kg, oil temperature 170°C, final moisture content 2% w.b.

Source: Data from B. Baumann and F. Escher, Mass and Heat Transfer During Deep-Fat Frying of Potato Slices, Rate of Drying and Oil Uptake, *Lebensmittel-Wissenschaft und Technologie*, Vol. 28, pp. 395–403, © 1995, Academic Press.

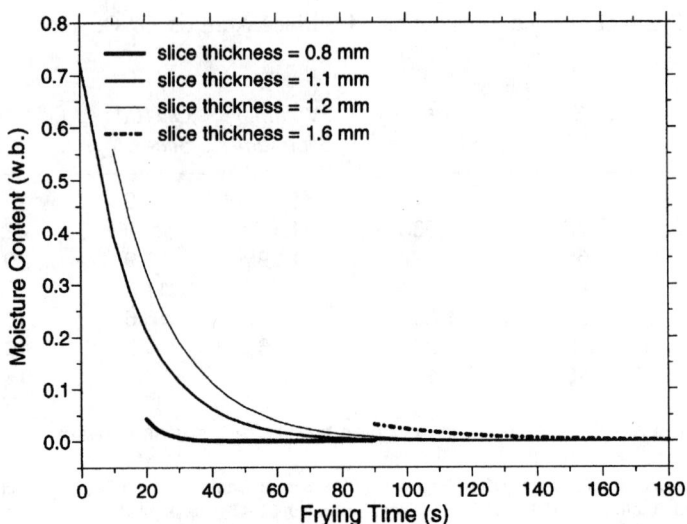

Figure 5–9 Effect of Potato Slice Thickness on Moisture Loss during Frying. Oil tempera-ture; 170°C, final moisture content, 2% w.b. *Source:* Adapted from *Lebensmittel-Wissenschaft und Technologie*, Vol. 28, B. Baumann and F. Escher, Mass and Heat Trans-fer During Deep-Fat Frying of Potato Slices, Rate of Drying and Oil Uptake, pp. 395–403, © 1995, by permission of the publisher Academic Press.

Table 5–9 Effect of Prefrying Conditions on French Fry Oil Content

Process Conditions*	Moisture Content before Frying (% w.b.)	Oil Content (% d.b.)
Blanched in water (75°C, 15 min)	81.5 ± 8.4	20.0 ± 1.4
Blanched and dried (80°C, 15 min)		
2% RH	80.7 ± 8.5	12.0 ± 1.0
30% RH	80.4 ± 8.4	15.0 ± 1.0
40% RH	81.3 ± 8.3	17.0 ± 2.0
Blanched and dried (80°C, 30 min)		
30% RH	79.7 ± 7.5	14.0 ± 2.0

*Fried at 180°C for 5 min

Note: RH, relative humidity.

Source: Adapted from *Lebensmittel-Wissenschaft und Technologie*, Vol. 23, I. Lamberg, B. Halstrom, and H. Olsson, Fat Uptake in a Potato Drying/Frying Process, p. 296, © 1990, by permission of the pub-lisher Academic Press.

Drying

The effect of prefrying treatment on the oil uptake in potato chips was studied by Gamble and Rice (1987). Microwave and hot-air treatment resulted in a reduction in the final oil content; freeze-drying resulted in an increase in the oil content (Table 5–10).

Microwave drying resulted in starch gelatinization, swelling prior to frying, highly heterogeneous structure, (with a modified moisture distribution), and large oil-free zones. Air predrying resulted in a chip with a lower oil content, moisture removed from the entire slice in a more regular pattern than with microwave drying, and oil more regularly distributed than in chips that have been only fried (which have most of the oil concentrated at the edges). Freeze-drying increased the final oil content of the product as freeze-drying time increased; freeze-dried slices showed no gross changes in structure, even distribution of oil, and the presence of ungelatinized starch, even after frying.

Baking

Table 5–11 presents the results of baking time on the final oil content of tortilla chips (Moreira et al., 1997). By increasing the baking time from 0 (unbaked chips) to 140 seconds (overbaked chips), the initial moisture content of tortilla chips decreased from 56.7% to 27.7% (w.b.), respectively. The tortilla chips were then fried for a constant time of 60 seconds. Initial moisture content significantly affected $(P<0.05)$ the final oil content of the chips.

When using the same frying time, the final moisture content will vary resulting in a different interpretation of the results. Therefore, to eliminate the effect of the final moisture content, it is more reasonable to consider the water removed during frying to compare different treatments (Pinthus et al., 1993). Table 5–11 shows the

---

Table 5–10 Effect of Prefry Drying Treatment on Oil Content of Potato Chips

Process Conditions	Final Moisture Content (% w.b.)	Oil Content (% w.b.)
Fried/control not predried[a]	0.9 ± 0.1	46.7 ± 0.4
Microwave predried[b]	2.4 ± 0.2	34.8 ± 0.6
Hot air (105°C) predried[c]	1.8 ± 0.1	33.5 ± 1.0
Freeze drying predried[d]	2.2 ± 0.3	45.4 ± 0.6

[a]Fried for 5 min at 165°C; [b]Drying time 5 min (frying time 1.5 min); [c]Drying time 35 min (frying time 1.5 min); [d]Drying time 5 min (frying time 1.5 min)

Source: Data from M.H. Gamble and P. Rice, Effect of Pre-Frying of Oil Uptake and Distribution in Potato Crisp Manufacture, *International Journal of Food Science and Technology*, Vol. 22, pp. 535–548, © 1987, Blackwell Science Ltd.

Table 5–11 Effects of Baking Time on Final Oil Content of Tortilla Chips Fried for 60 s

Samples	Temperature (°C)	Moisture Content (% w.b.)		Final Oil Content (% w.b.)	FOC/MR
		Initial	Final		
Unbaked	190 ± 5	56.7 ± 2.8	2.6[a]	30.8[a]	0.42[a]
Optimally baked	190 ± 5	44.1 ± 3.3	2.9[a,b]	26.4[b]	0.49[b]
Overbaked	190 ± 5	27.7 ± 0.8	2.6[a,b]	22.3[c]	0.83[c]

[a–c]Averages in the same column within the same data set (category) that are not followed by the same superscript letter are significantly different ($P<0.05$).

Note: FOC/MR, final oil content/moisture removed; w.b., wet basis.

Source: Adapted from *Journal of Food Engineering*, Vol. 31, No. 4, R.G. Moreira, X. Sun, and Y. Chen, Factors Affecting Oil Uptake in Tortilla Chips in Deep-Fat Frying, Pages No. 485–498, Copyright 1997, with permission from Elsevier Science.

relationship between oil uptake ratio (i.e., oil uptake/water removed) and initial moisture content at different frying temperatures. The oil uptake ratio was significantly ($P<0.05$) affected by the initial moisture content of the chips. Lower initial moisture content resulted in higher oil uptake ratio.

Product Composition

Tortilla chips prepared from finely ground masa showed excessive puffing and pillowing, resulting in higher oil absorption and lower porosity; tortilla chips made from coarse particles presented the lowest oil content and no puffing; and intermediate masa produced tortilla chips with final oil content closer to the control sample (Table 5–12). Final oil content/water removed ratio was lower for tortilla chips made from coarse flour than from fine flour.

The fine particles in the masa are responsible for most of the water uptake and viscosity development during mixing (Gomez et al., 1987). The function of the coarse particles is to produce fissures in the product that allow water to escape during frying and reducing oil absorption, and to reduce the extent of pillowing during baking and frying. Therefore, a moderately coarse masa is desirable for fried corn snacks.

Factors Affecting Fried Coating Characteristics

Coatings for fried foods are produced from batters in which wheat flour is a major ingredient. In the selection of flour suitable for coating applications, no definitive criteria exist. The flour functionality in a batter system is based on two main components, protein and starch. Because of the water-binding characteristics of the gluten protein fraction in wheat flours, hard wheat flours require more water than do soft wheat flours to produce the same batter consistency (viscosity). In a puff/tempura batter, gluten proteins help retain gases during leavening, resulting in the formation of a porous and crispy batter (Loewe, 1993).

Wheat starch is made up of linear (amylose) and branched (amylopectin) polymers of glucose. The ratio of these polymers has profound effects on starch functionality. High-amylose (70%) starches are useful in batter coatings that require a continuous membrane (e.g., for microwave reconstitution). During milling, starch granules can be damaged mechanically, thus altering its functionality. Hard wheat flour tends to contain a higher proportion of damaged starch than does soft wheat flour, resulting in higher water absorption during batter preparation (Loewe, 1993).

Olewnik and Kulp (1993) evaluated flour quality factors on various batter quality attributes (Table 5–13). A higher level of protein increased crispness of the fried product and produced a darker color. Drumsticks prepared with a soft wheat

Table 5–12 Oil Content, Porosity, and Bulk Density of Tortilla Chips as a Function of Particle Size Distribution

Sample	Final Oil Content (% w.b.)	Moisture Content (% w.b.)		Porosity	Bulk Density (kg/m³)	FOC/MR
		Initial	Final			
100% fine	31.2 ± 0.4	37.6 ± 1.5	1.4 ± 0.2	0.49 ± 0.1	717 ± 20	0.78
100% intermediate	23.7 ± 0.1	31.1 ± 1.5	1.0 ± 0.1	0.60 ± 0.1	603 ± 58	0.54
100% coarse	21.5 ± 0.4	37.2 ± 1.3	1.0 ± 0.1	0.56 ± 0.1	652 ± 43	0.48
Control sample: 33.3% fine 21.4% intermediate 45.3% large	25.1 ± 0.3	38.4 ± 2.3	1.3 ± 0.1	0.57 ± 0.1	662 ± 26	0.56

Note: Tortilla chips were optimally baked and fried in fresh oil for 60 s at 190°C. FOC/MR, final oil content/moisture removed.

Source: Adapted from *Journal of Food Engineering*, Vol. 32, No. 4, R.G. Moreira, X. Sun, and Y. Chen, Factors Affecting Oil Uptake in Tortilla Chips in Deep-Fat Frying, Pages No. 485–498, Copyright 1997, with permission from Elsevier Science.

Table 5–13 Organoleptic Characteristics of Batter-Coated Fried Chicken

Flour Type	Starch Damage (%)	Protein (Nx5.7) (%)	Color	Crispness	Greasiness	Adhesion
Soft-red (Chalange)	4.5	7.1	2.0	4.5	6.5	7.5
	6.4		4.5	7.0	8.5	6.5
	8.4		6.0	7.0	8.0	8.5
Soft-red (Emblem)	4.7	7.6	3.0	5.0	4.0	6.5
	6.5		4.5	7.0	6.0	10.0
	8.0		5.5	9.0	7.0	9.5
Hard-red (GM 10)	8.2	9.6	4.0	6.0	6.5	8.5
	12.4		5.5	8.0	6.5	8.0
	15.1		6.0	7.5	7.5	9.0
Hard-red (GM 11)	6.4	10.8	3.0	6.5	6.5	8.0
	9.6		4.0	6.0	6.5	7.0
	14.4		5.0	9.0	9.0	8.5
Hard-red (GM 12)	8.0	12.1	4.0	7.5	5.5	10.0
	16.4		4.0	8.0	8.0	9.5
	20.0		5.5	8.5	8.0	7.0

Note: Values are mean on a 0–10 scale.

Source: Adapted with permission from M. Olewnik and K. Kulp, Factors Influencing Wheat Flour Performance in Batter Systems, *Cereal World*, Vol. 38, No. 9, p. 683, © 1993, American Association of Cereal Chemists.

flour batter produced a more pancake-like coating when fried and the surface was smooth with a porous internal structure, resulting in a low crispness score. Adhesion, the measure of a coating's ability to stick to the chicken skin, increased in batters made with higher protein flour (12.1%). The level of protein did not have any effect on the perceived greasiness of the fried coatings, but oil content was higher for high-protein batters (Olewnik and Kulp, 1993).

In both soft and hard wheat, as the damaged starch level increased, the fried coatings became darker and crispier (Table 5–13). The increased crispness approached brittleness at high levels of damaged starch. The enhanced color formation was due to increased levels of reducing saccharides. Damaged starch had no effect on adhesion properties; however, high levels of starch damage caused the perception of a greasier product.

Batter prepared with fibrous ingredient, such as cellulose, reduces the oil uptake of fried products. Ang and Miller (1991) prepared batters containing powdered cellulose (PC) to coat fish and chicken and found that by adding 1% of fiber to the batters, oil content was reduced from 29.1% w.b. (control) to 25.5% w.b. for fried fish coated batters and from 25.3% w.b. (control) to 14.6% w.b. for fried chicken coated batter (Table 5–14). Moisture content was also increased significantly when 1% PC was added to the batters ($P \leq 0.05$). The same observations were made

Table 5–14 Effect of Powdered Cellulose on Oil Content of Coated Fried Fish, Chicken, and Cake-Type Doughnuts

Product	Oil Content (% w.b.)	Moisture Content (% w.b.)
Batter coated		
Fish		
Control	29.1	30.8
1% PC	25.6	33.6
Chicken		
Control	25.3	31.6
1% PC	14.6	43.1
Cake-type doughnuts		
Control	21.1	20.1
1% PC	19.2	21.1
2% PC	18.0	21.6
3% PC	16.8	22.4

Note: PC, powdered cellulose.

Source: Data from J.F. Ang and W.B. Miller, Multiple Functions of Powdered Cellulose as a Food Ingredient, *Cereal World*, Vol. 36, No. 7, pp. 558–564, © 1991, American Association of Cereal Chemists.

for cake-type doughnuts prepared with PC. Lower fat contents in batters and doughnuts containing cellulose were probably due to the strong hydrogen bonding between water molecules in the batter/doughnut samples.

Meat/Chicken Products

When meatballs are heated during frying, the components undergo physico-chemical and structural changes. Protein denaturation is related to temperature and frying time conditions. When product surface temperatures reach 100°C, the water evaporation zone recedes toward the center. The surface continues to lose water, the temperature rises, and structural and chemical changes in the protein result in crust formation.

Ateba and Mittal (1994) evaluated the kinetics of crust formation and quality changes in meatballs during frying. The crust depth increased linearly with frying time, becoming darker and harder. It took about 546 seconds for the center temperature of the sample to reach 70°C (accepted temperature for doneness) when the crust thickness was 1.84 mm. The increase in total crust color change was attributed to the high temperature and low moisture content of the product surface, which initiated browning reactions (Maillard and sugar caramelization) during the crust formation. The increase in hardness (peak force) was thought to be due to heat denaturation of protein, causing hardening of texture.

The effect of edible coating of chicken nuggets on oil content reduction in deep-fat frying processes was observed by Balasubramaniam et al. (1995). The frying time increased with increasing product fat content and the presence of edible film coating. Edible films function as moisture and oil barriers during frying, i.e., they reduce water loss and oil absorption (Table 5–15).

Frying chicken under pressure is known to yield more juicy and tender product than under atmospheric pressure. Mallikarjunan et al. (1995) fried chicken nuggets in a pressure fryer and observed that increasing the frying pressure and medium temperature decreased the frying time. Samples coated with edible films (high propyl methyl cellulose [HPMC]) had lower mass loss, compared with uncoated samples. Frying under pressure increased the moisture retention for both coated and uncoated samples but increased the fat uptake in the samples.

Oil Distribution and Structure of Fried Products

An understanding of the distribution of oil in finished fried products helps us better understand the mechanism of oil absorption. Several techniques have been used to visualize oil distribution in fried foods. Among them, the staining method, the magnetic resonance imaging (MRI) method, and the environmental electron microscopy method showed potential for oil distribution analysis in fried products.

Table 5-15 Change in Oil Content in Chicken Nuggets during Frying

Treatment	Initial Moisture Content (% d.b.)	Initial Oil Content (% d.b.)	Frying Time (s)[a]	Final Oil Content (% d.b.)		
				Surface	Midpoint	Center
Uncoated	62.7 ± 0.1	2.1 ± 0.2	340 ± 7	14.6 ± 0.1	3.9 ± 0.1	2.3 ± 0.0
Uncoated	58.9 ± 0.2	8.4 ± 0.2	386 ± 6	21.4 ± 0.1	12.7 ± 0.8	9.6 ± 1.2
Uncoated	57.5 ± 0.6	12.0 ± 0.5	390 ± 11	26.8 ± 0.1	14.8 ± 0.4	13.0 ± 1.4
Coated[b]	62.7 ± 0.1	2.1 ± 0.2	353 ± 6	12.5 ± 0.1	3.9 ± 0.1	2.5 ± 0.4
Coated	58.9 ± 0.2	8.4 ± 0.2	400 ± 7	16.8 ± 1.1	10.0 ± 0.3	9.4 ± 1.4
Coated	57.5 ± 0.6	12.0 ± 0.5	410 ± 4	21.8 ± 0.0	13.9 ± 0.7	12.9 ± 1.3

Note: Nuggets were fried at 175°C in peanut oil.
[a]Minimum frying time for center to reach 70°C; [b]Edible film—HPMC.

Source: Data from V.M Mallikarjunan, and M.S. Chinnan, Heat and Mass Transfer During Deep-Fat Frying of Chicken Nuggets Coated With Edible Film: Influence of Initial Fat Content, in *Advances in Food Engineering Proceedings of the 4th Conference in Food Engineering*, Narsimhan, Okos, and Lombardo, eds, pp. 103–106, © 1995.

Gamble et al. (1987) used the stain Oil Red O in a proportion of 0.5 g/L of oil to examine the oil distribution in fried products. Oil distribution pattern depended on the original moisture content and the structure of the food, which, in turn, influenced moisture loss.

Keller et al. (1986) also used the stain method to determine oil distribution in french fries. They used 1–3 g of Sudan Red B per liter of oil and fried potatoes (1 cm × 1 cm × 5 cm) at 180°C for 1–10 minutes. After frying, the products were then cut into 2- to 4-mm slices and visualized under a photomicrographic zoom system. Micrographic pictures showed that oil deposition was strictly limited to the crust—even after prolonged frying, no diffusion of oil toward the core occurred. Some cracks developed during frying over the fry cross-sectional area, due to shrinkage and heavy water loss. The overall changes in solid content and oil content are summarized in Table 5–16.

Moreira and Sun (1995) also used the stain method to determine oil distribution in tortilla chips fried in a soybean oil colored with Sudan Red 7B. During frying (0–60 seconds), the amount of oil absorbed was about 4.3% (w.b.) for the overbaked chips, 5.8% (w.b.) for optimally baked chips, and 7.8% (w.b.) for the unbaked chips. These values remained constant throughout the entire frying period, with exception of the unbaked chips, which showed a slight increase in oil content from 20 seconds of frying. These values represented about 20–24% of the final oil content in all cases (Color Plate 5). The unbaked tortilla chips became darker (absorbed more oil) during frying than did the optimally or overbaked chips.

Magnetic resonance imaging (MRI) has been proposed as an analytical tool for the visualization of oil and water concentration gradients within porous materials. The advantage of the MRI technique is that it is a noninvasive procedure for determining the dispersion of hydrogen nuclei in the form of water, oil, or other materials. Farkas et al. (1992) imaged the oil distribution in a cross-sectional area of a potato cylinder using MRI techniques. Results indicated a distinct oil-rich crust and water-rich core; their interface occurred at a depth of approximately 1–2 mm. Moreira et al. (1995) showed that oil distribution in tortilla chips concentrated in the puffed areas. Color Plate 6 shows an MRI of the cross-sectional area of three raw tortilla chips (A) and three tortilla chips fried for 60 seconds (B). Tortilla chip thickness increased about 40% after frying as a result of an expansion caused by an increase in porosity (Moreira et al., 1995).

Environmental scanning electron microscope (ESEM) is a technique that does not require food samples to be defatted or dehydrated before analysis, as is done in traditional light and scanning electron microscopy. The ESEM allows food samples to be viewed in their natural state with no modification before being analyzed. This avoids inducing artifacts into the samples and allows the location of oil in the sample to be determined. McDonough et al. (1993) examined the structural

Table 5–16 Influence of Deep-Fat Frying on the Oil Content of Fried French Fries

Samples	Total Solids Content (%)	Fat Content (% w.b.)
Whole French Fries		
Blanched	19.3	0.07
Fried at 180°C		
1 min	27.3	2.10
5 min	35.9	4.58
10 min	46.3	5.56
Crust		
Fried at 180°C		
1 min	46.9	8.91
5 min	49.7	9.02
10 min	90.7	11.18
Core		
Fried at 180°C		
1 min	19.6	0.19
5 min	19.6	0.27
10 min	16.7	0.30

Source: Adapted from *Lebensmittel-Wissenschaft und Technologie,* Vol. 19, Ch. Keller, F. Escher, and J. Solms, A Method for Localizing Fat Distribution in Deep-Fat Fried Potato Products, p. 348, © 1996, by permission of the publisher Academic Press.

changes that occurred during the frying process of tortilla chips using ESEM and observed that the immersion of tortilla chips in hot oil caused several structural transformations within the chip. As the steam exited the chip, the oil diffused into the chip's pores, filling the portions of the structure that once were occupied by moisture. Overfrying disrupted the uniformity of the pores, leading to large central air spaces and more oil absorption in the chip.

Pore size distribution affects oil absorption in fried products. Moreira et al. (1997) used ESEM to analyze the oil absorption patterns of tortilla chips prepared with three nixtamalized flours with different particle size distributions. The pore size distribution developed during frying was found to be the main cause for oil absorption during cooling. Small pores trapped more air during frying, resulting in higher capillary pressure during cooling and, therefore, higher final oil content.

Shrinkage in potato chips can be observed as soon as water loss occurs using transmission light microscope (Costa et al., 1998). The cells shrink faster at the surface and at higher temperatures. Potato volume decreases during frying but may increase at the last stages of frying. In french fries, little shrinkage was observed at the core; however, pore collapse and large shrinkage happened at the crust region. Crust formation is faster at higher temperatures. Potato chip volume reduced 40% and french fries 70% of their initial volume.

Crust Physical Properties

The formation of a crust influences the heat and mass transfer processes during the frying process. Fan et al. (1997) observed that starch gelatinization and subsequent swelling of the starch granules inhibited oil, thus resulting in a lower oil content in the core region of corn starch patties. The crust had most of the oil content.

Pinthus et al. (1995) demonstrated that crust yield strength and oil content were related to the modulus of deformation of the raw material. Lower modulus of deformation of restructured potato product resulted in thicker crust and higher oil content. Porosity and yield strength of the crust increased with increased raw material modulus of deformation. They observed that crust yield strength was more related to oil uptake than to porosity.

Porosity

The effect of porosity on oil uptake has been observed by Pinthus et al. (1995) for restructured potato product. A linear relationship was found between initial porosity (prior to frying) and oil uptake/water removed ratio. They concluded that initial porosity of the product determines the final oil uptake in fried products and that porosity and oil uptake increase during frying and are a function of each other.

REFERENCES

Ang, J.F., and Miller, W.B. 1991. Multiple functions of powdered cellulose as a food ingredient. *Cereal World*, 36(7):558–564.

Ateba, P., and Mittal, G.S. 1994. Dynamics of crust formation and kinetics of quality changes during frying of meatballs. *J Food Sci*, 59(6):1275–1278.

Balasubramaniam, V.M.; Mallikarjunan, P., and Chinnan, M.S. 1995. Heat and mass transfer during deep-fat frying of chicken nuggets coated with edible film: influence of initial fat content. In *Advances in food engineering*. Proceedings of the 4th Conference in Food Engineering, ed. Narsimhan, Okos, and Lombardo. West Lafayette, IN: Purdue University, 103–106.

Baumann, B., and Escher, F. 1995. Mass and heat transfer during deep-fat frying of potato slices. Rate of drying and oil uptake. *Lebensm-Wiss U-Technol*, 28:395–403.

Blumenthal, M.M. 1991. A new look at the chemistry and physics of deep-fat frying. *Food Technol*, 45(2):68–71, 94.

Blumenthal, M.M, and Stier, R.F. 1991. Optimization of deep-fat frying operations. Trends in *Food Sci Tech*, 144–148.

Costa, R.M.; Oliveira. F.A.R.; and Boutcheva G. 1998. *Structural changes during frying of potato slices*. IFT Annual Meeting. Atlanta, GA.

Diaz, A.; Totte, A.; Giroux, F.; Reynes, M.; and Raoult-Wack, A.L. 1996. Deep-fat frying of plantain. I. Characterization of control parameters. *Lebensm-Wiss U-Technol*, 29:489–497.

Du Pont, M.S.; Kirby, A.R.; and Smith, A.C. 1992. Instrumental and sensory tests of texture of cooked frozen french fries. *Intl J Food Sci Tech*, 27:285-295.

Fan, J.; Singh, R.P.; and Pinthus, E.J. 1997. Physicochemical changes in starch during deep-fat frying of a molded corn starch patty. *J Food Proc Preserv*, 21:443–460.

Farkas, B.E.; Singh, R.P.; and McCarthy, B. 1992. Measurement of oil/water interface in foods during frying. In *Advances in food engineering*, ed. Singh and Wirakartakusumah. Boca Raton, FL: CRC Press, Inc.

Gamble, M.H., and Rice, P. 1987. Effect of pre-frying drying of oil uptake and distribution in potato crisp manufacture. *Intl J Food Sci Tech*, 22:535–548.

Gamble, M.H., and Rice, P. 1988. The effect of slice thickness on potato crisp yield and composition. *J Food Engineering*, 8:31–46.

Gamble, M.H.; Rice, P; and Selman, J.D. 1987. Distribution and morphology of oil deposits in some deep fried products. *J Food Sci*, 52(6):1742–1745.

Gomez, M.H.; Rooney, L.W.; and Waniska, R.D. 1987. Dry corn masa flours for tortilla and snack foods. *Cereal World*, 32(5):370–377.

Keller, C.; Escher, F.; and Solms, J. 1986. A method for localizing fat distribution in deep-fat fried potato products. *Lebensm-Wiss U-Technol*, 19:346–348.

Kirkpatrick, M.E.; Heinze, P.H.; Croft, C.C.; Montjoy, B.M.; and Falatko, C.E. 1956. *Tech Bull*, 1142. Washington, DC: USDA, 1–46.

Lamberg, I.; Halström, B.; and Olsson, H. 1990. Fat uptake in a potato drying/frying process. *Lebensm-Wiss U-Technol*, 23:295–300.

Lesińska, G. and Leszczyński, W. 1989. *Potato science and technology*. New York: Elsevier Science.

Loewe, R. 1993. Role of ingredients in batter systems. *Cereal World*, 38(9):673–677.

Lulai, E.C., and Orr, P.H. 1979. Influence of potato specific gravity on yield and oil content of chips. *Am Potato J*, 56:379–390.

McDonough, C.; Gomez, M.H.; Lee, J.K.; Waniska, R.D.; and Rooney, L.W. 1993. Environmental scanning electron microscopy evaluation of tortilla chip microstructure during deep-fat frying. *J Food Sci*, 58:199-203.

Mallikarjunan, P.; Chinnan, M.S.; and Balasubramaniam, V.M. 1995. Mass transfer in edible film coated chicken nuggets: influence of frying temperature and pressure. In *Advances in food engineering proceedings of the 4th Conference in Food Engineering*, ed. Narsimhan, Okos, and Lombardo, 107–111.

Moreira, R.G.; Palau, J; and Sun, X. 1995. Deep-fat frying of tortilla chips: an engineering approach. *Food Technol*, 49(4):146–150.

Moreira, R.G., and Sun, X. 1995. *Oil distribution in tortilla chips using the stain method*. Internal Report. Department of Agricultural Engineering. College Station, TX: Texas A&M University.

Moreira, R.G.; Sun, X.; and Chen, Y. 1997. Factors affecting oil uptake in tortilla chips in deep-fat frying. *J Food Eng*, 31(4):485–498.

Olewnik, M., and Kulp, K. 1993. Factors influencing wheat flour performance in batter systems. *Cereal World*, 38(9):679–684.

Pinthus, E. J.; Weinberg, P.; and Saguy, I.S. 1993. Criterion for oil uptake during deep-fat frying. *J Food Sci*, 58(1):204–205, 222.

Pinthus, E.J.; Weinberg, P.; and Saguy, I.S. 1995. Oil uptake in deep fat frying as affected by porosity. *J Food Sci*, 60(4):767–769.

Totte, A.; Diaz, A.; Marouze, C.; and Raoult-Wack, A.L. 1996. Deep-fat frying of plantain. II. Experimental study of solid/liquid phase contacting systems. *Lebensm-Wiss U-Technol*, 29:599–605.

Varela, G.; Bender, A.E.; and Morton, I.D., eds. 1988. Frying of foods: principles, changes, new approaches. New York: VCH Publishers.

Zak, J., and Holt, C. 1973. Effect of finish-frying conditions on the quality of french fried potatoes. *J Food Sci*, 38:92–95.

CHAPTER 6

Theory and Simulation of Frying

Heat transfer in deep-fat frying of foods is coupled to the transport of mass. Heat is transferred from the hot oil to the product surface by convection and from the surface to the center of the food by conduction. The liquid water moves from the inside of the product to the evaporation zone, leaving the surface as vapor.

The simultaneous aspect of both heat and mass transfer, in addition to the fact that the physical properties of the material vary with changing temperature and moisture content, make theoretical treatment more complicated, and a complete analytical solution to deep-fat frying processes is not available. Several simplified models have been developed, each of which are specific with regard to the material and boundary conditions. Review of these models is the principal subject of this chapter.

PHYSICAL PROPERTIES

Some properties (often called *engineering properties*) influence heat transport in food products. Such properties are density, specific heat, and thermal conductivity. Knowledge of these properties is needed in calculation and design of processes and equipment. All of these properties are influenced to a greater or lesser extent by the temperature during heat treatment. These changes may be of importance for the quality of the product but may also affect the basic properties important in heat transport. Some of the engineering properties relevant to frying are discussed below.

Density and Porosity

Bulk density, ρ_b (kg/m^3), is defined as the mass per unit bulk volume. *Porosity, ϕ*, is defined as:

$$\phi = 1 - \frac{\rho_b}{\rho_s} = 1 - \frac{V_s}{V_b} \qquad [1]$$

where ρ_s is the density of the dry matter (or true density). Most food products contain gas that is, in general, a mixture of air and water vapor. The gas is contained in capillaries, which can be open or completely closed. If the diameter of some pores is less than 10^{-7} m, the material is said to be *capillary-porous* (Hallström et al., 1988).

Solid density, ρ_s, is the weight of the material per unit of solid volume. Generally, solid volume is considered the volume of solid material (including water) from which the volume of all open and closed pores has been excluded.

In the case of foods with a regular shape, bulk volume can be calculated from the outside geometric characteristics of the food (Karathanos et al., 1996). The bulk volume of irregularly shaped samples may be determined by volumetric displacement of glass beads (Marousis and Saravacos, 1990). Pinthus et al. (1995) measured the bulk volume of restructured fried potato using the displacement method with 100-μm diameter glass beads.

Total volume, or bulk volume, of foodstuffs has also been measured using the liquid displacement technique with toluene (Wang and Brennan, 1995; Lozano et al., 1983) or n-heptane (Zogzas et al., 1994). The method consists of immersing the sample in the compartment (A) of an apparatus as the one depicted in Figure 6-1, then closing it hermetically with the lid (C). The apparatus is filled with the liquid, and the volume displacement is measured by turning it upside down twice—once without and once with the sample immersed in the liquid. The accuracy of the measurement depends on the accuracy of the burette (B).

The solid density is generally measured using a pycnometer (stereopycnometer, multipycnometer or micropycnometer). This instrument uses a gas displacement method in which the gas is capable of penetrating all open pores up to the diameter of the gas molecule. The remainder of the sample that is not penetrated by the gas is considered as the volume of solids.

The stereopycnometer method (Karathanos and Saravacos, 1993) consists of preweighing a sample and placing it in the sample holder of known volume, V_1. The sample is outgassed by pressurization-depressurization cycles with helium gas prior to analysis. Operating conditions are ambient temperature and pressure up to 136 kPa. Helium is then introduced into the chamber and the pressure P_1 is recorded. Then, with the helium gas tank closed, a valve is opened and the helium gas enters a second chamber of known volume, V_2. The recorded pressure P_2 (equal in both chambers), is lower than P_1. From the gas law, the volume of the solid, V_s, is then calculated by:

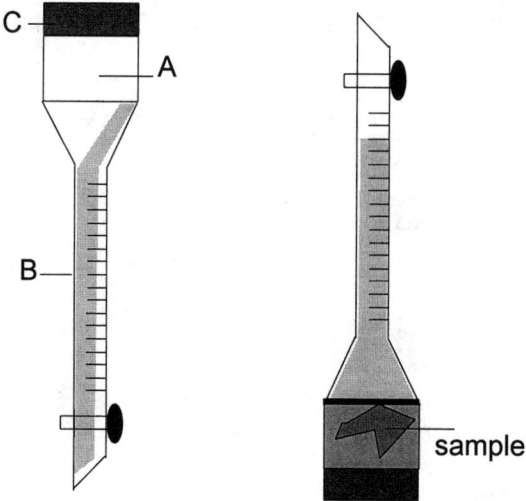

Figure 6–1 Volume Displacement Apparatus to Measure Total Volume of Food-stuffs

$$P_1(V_1 - V_s) = P_2(V_1 + V_2 - V_s) \qquad [2]$$

Moreira et al. (1995a) measured the bulk density of tortilla chips using the volume displacement method with amaranth seeds and the dry matter density with a helium multipycnometer (Quantachrome Co., Florida). Table 6–1 presents the results of bulk density, true density, and porosity of tortilla chips as a function of frying time. The bulk density decreased, the true density remained the same, and porosity increased during frying. The bulk density of raw tortilla chips was 880 ± 3 kg/m³, true density about 1,300 ± 11 kg/m³, and porosity 0.32 ± 0.02. The bulk and true densities of tortilla chips fried for 60 seconds were 579 ± 2 kg/m³ and 1,288 ± 2 kg/m³, respectively, and porosity 0.55 ± 0.04.

Pore size distribution in food products have been measured using a mercury porosimeter (Karathanos et al., 1996; Karathanos and Saravacos, 1993), stereomicroscopy (Marousis and Saravacos, 1990), and scanning electron microscopy (Lujan-Acosta, 1996). The mercury porosimeter is capable of measuring very small pores with a down limit of 0.0018 μm using a maximum pressure of 408 mPa (Quantachrome Poremaster-60, Quantachrome Co., Florida. The principle of the mercury porosimeter is based on the capillary law. The relationship between capillary pressure and the pore radius is expressed as:

Table 6–1 Bulk Density, True Density, and Porosity of Tortilla Chips As a Function of Frying Time

Frying Time (s)	Bulk Density (kg/m³)	True Density (kg/m³)	Porosity*
0	880 ± 3	1300 ± 10	0.32 ± 0.02
5	880 ± 3	1259 ± 10	0.31 ± 0.03
10	794 ± 3	1282 ± 10	0.45 ± 0.03
15	646 ± 3	1293 ± 9	0.50 ± 0.01
20	656 ± 2	1312 ± 7	0.50 ± 0.01
25	637 ± 2	1301 ± 4	0.51 ± 0.01
30	547 ± 2	1303 ± 2	0.58 ± 0.01
45	520 ± 2	1301 ± 2	0.60 ± 0.01
60	579 ± 2	1288 ± 2	0.55 ± 0.01
AVG	682 ± 3	1293 ± 6	0.42 ± 0.01

*See Equation 1.

Source: Adapted with permission from R.G. Moreira et al., Thermal and Physical Properties of Tortilla Chips as a Function of Frying Time, *Journal of Food Processing and Preservation*, Vol. 19, No. 3, pp. 175–190, © 1995, Food and Nutrition Press, Inc.

$$P = \frac{2 \cdot \gamma \cdot \cos \Theta}{r} \qquad [3]$$

where P is the applied pressure to force the mercury into the sample pores, r the sample pore radius, γ the surface tension of mercury (0.48 N/m), and Θ the contact angle of mercury on foods ($\Theta = 140°$). Based on Eq.(3), the principle of pore size distribution measurement shows that as the applied pressure increases, it causes the next smaller group of pores to be filled, with a simultaneous increase in total volume of mercury penetrated into the solid pores. From the volume versus pore radius data, the pore volume distribution and the specific surface area can then be calculated. The distribution function, $D_v(r)$, defined as:

$$D_v(r) = \left(\frac{P}{r} \right) \left(\frac{dV}{dP} \right) \qquad [4]$$

is then related to the slope of the intruded volume versus pressure curve (which is determined numerically, i.e., $dV/dP = (V_{i+1}-V_i)/(P_{i+1}-P_i)$). The specific surface area of the pores is then given by:

$$S = \left[\frac{1}{\gamma \cos\Theta}\right] \int_0^{V_F} P_i dV = \left[\frac{1}{\gamma \cos\Theta}\right] \Sigma_i P_i \Delta V_i \qquad [5]$$

where ΔV_i is the volume parameter $(V_{i+1} - Vi)$ between two pressures P_i and P_i+1. The total specific surface area corresponds to the highest volume of intruded mercury, V_F, or the highest intrusion pressure P_F (Karathanos and Saravacos, 1993).

Thermal Conductivity

Thermal conductivity (k) of a product gives in qualitative terms the rate of heat that will be conducted through a unit thickness of the material if a unit temperature gradient exists across that thickness. For porous materials consisting of a solid phase and a gas phase (two-phase system), the measured thermal conductivity is an apparent one, usually called the *effective thermal conductivity* (k_{eff}). It is an overall thermal transport property, assuming that heat is transferred by conduction through the solid and the porous phase.

Several methods for measurement of effective thermal conductivity of foods have been reviewed by Mohsenin (1980). Among these methods, the heated probe method is preferred for food materials (Sweat, 1986). Predictive models have also been used to estimate the effective thermal conductivity of foods. A number of models exist in the literature (Murakami and Okos, 1989), but many of them contain empirical factors and product-specific information. In addition, prediction of the effective thermal conductivities of heterogeneous materials depends on the heat flow direction, i.e., parallel or perpendicular to the direction of the material layers (Hallström et al., 1988).

In the case when the orientation of the material is completely random, the isotropic model developed by Kopelman (1966) and modified by Sastry (1992) can be used to predict the effective thermal conductivity of food materials. The model describes the thermal conductivity of a composite material as the combination of continuous and discontinuous phases. The approach chosen by Sastry (1992) consisted of successive determination of the thermal conductivity of two component systems, starting with water as continuous and carbohydrate as discontinuous; then using water-carbohydrate as continuous, protein as discontinuous, and continuing throughout all phases. Using the order water (phase no. 1), carbohydrate (2), protein (3), fat (4), ash (5), air (6), the following algorithm was used to evaluate the thermal conductivity of a system of $i+1$ components:

$$k_{o,i+1} = \frac{k_i\left[1 - Q_{i+1}\right]}{1 - Q_{i+1}\left[1 - (X_{d,i+1}^v)^{1/3}\right]} \qquad [6]$$

where $k_{o,i+1}$ is the composite thermal conductivity and the following definitions apply:

$$Q_{o,i+1} = \left[X_{d,i+1}^v\right]^{2/3}\left[1 - \frac{k_{i+1}}{k_i}\right] \qquad [7]$$

$$X_{d,i+1}^v = \frac{V_{i+1}}{\sum\limits_{1}^{i+1} V_i} \qquad [8]$$

where V_{i+1} is the volume of the solid phase, i.e., for $i=1$ (phase no. 1 = water) $V_{i+1} = V_2$ (volume of carbohydrate, phase no. 2) and $\sum\limits_{1}^{i+1} V_i = V_{water} + V_{carb}$. Thus, for an N-component system, the composite conductivity, $k_{o,N}$ is given by:

$$k_{o,N} = \frac{k_{N-1}\left[1 - Q_N\right]}{1 - Q_N\left[1 - (X_{d,N}^v)^{1/3}\right]} \qquad [9]$$

A procedure has been developed by Moreira et al. (1995a) to measure the thermal conductivity of tortilla chips at room temperature for different frying times using a line heat source thermal conductivity probe. Normally, when measuring thermal conductivity of a sample with the heat line source probe, the probe should be surrounded by the homogenous sample at a recommended radius of at least 0.5 cm to minimize error due to the finite sample diameter. In the case of tortilla chips, this is impossible, but an attempt was made to enclose the probe with the sample approximating the ideal case. Eight tortilla chips were "sandwiched" around the probe and pressed together with a clamp to minimize the void spaces around the probe. The force on the clamp was sufficient to flatten the chips but not enough to pulverize them. Aluminum foil was wrapped around the sample "sandwich" to prevent change in water content of the chips.

The values of thermal conductivity decreased with frying time from 0.23±0.02 W/m°C for raw tortilla chips to 0.09±0.01 W/m°C for tortilla chips fried for 60 seconds in fresh soybean oil (Table 6–2). Moisture content ranged from 35.6% (w.b.) for raw tortilla to 1.4% (w.b.) for tortilla chips fried for 60 seconds. Kopelman's equation (Eq. 9) was shown to predict well the thermal conductivity values of fried tortilla chips measured at 25°C (Moreira et al., 1995a). Using this equation to predict k_{eff} at 190°C resulted in a change in k_{eff} from 0.37 to 0.11 W/m°C as frying time increased (Table 6–2).

Elansari and Singh (1996) determined the thermal conductivity and thermal diffusivity of the crust and inner core of cylindrically shaped fried potato-based products by using DSC. The technique is based on linearly increasing heat input with time on one face of the slab while the other face and sides are insulated. A needle probe with a 21-gauge type-T was used to measure the temperature in the product. The potato was fried in a 180°C canola oil bath for 10 and 20 minutes. After frying, the crust and the core were separated and the thermal properties determined using the least squares method for the analytical solution of the heat transfer equation. The average thermal conductivity of the crust (6.25% w.b. moisture) and core (51.35% w.b.) were 0.162 and 0.545 W/m K, respectively. The average values for the thermal diffusivity were 1.05×10^{-7} m²/s for the crust and 1.265×10^{-7} m²/s for the core.

The line heat source probe accuracy is around ±2% for nonviscous fluids. In the case of fried products, a larger error is expected, due to the difficulty of placing a uniform sample around the probe. In addition, there is not a good standard to calibrate the probe in the range below 0.3 W/m°C (Sweat, 1986). However, the values obtained with the heat source probe approximate to the values obtained with DSC (Elansari and Singh, 1996).

Table 6–2 Thermal Conductivity and Specific Heat of Tortilla Chips as a Function of Frying Time

Frying Time (s)	Measured k_{eff}[a] (W/m°C) 25°C	Predicted k_{eff} (W/m°C) 190°C	Measured C_p[b] (kJ/kg°C) 190°C
0	0.23 ± 0.02	0.37	3.36 ± 0.10
5	0.20 ± 0.02	0.26	2.82 ± 0.10
10	0.13 ± 0.03	0.17	2.60 ± 0.10
15	0.12 ± 0.02	0.15	2.50 ± 0.10
20	0.11 ± 0.01	0.13	2.51 ± 0.08
25	0.11 ± 0.01	0.13	2.52 ± 0.05
30	0.10 ± 0.01	0.11	2.27 ± 0.03
45	0.10 ± 0.01	0.10	2.19 ± 0.03
60	0.09 ± 0.01	0.11	2.31 ± 0.02
AVG	0.13 ± 0.02		2.56 ± 0.10

[a]line-heat source thermal conductivity probe method; [b]DSC method.

Source: Adapted with permission from R.G. Moreira et al., Thermal and Physical Properties of Tortilla Chips as a Function of Frying Time, *Journal of Food Processing and Preservation*, Vol. 19, No. 3, pp. 175–190, © 1995, Food and Nutrition Press, Inc.

Specific Heat

Specific heat (C_p) is defined as the amount of heat needed to raise the temperature of 1 kg of material by 1 degree K. The unit is, therefore kJ/kg K. The specific heat of food materials depends very much on the composition of the material. The specific heat decreases as moisture content in the material decreases. Choi and Okos (1984) suggested the following model to estimate the specific heat of food materials based on composition:

$$C_p = 1.547 m_c + 1.711 m_p + 1.928 m_f + 0.908 m_a + 4.180 m_m \qquad [10]$$

where m is mass fraction; the subscripts are c, carbohydrate; p, protein; f, fat; a, ash; and m, moisture.

Moreira et al. (1995a) measured the specific heat of tortilla chips using DSC (Perkin-Elmer DSC-4) at 190°C. Values of C_p varied from 3.36±0.10 to 2.19±0.03 kJ/kg°C as moisture content decreased during frying (Table 6–2).

In comparing the two methods for determining the specific heat of food products, DSC is the recommended one (Sweat, 1986). The main disadvantages of this method are: (1) it requires a small sample size (5–15 mg), which may be not homogenous or representative of the product being tested, (2) it is expensive, and (3) it is time-consuming.

On the other hand, Eq. (10) should be accurate as it includes the primary components of a food product. The values of the specific heat for each component were determined statistically (Choi and Okos, 1984) for a broad range of food materials, so these tend to be averaged for all types of protein, fat, carbohydrate, and ash. The only limitation is that this model neglects interaction between components, which may be significant in some cases, but the model is probably well below an accuracy of 2–5% (Sweat, 1986).

MASS DIFFUSION COEFFICIENT

The drying behavior of biological materials can be classified into constant-rate drying period and falling-rate drying period.

Constant-drying rate period is observed in products for which the internal resistance to moisture transport is much less than the external resistance to water vapor removal from the product surfaces. The product behaves as if a thin water layer covers the surface. The moisture transport is then convection controlled.

During the falling-rate drying period, the internal resistance to moisture transport is greater than the external resistance. In this case, diffusion is the controlling mechanism for mass transport. The drying rate of food materials during the falling-rate drying period can be estimated by the unsteady-state diffusion equation (Brooker et al., 1992):

$$\frac{\partial M}{\partial t} = D\nabla^2(M).$$ [11]

The initial and boundary conditions to solve Eq.(11) are:

$$M(x,0) = M_o \text{ for } x < L/2$$
$$M(x_o,t) = M_e \text{ for } t > 0,$$ [12]

where x_o indicates conditions at the surface and $L/2$ at the center of the product; L is the product thickness, M_o the initial moisture content, M_e the equilibrium moisture content, and t the time. The analytical solution for the average moisture content for different shapes can be obtained directly from texts on heat transfer and diffusion (Crank, 1975). The analytical solution of Eq.(11) for an infinite slab with initial and boundary conditions (Eq.12) is:

$$\frac{\overline{M}(t) - M_e}{M_o - M_e} = \frac{8}{\pi^2} \sum_{n=0}^{\infty} \frac{1}{(2n+1)^2} \exp\left[-\frac{(2n+1)^2 \pi^2 Dt}{L^2}\right].$$ [13]

Simplification of the above equation has been used frequently to predict the drying of foods. Instead of an infinite number of terms, only the first term of Eq.(13) is employed to calculate the drying rate. The following semi-empirical expression results:

$$\frac{\overline{M}(t) - M_e}{M_o - M_e} = \frac{8}{\pi^2} \exp\left[-\frac{D\pi^2 t}{L^2}\right] = \frac{8}{\pi^2} \exp(-k_1 t)$$ [14]

Eqs.(13) and (14) give significantly different results only for small values of t.

A relationship similar to Eq.(14) is often used in drying analysis. It is assumed that the rate of moisture loss is proportional to the difference between the product moisture and its equilibrium moisture content:

$$\frac{\overline{M}(t) - M_e}{M_o - M_e} = \exp(-k_2 t)$$ [15]

where k_1 and k_2 are the drying constants.

Purely empirical drying equations are frequently used for foods because they represent the drying behavior better than do the theoretic and semi-empirical equations. An example is the equation proposed by Rice and Gamble (1989) for potato chips:

$$t = A + BM^2 \qquad [16]$$

where A and B are product-specific and A is expressed as a function of temperature. Baumann and Escher (1995) used the one-term exponential equation to describe the drying behavior of potato chips:

$$M(t) = a \exp(-bt) \qquad [17]$$

where a and b are the product constants.

Diffusion Coefficient and Drying Constant

The relationship between the diffusion coefficient/drying constant and the product temperature is often of the Arrhenius type:

$$D = C_1 \exp\left(-\frac{C_2}{\theta_{abs}}\right), \qquad [18]$$

where θ_{abs} is the product absolute temperature.

As examples of type of equations found in the literature for frying foods, four different equations are presented. The diffusion coefficient for french fries, based on Kozempel et al. (1991), is:

$$D = 1.39 \times 10^{-9} \exp\left(-\frac{751}{\theta_{abs}}\right) \qquad [19]$$

for temperature ranging from 458 K to 474 K. For potato chips, Rice and Gamble (1989) found the following constants for the temperature range of 418 K to 458 K:

$$D = 11.04 \exp\left(-\frac{2911}{\theta_{abs}}\right) \qquad [20]$$

Moisture diffusivity for the muscle part of chicken drums was estimated as (Ngadi et al., 1997):

$$D = 8.35 \times 10^{-6} \exp\left(-\frac{2930}{T_\infty} - 0.56 M_d + 0.092 M_d^2 \right) \quad [21]$$

where M_d is the moisture content in decimal dry basis (d.b.). For the drying Eq. (15), the value of k_2 for tortilla chips is (Moreira et al., 1997):

$$k_2 = 78.72592 \exp\left(-\frac{3074}{\theta_{abs}} \right) \quad [22]$$

for θ_{abs} in the range of 403 K to 463 K. The constant A in Eq.(16) is used to determine the drying rate of potato chips in the temperature range of 145°C to 185°C as:

$$A = 40\theta - 6036 \quad [23]$$

CONVECTIVE HEAT TRANSFER COEFFICIENT

Knowledge of the heat transfer coefficient (h') during frying is very important in modeling and design of frying systems for foods. It is defined as the rate of heat that will be convected at the product surface-fluid interaction through a unit surface area of the material if a unit temperature gradient exists between the product surface and the surrounding fluid. The methods reported in the literature for measuring heat transfer coefficients during frying have the same limitations as discussed by Hallström et al. (1988), that is:

• there is lack of a standard measurement technique
• mass transfer is seldom taken into consideration
• effects of packing are neglected
• there is lack of knowledge of local heat transfer coefficient (i.e., function of position of the material)
• there is lack of uniformity in estimating and reporting errors in measurement.

Methods of measuring the convective heat transfer coefficient can be divided into three classes (Hallström et al., 1988):

- *Steady-state measurement of surface temperature:* Based on Newton's law of cooling, this method requires constant conditions during the test and no mass transfer. It is limited to food products because physical properties of food-stuffs change continuously during processing.
- *Transient measurement of temperature:* This technique is based on transient temperature measurements. The heat transfer coefficient can be determined by using the lumped capacity method, which essentially assumes that the temperature is uniform in the object during heating. This is accomplished by having a material with a thermal conductivity value much larger than the convective heat transfer coefficient, so that the Biot number ($Bi = [(hd)/k]$) will be lower than 0.1.

Figure 6–2 illustrates the temperature change with time of a copper sphere ($k = 401$ W/m K) transducer submerged in soybean oil heated to 190°C. Based on the lumped capacity method (Incropera and DeWitt, 1996), energy balance on the sphere results in:

$$\frac{\theta(t) - T_\infty}{\theta_o - T_\infty} = \exp\left[-\left(\frac{h' A_s}{\rho_p V C_p}\right) t\right] \qquad [24]$$

so that h' can be calculated by fitting Eq.(25) to the experimental data using nonlinear regression:

$$\theta(t) = (\theta_o - T_\infty) \exp\left[-\left(\frac{3h'}{\rho r C_p}\right) t\right] + T_\infty \qquad [25]$$

where T_∞ is the oil temperature (190°C), θ_o the initial temperature of the copper sphere (25°C), A_s the surface area of the sphere (with a 25.4-mm diameter), V the object volume (for a sphere $A_s / V = 3/r$, where r is the radius of the sphere), ρ_p the copper sphere density (8954 kg/m³), C_p the sphere thermal conductivity (383 J/kg K). The value of h' obtained from the data shown in Figure 6–2 is 282.1 W/m² K.
- *Heat flux measurement of surface temperature:* This method requires using a sensor that can detect heat flux at the surface (and change in temperature at the surface with time). This method would account for the mass transfer, but the presence of the device would alter the surface of the body, making the readings inaccurate.

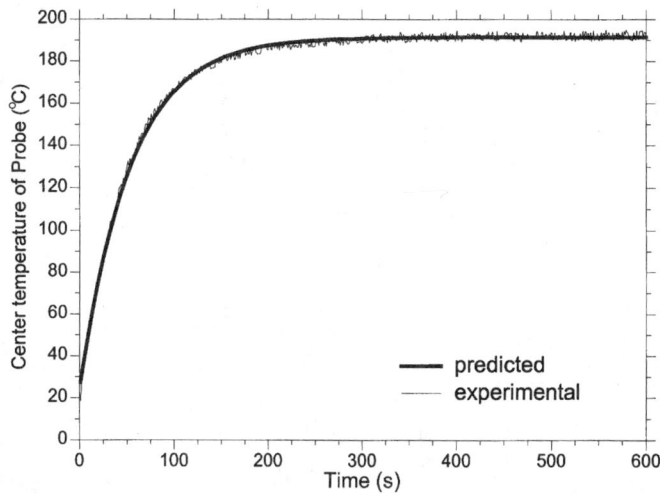

Figure 6–2 Temperature History of Copper Sphere Transducer Immersed in Soybean Oil Heated to 190°C

Table 6–3 presents a summary of the techniques that have been used to measure the convective heat transfer coefficient in deep-fat frying systems. Dagerskog and Sörenfors (1978) examined the heat transfer in deep-fat fried meat patties at 160°C. Temperatures at the center of the patties were recorded using a thin copper/constantan thermocouple (2-mm diameter), and surface temperature was assumed to be 135°C (close to the temperature obtained under the same conditions in convection heating). The convective heat transfer coefficient can then be estimated by the following relationship:

$$h'S(T_\infty - \theta_s) = (\rho_p C_p + \overline{M}\rho_w C_{pw})\frac{d\theta}{dt} + h_{fg}\rho_p \frac{d\overline{M}}{dt} \qquad [26]$$

where S is the specific surface area, ρ_w the water vapor density, M the average moisture content, C_{pw} the specific heat of water vapor, and h_{fg} the latent heat of vaporization. Hallström (1979) reported that during frying of meat patties, two periods could be observed. Before water evaporation starts, the heat transfer coefficient was found to be 200–300 W/m² K. During water evaporation, the turbulence close to the surface increases the heat transfer coefficient to values up to 978 W/m²°C at 190°C.

Mittelman et al. (1984) used the lumped capacity method to determine the h' value of a copper sphere submerged in soybean oil at 135–175°C. They found the

Table 6–3 Methods used to Determine the Convective Heat Transfer Coefficient (*h*) of the Oil Film During Frying

Method	Probe/Product	Temperature (°C)	Oil Type*	h' (W/m²°C)	Reference
Transient temperature and mass transfer	meat patty	160	hydrogenated	500 ± 200	Dagerskog and Sörenfors (1978)
Transient temperature and mass transfer	meat patty	130–190	hydrogenated	600 ± 300	Hallström (1979)
Lumped capacity	solid copper sphere (70 mm diam T-type thermocouple)	135–175	soybean	180	Mittelman et al. (1984)
Lumped capacity	solid aluminum sphere (12.2 mm)	170–190	canola soybean palm	251–264 261–276 249–271	Miller et al. (1994)
Lumped capacity	solid copper sphere (25.4 mm diam E-type thermocouple)	190	soybean	279	Tseng et al. (1996)
Transient temperature and mass loss	aluminum wool filled with potato mash (50x50x3 mm)	150–190	sunflower	80–180	Sahin and Sastry (1997)
Transient temperature and mass loss	potato (20x80 mm cylindrical shape)	120–180	canola	200–3500	Hubbard and Farkas (1998)

*Fresh oils.

value to be $180 W/m^2 °C$ and not affected by the oil temperature in the range studied. Miller et al. (1994) measured the heat transfer coefficients of canola, soybean, and palm oils at temperatures ranging from 170°C to 190°C. A spherical aluminum transducer was used to determine the convective heat transfer coefficient. Values for h' ranged from 251 to 276 $W/m^2 °C$. Palm oil was the most affected by temperature. Tseng et al. (1996), using a copper sphere, found the convective heat transfer coefficient to be 279 $W/m^2 °C$ for soybean oil heated at 190°C, close to the value found by Miller et al. (1994).

Sahin and Sastry (1997) used aluminum wool filled with potato mash to determine the convective heat transfer coefficient of sunflower oil film heated from 150°C to 190°C. Temperature at the center of the samples was measured and the heat transfer calculated by solving the heat transfer equation. Hubbard and Farkas (1998) measured the temperature and mass loss of potato cylinders continuously on line during frying in canola oil for 10 minutes at 120°C, 150°C, and 180°C. The convective heat transfer coefficient was calculated based on a single equation similar to Eq.(26). The h' values varied during frying and increased with oil temperature. Maximum values of h' were measured at 1700 W/m^2 K, 2300 W/m^2 K, and 3500 W/m^2 K for oil temperatures of 120°C, 150°C, and 180°C, respectively.

SINGLE-PIECE FRYING

Many physical, chemical, and nutritional changes occur in foods during deep-fat frying. Many of these changes are functions of oil temperature and quality, product moisture and oil contents, and product residence time in the fryer. Undesirable effects could be minimized and the process could be better controlled if temperature, moisture, and oil distributions in food with respect to time could be accurately predicted.

In general, the frying process of a single product such as french fries can be divided into four stages (Farkas et al., 1996a):

- *Initial Heating:* During the initial heating stage, usually lasting a few seconds, the food submerges in the heated oil and its surface is heated to the boiling point of water (around 103°C). The mode of heat transfer between the oil and the food product during this short period of time is due to natural convection, and no evaporation of water occurs from the surface of the food.
- *Surface Boiling:* The vaporization process signals the start of the surface boiling stage. The convection mode of heat transfer changes from natural convection to forced convection because of the presence of considerable turbulence in the oil surrounding the food. During this stage, the crust begins to form at the surface of the food.

- *Falling Rate:* During this stage, more internal moisture leaves the food, and the internal core temperature rises to the boiling point. This stage is similar to the falling-rate period observed in food dehydration processes. The crust layer continues to increase in thickness, and after sufficient time and more removal of moisture, the vapor transfer to the surface decreases.
- *Bubble End Point:* If the food is fried for a considerably long period of time, the rate of moisture removal diminishes, and no more bubbles are seen escaping the surface of the food. As the frying process proceeds, the crust layer continues to increase in thickness.

The extent to which each of these stages exists depends on initial conditions, process parameters, and product characteristics (shape, size, composition).

Many attempts have been made to combine heat and mass transfer principles to describe the temperature and moisture content profiles in a product in deep-fat frying processes (Ateba and Mittal, 1994; Moreira et al., 1995b; Farkas et al., 1996a,b).

The single-piece frying models can be divided into two types, namely, heat and mass balance (semi-empirical) and DE (differential equation) models. The first type is treated partially for its historic interest and partly to place it in the context of a more general model. A number of DE models, one for meatballs (crustless), one for tortilla chips (all crust), and one for potato slabs (crust and crumb), is presented and given a much more detailed treatment in view of greater accuracy and applicability over a wider range of frying problems.

Heat and Mass Balance Models

Mittelman et al. (1984) developed frying models for crustless (using a rigid sponge as a model system) and crust-forming products (french fries). The rigid sponge model was supposed to represent porous rigid foods that would not form a crust of low water-vapor permeability during frying. Examples of such food products are: bread croutons, extruder puffed cereal snacks, soy protein, and coarse breading. Moisture content during frying was found to be proportional to the square root of frying time and the difference between oil temperature and boiling temperature of water. For french fries, an ordinary differential equation similar to Plank's equation for freezing was proposed to predict the crust thickness of the product during frying.

The limitations of the models are evident: (1) the procedure assumes constant physical and thermal properties of the product during frying; (2) the model neglects oil absorption during the process; (3) the models use only the initial boiling point of water in the computation equation and neglect the time required to add sensible heat below evaporation point; and (4) the models did not address the transient temperature and moisture profiles in the crumb. Keller and Escher (1989) developed a mathematic model for the frying of potato sticks similar to the model of Mittelman et al. (1984) with the addition of a term for the sensible heat required to heat the dry crust region from the boiling point of water to the oil temperature.

Differential Equation Models

DE frying models are based on the laws of heat and mass transfer and lead to rather complicated systems of equations. The models can be solved only with the aid of digital computers.

General Model

During frying, two regions are formed in the product. The outer layer, or *crust*, is characterized by having a temperature higher than the boiling point of water and a low moisture content. The core or *crumb* region consists basically of the interior part of the product. The crust/crumb interface moves toward the center of the product during frying. At the interface, the temperature remains at the water boiling point for a short period of time to allow for the water present in that region to evaporate.

Figure 6–3 shows a schematic of an infinite slab-shaped material (such as chips, french fries, etc.) undergoing frying. Heat (q) is transferred by convection from the oil to the surface of the chip and by conduction to the center of the chip. There is, however, a certain transfer of heat coupled to the transfer of water (M) or vapor that is the energy carried by the water vapor. Most of the water escapes from the product in the form of vapor during frying, and a small percentage of the frying oil also diffuses into the material. Diffusion of moisture and diffusion of oil are in two opposite directions. The moisture content decreases, oil content increases, and the product becomes more porous during frying.

Thermal and physical properties of the product change greatly during frying. The bulk density of the food material decreases, and the food becomes more porous. The thermal conductivity decreases as the porosity increases, the specific heat decreases as moisture content decreases, and oil content increases during frying (Moreira et al., 1995a).

When a product is taken out of the fryer, it is covered with a thin layer of oil. As the temperature of the product decreases by natural convection with ambient air, the vapor pressure within the pores of the product decreases, forcing the surface oil to flow into the chips. Moreira et al. (1997) observed that almost 80% of the oil is absorbed by the tortilla chips during the cooling period. Investigation of the cooling process is very important to fully understand the frying mechanism. The mechanism of oil absorption during cooling is described in Chapter 7.

The following simplified assumptions can be made in the development of the DE frying models:

1. The product to be fried is homogeneous and isotropic.

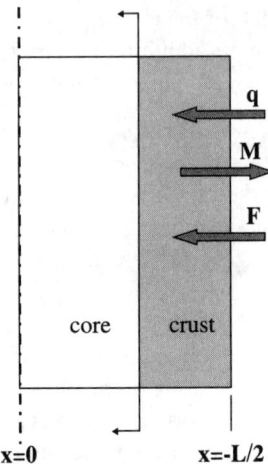

Figure 6–3 Schematic of an Infinite Slab during Deep-Fat Frying. *Source:* Adapted with permission from Y. Chen and R.G. Moreira, Modelling of a Batch Deep-Fat Frying Process for Tortilla Chips, *Transactions of the Institution of Chemical Engineers*, Vol. 75(C), pp. 181–190, © 1997, Institute of Chemical Engineers.

2. Initial moisture and temperature in the product are uniformly distributed.
3. The heat required for chemical reactions (i.e., starch gelatinization, protein denaturation) is small compared with the heat required to evaporate the water.
4. Thermal and physical properties are functions of local temperature and moisture content during the frying process.
5. A "microscopically uniform" porous medium is formed after frying. The surface of the chip is covered with a uniform layer of oil after frying, and most of the oil diffuses into the product during the cooling period.

Governing equations. Because of its symmetry, the computational domain may be simplified to a half section of a product. The temperature (θ), moisture content (M), and oil content (F) in the product change dramatically during frying. Energy and mass balances are written on a differential volume (Figure 6–3). The basic one-dimensional heat and mass diffusion equations are employed. Thus, three balances result in four equations.

- The governing differential equation describing the temperature change in the product during frying is:

$$\frac{\partial}{\partial x}\left(k\frac{\partial \theta}{\partial x}\right) - \frac{\partial(\Gamma C_{pw}\theta)}{\partial x} = \frac{\partial(\rho_b C_p \theta)}{\partial t} \qquad [27]$$

The second term in the left side of the equation represents the heat transfer caused by diffusion of water vapor, where Γ is the water vapor flux:

$$\Gamma = -\frac{\partial(\rho_b D_w M)}{\partial x} \qquad [28]$$

- Fick's law of diffusion is used to calculate the mass transfer rate in the product in two different directions: moisture (water vapor) diffuses from the chip to the oil, and the oil diffuses from the surface to the center of the chip:

$$\frac{\partial}{\partial x}\left(D_w \rho_b \frac{\partial M}{\partial x}\right) = \frac{\partial(\rho_b M)}{\partial t} \qquad [29]$$

$$\frac{\partial}{\partial x}\left(D_f \rho_b \frac{\partial F}{\partial x}\right) = \frac{\partial(\rho_b E)}{\partial t} \qquad [30]$$

Eqs.(27)–(30) describe temperature and moisture changes at all points in the chip (i.e., crumb and crust). Inside the crust, there is an evaporation zone that moves toward the center of the product. In the evaporation zone, the temperature is constant, and the energy is mainly used to evaporate the water. The duration of this constant period depends on the water content in that location. As the water content is reduced at the level corresponding to M_e, the temperature increases rapidly, and this part of the product becomes part of the crust. Different thermal properties are used for the two different zones in the product. When the temperature of the product is higher than the boiling temperature of water, the properties of vapor instead of liquid water are used to predict temperature and moisture changes in the product. The crust has thermal and physical properties of an insulating material. Its low thermal conductivity and porosity slow down the rate of heat transfer and, therefore, the rate at which the product cooks and water vaporizes.

Two boundary conditions and one initial condition are needed for each governing equation. At $x=L/2$ (center line) for any time, no temperature, moisture, or oil gradient exist at the center of the product,

$$\frac{\partial \theta}{\partial x} = 0 \;;\; \frac{\partial M}{\partial x} = 0 \;;\; \frac{\partial F}{\partial x} = 0 \qquad [31]$$

At the surface ($x_o=0$). for any time, the energy transferred by convection from the oil to the product's surface is equal to the sum of energy required for transferring heat to the center of the product by conduction, for evaporating water from the product, and for heating the water vapor evaporated from the product at temperature θ to the oil temperature T:

$$h'(\theta_{sur} - T) = -k\frac{\partial\theta}{\partial x} + h_{fg}\Gamma + \Gamma C_{pw}(\theta_{sur} - T) \qquad [32]$$

The second term at the right side of Eq.(32) is eliminated when the temperature of the product is above the boiling point of water, with two mass transfer boundary conditions:

$$k_d \rho_b (M_{sur} - M_\infty) = -\Gamma \qquad [33]$$

where $M_\infty = M_e \simeq 0$, i.e., the moisture content of the surrounding medium; and:

$$k_f (F_{sur} - F_\infty) = -D_f \frac{\partial(F)}{\partial x} \qquad [34]$$

where $F_\infty = F_e \simeq 1$, i.e., oil content of the surrounding medium.

The initial conditions for any location x in the product at time zero are the following:

$$\theta(x,0) = \theta_o ; \ M(x,0) = M_o ; \ F(x,0) = F_o \qquad [35]$$

The above three governing equations and nine boundary conditions can be solved simultaneously by using finite difference techniques. Heat and mass transfer equations are coupled by the transport properties and thermal properties, which are functions of moisture content and temperature (Chen and Moreira, 1997). Change in bulk density was calculated as function of moisture content during the process as a means of accounting for change in the product porosity.

Solution of the mathematic model. A control-volume formulation to discretize the governing equations, initial conditions, and boundary conditions can be used to solve those sets of DE. The continuous physical space (infinite slab) is divided into a number of nonoverlapping control volumes in the x-direction, such that there is one control volume surrounding each node. The differential equations are integrated over each control volume. The control volume formulation method obtains the finite difference equation by applying conservation of energy to a control volume around each node. The most attractive feature of this method is that the resulting solution would imply that the integral conservation of energy is exactly satisfied over any group of control volumes and over the entire calculation domain (Chen and Moreira, 1997).

After the discretization of heat and mass transfer equations for every node, a set of simultaneous algebraic relationships are obtained and solved to obtain the transient temperature, moisture content, and oil content at each node.

Grid sensitivity and stability. From the mathematical point of view, the development of the mathematical model cannot end at this step. Even when the finite difference equations have been properly formulated and solved, the results might still represent a coarse approximation to the exact solution. A numeric simulation enables determination of the temperature at only discrete points, which represents the average value of the surrounding region. However, the finite difference approximations can be made more accurate as the nodal network is refined. The cost is that the computer takes longer CPU time to complete the iterations. Grid sensitivity study should be performed to define when the computed results no longer depend on the choice of Δx and Δt.

The finite difference technique can be used to solve the previous equations. The temperature, moisture content, and oil content of any node at $t+\Delta t$ can be calculated from the knowledge of temperatures, moisture contents, and oil contents at the same and neighboring nodes for the preceding time t. Thus, determination of a nodal temperature and moisture/oil content at some time is independent of temperatures and moisture/oil content at other nodes of the same time. In this method, the choice of Δx is based on a compromise between accuracy and computational time requirements, as mentioned above. Once this selection has been made, however, the value of Δt may not be chosen independently. Instead, it is determined by stability requirements. For a one-dimensional node, the following criterion should be used to select the maximum allowable value of Fo, and, thus, dt, to be used in the calculation:

$$Fo(1+Bi) \leq \frac{1}{2}$$ [36]

where Fo is Fourier number, $Fo = 4at/L_2$; the thermal diffusivity, $a = \dfrac{k}{\rho C_\rho}$; L the thickness; t the time; and Bi the Biot number, $Bi = h'L/2k$ (Chen and Moreira, 1997).

Particular Models

Ateba and Mittal (1994) were among the first to present a DE model for the simulation of simultaneous heat and mass transfer in deep-fat frying. They developed a model to describe the frying of meatballs by dividing the process into two fat transfer periods: (1) the fat absorption period, consisting of diffusion of the frying fat from the product, and (2) the fat desorption period, consisting of the migration of the fat from the product, to the frying oil by capillary flow. They made the assumptions that properties were constant during frying and neglected the effects of crust formation on physical properties. Moisture was assumed to diffuse in liquid state throughout the product and as vapor at the surface only. In addition, the product surface moisture and fat content reaches the equilibrium values instantaneously. During the desorption period, they assumed that the fat content in the crumb remains constant and increases at the product's surface during the process. The desorption of fat is then described by:

$$V \frac{dF}{dt} = -k_f A_s (F - F'_e)$$ [37]

where k_f is the fat conductivity, V the product volume, and F'_e the equilibrium fat content (minimum fat content below which there is no transfer of fat).

Using a similar approach, Moreira et al. (1995b) simulated the deep-fat frying of tortilla chips. The main differences were that: (1) oil content was assumed to be mainly absorbed during the cooling period and (2) the diffusion equation was solved with a convective-type boundary condition. The tortilla chips model has been improved by considering changes in physical properties of the material with temperature and moisture content (Chen and Moreira, 1997) and is described by Eqs.(27)–(35).

Farkas et al. (1996a,b) developed a mathematic model to describe the frying of a potato product (infinite slab). The main differences were: (1) the process was viewed as a moving boundary problem similar to the freezing and freeze-drying situations; (2) the product was divided into crust and crumb (same as Figure 6–3) and mass and energy equations were developed for each region; (3) the vapor moves in crust region due to pressure differences, and (4) the mass fraction of oil in the fried product was assumed to be negligible with negligible effect on other mass and energy fluxes. In this model, shrinkage of the material was assumed negligible. The resulting set of equations are the following:

Crumb region 0<x<X(t)

$$k^{II} \frac{\partial^2 \theta}{\partial x^2} + N_{\beta x} C_{p\beta} \frac{\partial \theta}{\partial x} = (\varepsilon_\beta \rho_\beta + \varepsilon_\sigma \rho_\sigma C_{p\sigma}) \frac{\partial \theta}{\partial t}$$ [38]

$$N_{\beta x} = -D_{\beta\sigma} \frac{\partial c_\beta}{\partial x} \tag{39}$$

$$N_{yx} = -\frac{\rho_\gamma K_\gamma}{\mu_\gamma} \frac{\partial P_\gamma}{\partial x} \tag{40}$$

$$\frac{\partial c_\beta}{\partial t} = D_{\beta\sigma} \frac{\partial^2 c_\beta}{\partial x_2} \tag{41}$$

Crust region X(t)<x<L

$$(\varepsilon_\gamma \rho_\beta C_{p\gamma} + \varepsilon_\sigma \rho_\sigma C_{p\sigma}) \frac{\partial \theta}{\partial t} = k^I \frac{\partial^2 \theta}{\partial x^2} + N_{yx} C_{p\gamma} \frac{\partial \theta}{\partial x} \tag{42}$$

$$\frac{\partial}{\partial x}\left[\rho\gamma \frac{\partial P_\gamma}{\partial x} \right] = 0 \tag{43}$$

where the subscripts β, γ, and σ refer to liquid water, water vapor, and solid, respectively; superscripts I and II are crust and crumb regions, c is the concentration, K the permeability, μ the viscosity, P the pressure, and N the mass fluxes in the x-direction. Sets of 8 boundary conditions and 3 initial conditions are specified to solve these equations. The boundary condition at the crust-crumb interface is defined as:

$$-k^I \frac{\partial T}{\partial x} + k^{II} \frac{\partial T}{\partial x} - \frac{\rho_\gamma K_\gamma}{\mu_\gamma} \frac{\partial \rho_\gamma}{\partial x}\left(h'_\beta - H_\gamma \right) = \left[\varepsilon_\sigma \rho_\sigma \left(h'^I_\sigma - h'^{II}_\sigma \right) + \varepsilon_\gamma \rho_\gamma \left(H_\gamma - h'_\beta \right) \right] \frac{DX}{dt} \tag{44}$$

and the interfacial mass balance:

$$\frac{dX}{dt} = \frac{N_{\beta x} - N_{yx}}{\varepsilon_\gamma \rho_\gamma - \varepsilon_\beta \rho_\beta} \tag{45}$$

Ni and Datta (1998) presented a mechanistic model to describe the frying of potato based on the volume average approach of a multiphase porous media, as described by Whitaker (1977). The model considers the transport of liquid water, vapor, air, and oil separately. Three modes of mass transport are used in the model: diffusional, capillary, and pressure-driven (Darcy) flow. The model predicts that most of the oil concentrates at the crust region and is all absorbed during frying. Shrinkage was ignored in this model (the food product was assumed non-hygroscopic).

BATCH FRYING

Isothermal frying is possible only when frying a single piece of a product. Depending on the fryer size, oil volume, batch size, and water content of the product, a temperature drop of 30°C to 45°C of the frying oil can be observed in industrial operations (Benson et al., 1992). Thus, the practical importance of this information is limited because foods are seldom fried as individual pieces. Instead, tortilla and potato chips, french fries, etc., are fried either in a stationary (batch fryer) or moving bed (continuous fryer).

Batch frying is quite different from the single-chip frying process. For single-chip frying, the temperature of the oil is assumed to be constant, whereas, for a batch process, changes in temperature of the oil depend on how many chips are fried at the same time. The temperature of the oil decreases greatly at the moment the chips are placed into the oil, then increase gradually to the setting temperature, as observed in Figure 6–4.

Chen and Moreira (1997) presented a model to describe the frying of a batch of tortilla chips. In addition to the assumptions presented by the general model, the following was considered:

> The volume of frying oil in the fryer kettle to the volume of chips ratio is about 15, i.e., about 20–50 chips are placed in a 5×10^{-4} m³ basket, then into a fryer kettle containing 7.5×10^{-3} m³ of oil. It was previously observed that only the oil that was in direct contact with the chips (in between the chips) was affected by the chip moisture loss during frying, whereas the change in temperature of the rest of the oil in the kettle was negligible. Therefore, it was assumed that the fryer consists of two oil zones, and the exchange of energy between these two zones is by convection only. This would not be true for a small ratio of volume oil/ volume chips.

Figure 6–5 shows a schematic of the batch of chips surrounded by the oil.

The temperature of the oil decreases significantly during the first seconds of frying when tortilla chips are dropped into the fryer. The change in enthalpy of the oil with respect to time in the void space (between chips) is equal to the sum of energy required for: (a) heating the product; (b) evaporating water from the chips;

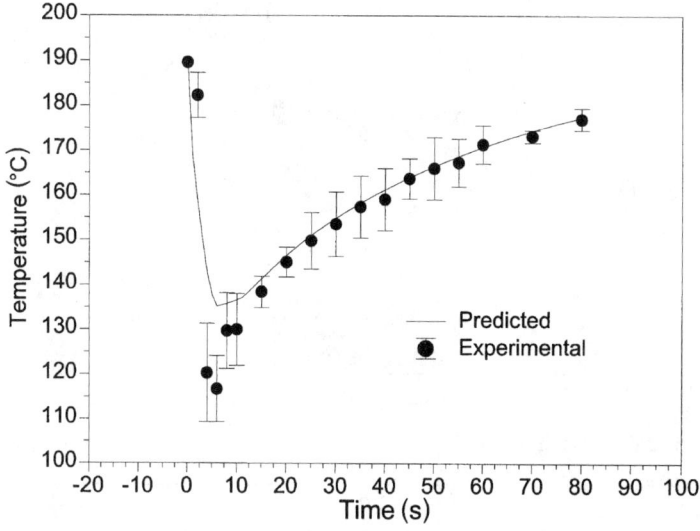

Figure 6–4 Predicted and Experimental Results of Tortilla Chip Average Moisture Content during Frying as Affected by Frying Oil Temperature. Initial moisture content, 44% w.b.; chip thickness, 1.6 mm; batch frying of 20 chips. *Source:* Adapted with permission from Y. Chen and R.G. Moreira, Modelling of a Batch Deep-Fat Frying Process for Tortilla Chips, *Transactions of the Institution of Chemical Engineers*, Vol. 75(C), pp. 181–190, © 1997, Institute of Chemical Engineers.

(c) heating the water vapor evaporated from the chips; and (d) exchanging energy to the surrounding oil. The changes in the *temperature of the oil* are described by:

$$\rho_{oil}C_{pf} = \frac{\partial T}{\partial t} = \left[-k\frac{\partial \theta}{\partial x} + h_{fg}D_w\frac{\partial(\rho_b M)}{\partial x} + \Gamma C_{pw}(\theta_{sur} - T)\right]\frac{A}{(1-\Phi)} + h_s'S(T_{fo} - T) \quad \textbf{[46]}$$

where Φ is the porosity of the bed of tortilla chips.

Eq.(46) is used to calculate the temperature of the oil between the chips at each time step, and this value is then used in Eqs.(27)–(30) to obtain the temperature, moisture content, and oil content profiles in the chip during batch frying process.

The frying of a batch of tortilla chips was simulated with an IBM-compatible computer with a time step of 0.01 seconds and a distance step 0.1 mm. The computer program to solve the above problem was written using MATLAB (The MathWorks, Natick, Massachusetts).

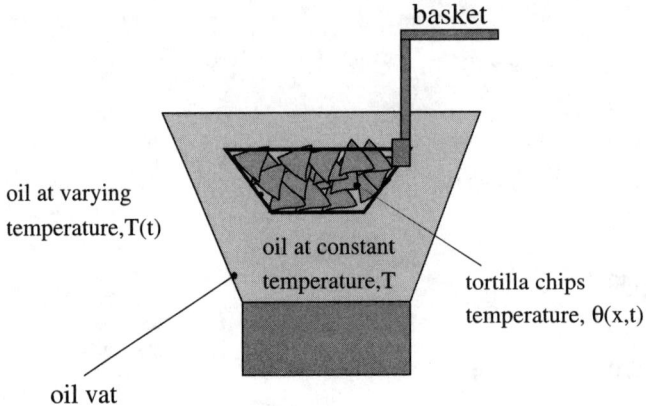

Figure 6–5 Batch Frying of Tortilla Chips

Analysis of the Frying Process

Frying Oil Temperature

This is an important parameter in frying because the oil serves as heating medium in the process. Figure 6–6 shows the temperature history at the center of the chip using different frying oil temperature set points (160°C and 190°C). This is a sensible heating period at the beginning of drying, while the temperature of the chips increases up to the boiling temperature of water. The chip temperature then remains at the boiling point for several (3–5) seconds. During this period, all the heat has been used to evaporate the water. When most of the free water is lost, the chip temperature begins to increase up to the temperature of the frying oil (second sensible heating period). During this period, the center temperature in the chip increases much faster for the chip fried in the oil at 190°C than for the one fried in the oil at 160°C. However, in both cases, the temperature of the chip reaches the temperature of the frying oil after 60 seconds of frying. A similar phenomenon was observed by Farkas et al. (1996b), who suggested that the temperature profile in the crust region was mainly a function of oil temperature whereas the temperature profile in the core region was unaffected by the oil temperature.

The rate of moisture content decrease of the chips is higher at a higher oil temperature (Figure 6–6). The higher the oil temperature, the higher will be the diffusion coefficient and, thus, the higher the mass transfer of water vapor. To achieve the same final moisture content (2% w.b.), chips fried at 190°C oil temperature need only 60 seconds of frying whereas 90 seconds is required to fry the chips in the frying oil maintained at 160°C.

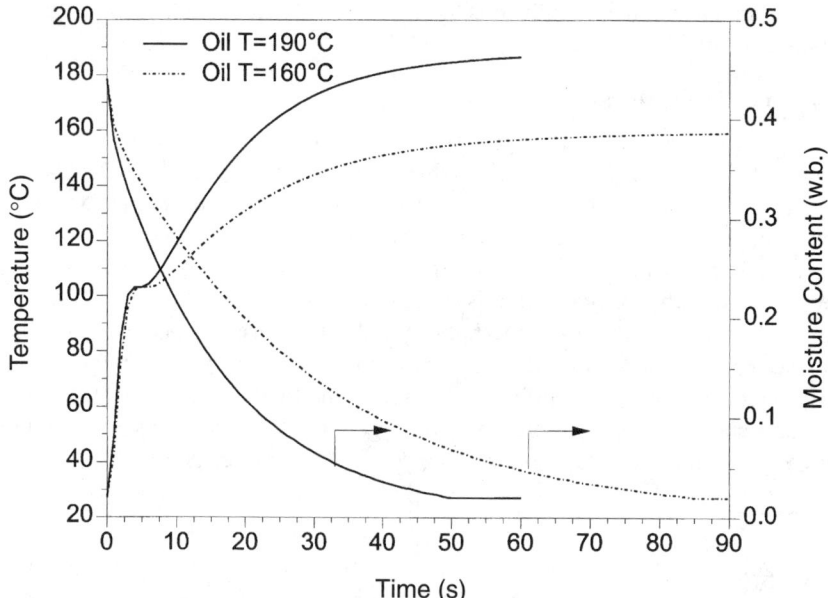

Figure 6–6 Effect of Oil Temperature on the Center Temperature and Average Moisture Content of Tortilla Chips during Frying. Initial moisture content, 44% w.b.; chip thickness, 1.6 mm; batch frying of 20 chips. *Source:* Adapted with permission from Y. Chen and R.G. Moreira, Modelling of a Batch Deep-Fat Frying Process for Tortilla Chips, *Transactions of the Institution of Chemical Engineers*, Vol. 75(C), pp. 181–190, © 1997, Institute of Chemical Engineers.

It was observed (Moreira et al., 1995b) that as the water evaporates from the chip, during the first 10 seconds of frying, the product becomes harder (crust formation), due to faster water evaporation, resulting in the formation of a structure with a number of small pores. As frying continues, the pores start to enlarge (due to vapor expansion), the material expands and becomes crispier (decreased hardness). In the last stage of frying (30–60 seconds), the pores stop expanding, due to the increase in the matrix complex viscosity, as the amount of plasticizing water has been greatly reduced. Similar to drying, it is believed that the average T_g of the material elevates above the frying temperature, thus stopping expansion rates and increasing porosity. The bulk porosity is then formed only at the end of the frying process.

The results showed that temperature of the oil must be kept as uniform as possible to reduce the frying time. This phenomenon is more pronounced in industrial operation, where the mass of product per volume of oil is larger. Larger temperature variations within the continuous fryers will result in longer frying time than in isothermal situations.

Tortilla Chip Thickness

Figure 6–7 shows the effect of chip thickness on the center temperature. As expected, the thinner chip reaches a higher temperature more quickly. The plateau for the thicker chip is much longer because the thermal resistance increases significantly, due to the low thermal conductivity of the frying product.

Moisture removal rate is slower for the chips with a thickness of 2.6 mm than for the chips with 1.6-mm thickness (Figure 6-7). More than 100 seconds are required for the thicker chips to reach the equilibrium moisture content (2% w.b.). Thicker chips result in a fried product with lower oil content. A previous study also suggested that higher surface/mass ratio of the food increases oil absorption (Gamble and Rice, 1988). Therefore, it is suggested that thicker chips will result in less oil content but will take longer to cook to the desired crispness.

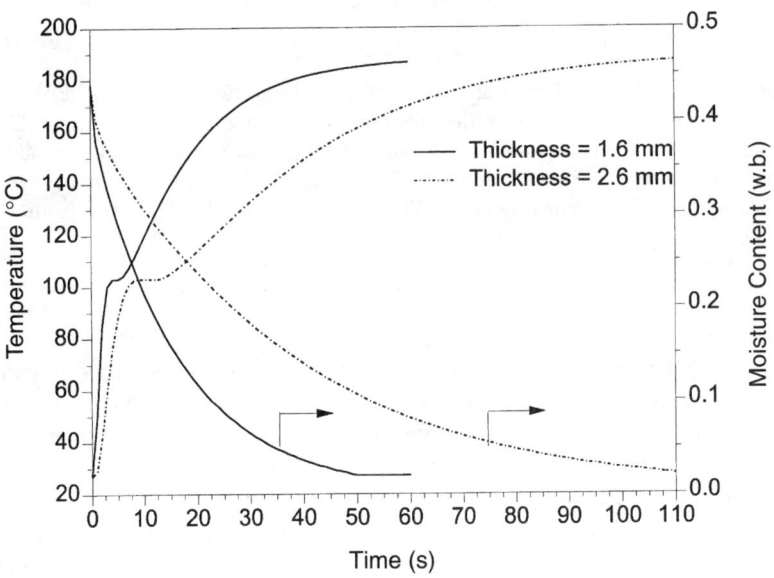

Figure 6–7 Effect of Thickness on the Temperature and Average Moisture Content of Tortilla Chips during Frying. Oil temperature, 190°C; initial moisture content, 44% w.b.; chip thickness, 1.6 mm; batch frying of 20 chips. *Source:* Adapted with permission from Y. Chen and R.G. Moreira, Modelling of a Batch Deep-Fat Frying Process for Tortilla Chips, *Transactions of the Institution of Chemical Engineers*, Vol. 75(C), pp. 181–190, © 1997, Institute of Chemical Engineers.

Initial Moisture Content

Figure 6–8 shows the temperature at the center of the tortilla chip. The temperature of the chip with higher initial moisture content (54% w.b.) is a little lower at the beginning of the frying process and soon reaches the same temperature of the chip with lower initial moisture content (27% w.b.). The temperature profile is not greatly affected by the chip's initial moisture content during frying. The higher the initial moisture content, the higher will be the diffusion coefficient and, thus, the higher the mass transfer of water vapor.

Moisture loss rate is faster for the tortilla chips having higher initial moisture content (Figure 6–8). The equilibrium moisture content is reached in 50 seconds of frying for both cases. It is known from a previous study (Moreira et al., 1997) that chips with higher initial moisture content will have higher final oil content. This is related to the effective porosity of the chip. Chips with higher initial moisture content have more space available for oil absorption after the moisture is removed during frying.

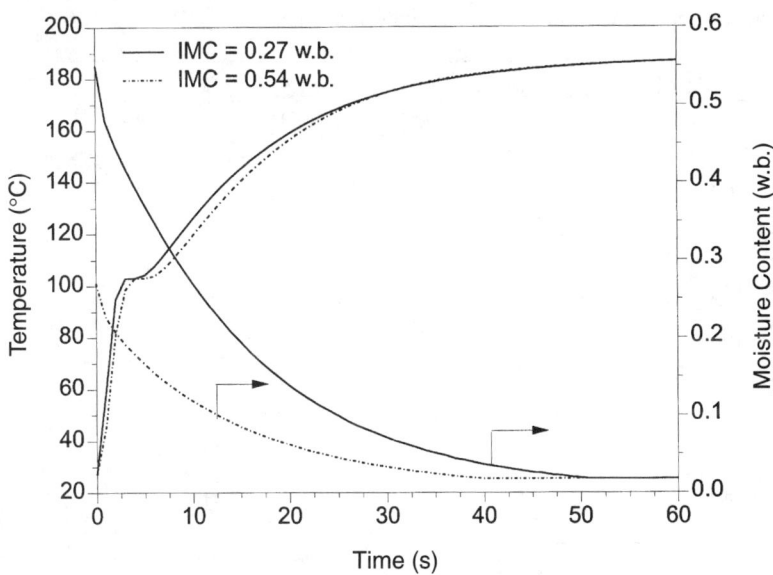

Figure 6–8 Effect of Initial Moisture Content on the Temperature and Average Moisture Content of Tortilla Chips during Frying. Oil temperature, 190°C; chip thickness, 1.6 mm; batch frying of 20 chips. *Source:* Adapted with permission from Y. Chen and R.G. Moreira, Modelling of a Batch Deep-Fat Frying Process for Tortilla Chips, *Transactions of the Institution of Chemical Engineers*, Vol. 75(C), pp. 181–190, © 1997, Institute of Chemical Engineers.

Although initial moisture content has little effect on the frying time, it will affect the oil content of the product substantially. These results are discussed in detail in Chapter 7. The crispness of the chips is also affected by the initial moisture content, i.e., high initial moisture chips are more brittle than are lower-moisture chips (Moreira et al., 1997).

Batch Frying Process

Figure 6–9 shows the center temperature of the chips during batch frying. The temperature of the chip increases slowly when more chips (about 60) are fried at the same time. The temperature of the oil drops more dramatically (volume of oil/volume chips decreased) when more chips are fried at the same time, as shown in Figure 6–10.

The results clearly show that, as in typical commercial operations, a temperature gradient within the frying oil can affect the residence time of the product in the fryer and the quality of the final product. Ways of minimizing temperature differences in oil batches will result in a more efficient process. As discussed in Chapter 5, stirring and forced immersion systems are examples of techniques used to produce better products in commercial operations.

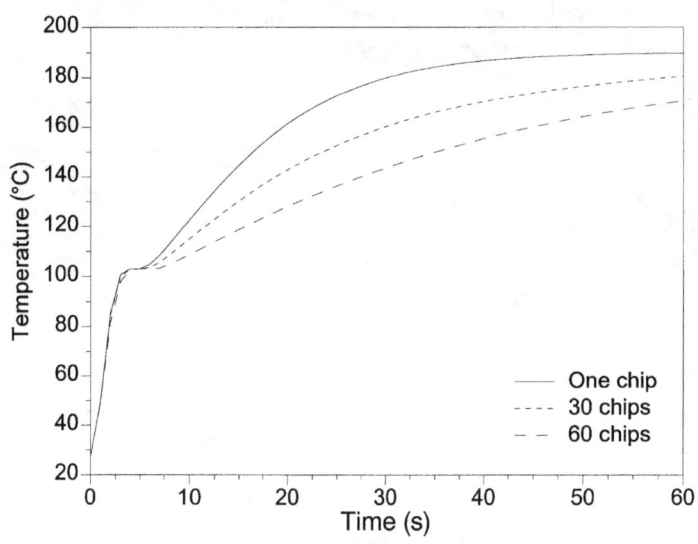

Figure 6–9 Effect of Batch Frying Process on the Center Temperature of Tortilla Chips during Frying. Oil temperature, 190°C; chip thickness, 1.6 mm; initial moisture content, 44% w.b. *Source:* Adapted with permission from Y. Chen and R.G. Moreira, Modelling of a Batch Deep-Fat Frying Process for Tortilla Chips, *Transactions of the Institution of Chemical Engineers*, Vol. 75(C), pp. 181–190, © 1997, Institute of Chemical Engineers.

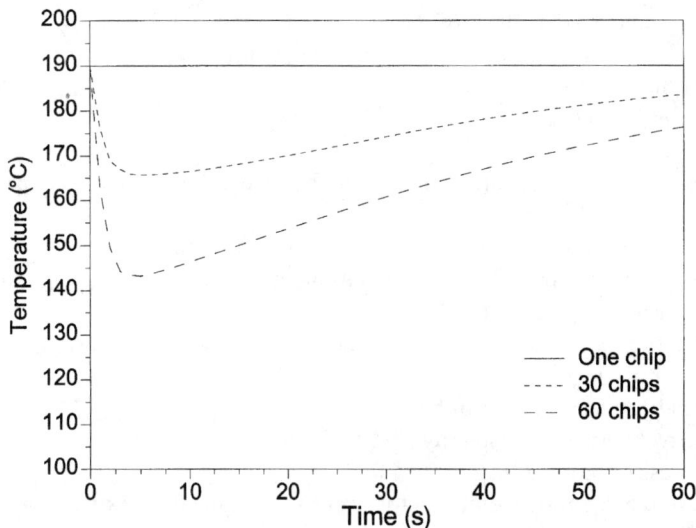

Figure 6–10 Effect of Batch Frying Process on the Frying Oil Temperature during Frying. Settling Temperature of the oil, 190°C; chip thickness, 1.6 mm; initial moisture content, 44% w.b. *Source:* Adapted with permission from Y. Chen and R.G. Moreira, Modelling of a Batch Deep-Fat Frying Process for Tortilla Chips, *Transactions of the Institution of Chemical Engineers*, Vol. 75(C), pp. 181–190, © 1997, Institute of Chemical Engineers.

NOTATIONS

A   constant in Eq.(16)

A_s surface area [m²]

a   constant in Eq.(17)

Bi   Biot number [ ]

C_1 constant in Eq.(18)

C_2 constant in Eq.(18)

C_p specific heat of chip [J/kg K]

C_{pf} specific heat of oil [J/kg K]

C_{pw} specific heat of water vapor [J/kg K]

D_f mass diffusivity of oil [m²/s]

D_w mass diffusivity of moisture [m²/s]

F   oil content of the product in wet basis, defined as: (kg oil absorbed)/(kg product) [% w.b.]

Fe   final oil content at surface interface between batch and the bath oils [% w.b.]

Fo   Fourier number

F_∞ oil content of the surrounding oil [decimal, w.b.]

F_o initial value of oil content of the product [decimal, w.b.]

F_{sur} surface oil content of the product [decimal, w.b.]

K   permeability [m²]

k_1 constant in Eq.(14)

k_2 constant in Eq.(15)

h' convection heat transfer coefficient between the oil and the product [W/m² K]

k_d convection mass transfer coefficient of water vapor [m/s]

k_f convection mass transfer coefficient of oil at the surface of chips [m/s]

h_{fg} latent heat of vaporization [J/kg]

h'_s convection heat transfer coefficient between frying oil and surrounding oil [W/m² K]

k   thermal conductivity [W/m K]

k_{eff} effective thermal conductivity [W/m K]

L   thickness of product [m]

M   moisture content of the product in wet basis, defined as: (kg water evaporated)/(kg product) [w.b.]

$\overline{M}$ average moisture content [w.b.]

M_d moisture content of the product in dry basis, defined as: (kg water evaporated)/(kg product – evaporated water) [d.b.]

Me   equilibrium moisture content [w.b.]

M_∞ moisture content of the surrounding oil [w.b.]

M_o initial value of moisture content [w.b.]

M_{sur} surface moisture content of the product [w.b.]

P   pressure [Pa]

r   radius [m]

S   specific surface area [m²/m³]

t   time [s]

T_∞ temperature of the frying oil [°C]

T_{abs} absolute temperature of the frying oil between the product [K]

T_{fo} set temperature of the frying oil [°C]

T_{fo} initial value of oil temperature [°C]

V   volume [m³]

V_s volume of the solid [m³]

V_b volume of the bulk [m³]

x   product thickness [m]

X   crust thickness [m]

α   thermal diffusivity [m²/s]

ε   volume fraction [m³/m³]

φ   porosity of the bed of product [ ]

γ   surface tension [N/m]

μ   viscosity [Pa.s]
ρ_p density of the product [kg/m³]
ρ_b bulk density of the product [kg/m³]
ρ_{oil} density of the oil [kg/m³]
ρ_s density of the solid [kg/m³]
Θ   contact angle [°]
θ   temperature of the product [°C]
θ_0 initial values of temperature [°C]
θ_{sur} surface temperature of the product [°C]
θ_{abs} absolute temperature of the product [K]

REFERENCES

Ateba, P., and Mittal, G.S. 1994. Modelling the deep-fat frying of beef meatballs. *Intl J Food Sci Technol*, 29:429–440.

Baumann, B., and Escher, F. 1995. Mass and heat transfer during deep-fat frying of potato slices. I: Rate of drying and oil uptake. *Lebensm-Wiss U-Technol*, 28(4):395–403.

Benson, C.K.; Caradis, A.A.; and Klein, L.F. 1992. *Continuous food processing methods*. United States Patent number 5,137,740.

Brooker, D.B.; Bakker-Arkema, F.W.; and Hall, C.W. 1992. *Drying and storage of grains and oilseeds*. New York: Van Nostrand Reinhold.

Chen, Y., and Moreira, R.G. 1997. Modelling of a batch deep-fat frying process for tortilla chips. *Trans Chem Engr*, 75(C):181–190.

Choi, Y., and Okos, M.R. 1984. *The thermal properties of liquid foods—review*. Presented at the 1983 Winter Meeting of the American Society of Agricultural Engineers, paper no. 83-6517.

Crank, J. 1975. *The mathematics of diffusion*. Oxford: Claredon Press.

Dagerskog, M., and Sörenfors, P. 1978. A comparison between four different methods of frying meat patties: heat transfer, yield, and crust formation. *Lebensm-Wiss U-Technol*, 11:306–311.

Elansari, A.M., and Singh, R.P. 1996. Determination of thermal properties of fried foods using the differential scanning calorimeter. IFT Annual meeting. Orlando, FL. June 14–18.

Farkas, B.E.; Singh, R.P.; and Rumsey, T.R. 1996a. Modeling heat and mass transfer in immersion frying. I. Model development. *J Food Eng*, 29:211–226.

Farkas, B.E.; Singh, R.P.; and Rumsey, T.R. 1996b. Modeling heat and mass transfer in immersion frying. II. Solution and verification. *J Food Eng*, 29:227–248.

Gamble, M.H., and Rice, P. 1988. The effect of slice thickness on potato crisp yield and composition. *J Food Eng*, 8:31–46.

Hallström, B. 1979. Heat and mass transfer in industrial cooking. In *Food process engineering*, Vol 1, ed. Linko, Malkki, Olkku, and Larinkari. London: Applied Science Publishers Ltd., 457–465.

Hallström, B.; Skjöldebrand, C.; and Trägårdh, C. 1988. *Heat transfer and food products*. New York: Elsevier Science.

Hubbard, J.J., and Farkas, B.E. 1998. *Oil temperature effects on the convective heat transfer coefficient during immersion frying*. Annual meeting. Atlanta, GA. June 20–24.

Incropera, F.P., and DeWitt, D.P. 1996. *Fundamentals of heat and mass transfer*, 3rd. ed. New York: John Wiley and Sons.

Karathanos, V.T.; Kanellopoulos, N.K.; and Belessiotis, V.G. 1996. Development of porous structure during air drying of agricultural plant products. *J Food Eng*, 29:167–183.

Karathanos, V.T., and Saravacos, G.D. 1993. Porosity and pore size distribution of starch materials. *J Food Eng*, 18:259–280.

Keller, C., and Escher, F. 1989. Heat and mass transfer during deep-fat frying of potato products. Intl. Congress on Engineering and Food 5. Cologne, Germany. May 25–June 1.

Kopelman, I.J. 1966. *Transient heat transfer and thermal properties in food systems*. PhD diss. W. Lafayette, IN: Purdue University.

Kozempel, M.; Tomasula, P.M.; and Craig, J., Jr. 1991. Correlation of moisture and oil concentration in french fries. *Lebensm-Wiss U-Technol*, 24:445–448.

Lozano, J.E.; Rotstein, E.; and Urbicain, M.J. 1983. Shrinkage, porosity and bulk density of foodstuffs at changing moisture content. *J Food Sci*, 48(5):1497–1502, 1553.

Lujan-Acosta, F.J. 1996. *Production of low-fat tortilla chips using alternative methods of drying before frying*. Thesis. College Station, TX: Texas A&M University.

Marousis, S.N., and Saravacos, G.D. 1990. Density and porosity of drying starch materials. *J Food Sci*, 55(5):1367–1372.

Miller, K.S.; Singh, R.P.; and Farkas, B.E. 1994. Viscosity and heat transfer coefficients for canola, corn, palm, and soybean oil. *J Food Process Preserv*, 18:461–472.

Mittelman, N.; Mizrahi, S.; and Berk, Z. 1984. Heat and mass transfer in frying. In *Engineering and food, Vol. 1: Engineering sciences in the food industry*, ed. B. M. McKenna. New York: Elsevier Applied Science Publishers.

Mohsenin, N.N. 1980. *Thermal properties of foods and agricultural materials*. New York: Gordon and Breach Science Publishers.

Moreira, R.G.; Palau, J.E.; Sweat, V.E.; and Sun, X. 1995a. Thermal and physical properties of tortilla chips as a function of frying time. *J Food Process Preserv*, 19:175–189.

Moreira, R.G.; Palau, J.E.; and Sun, X. 1995b. Deep-fat frying of tortilla chips: an engineering approach. *Food Technol*, 49(4):146–152.

Moreira, R.G.; Sun, X; and Chen, Y., 1997. Factors affecting oil uptake in tortilla chips in deep fat frying. *J Food Eng*, 31:485–498.

Murakami, E.G., and Okos, M.R. 1989. Measurement and prediction of thermal properties of foods. In *Food properties and computer-aided engineering of food processing systems*. San Diego: Academic Press.

Ngadi, M.O.; Watts, K.C.; and Correa, L.R. 1997. Finite element method modelling of moisture transfer in chicken drum during deep-fat frying. *J Food Eng*, 32:11–27.

Ni, H., and Datta, A.K. 1998. *Moisture, oil and energy transport during deep fat frying of food materials*. Submitted to the transactions of the Institute of Chemical Engineers.

Pinthus, E.J.; Weinberg, P.; and Saguy, I.S. 1995. Oil uptake in deep-fat frying as affected by porosity. *J Food Sci*, 60: 767–772.

Rice, P., and Gamble, M.H. 1989. Technical note: modelling moisture loss during potato slice frying. *Intl J Food Sci Technol*, 24:183–187.

Sahin, S., and Sastry, S.K. 1997. *The determination of convective heat transfer coefficient during frying*. IFT annual meeting. Orlando, Florida. June 14–18.

Sastry, S.K. 1992. *Modeling thermal conductivity of foods*. Personal communication.

Sweat, V.E. 1986. Thermal properties of foods. In *Engineering properties of foods*, ed. Rao and Rizvi. New York: Marcel Dekker.

Tseng, Y.; Moreira, R.G.; and Sun, X. 1996. Total frying-use time effects on soybean oil deterioration and on tortilla chip quality. *Intl J Food Sci Technol*, 31:287–294.

Wang, N., and Brennan, J.G. 1995. Changes in structure, density and porosity of potato during dehydration. *J Food Eng*, 24:61–76.

Whitaker, S. 1977. Simultaneous heat, mass and momentum transfer in morous media: a theory of drying. In *Advances in Heat Transfer*, 13:119–203.

Zogzas, N.; Maroulis, Z.B.; Marinos-Kouris, D.; and Saravacos, G.D. 1994. Densities, shrinkage and porosity of some vegetables during air drying. In *Drying 94*, Vol. B, eds. V. Rudolph and R.B. Keey. Proceeding of the 9th International Drying Symposium, 863–870.

CHAPTER 7

Oil Absorption in Fried Foods

Deep-fat frying is a complex process that involves heat and mass transfer mechanisms and a variety of physical and chemical changes in both food and frying oil. Understanding the mechanism of oil uptake is crucial to product quality control as oil content is one of the most important quality factors of fried foods in deep-fat frying processes. Most of the research in oil absorption during deep-fat frying has been based on total oil content related to process conditions and moisture loss. Recently, quantitative results have been presented in the literature regarding oil distribution in foods during frying. An interesting finding was the importance of the cooling period on oil uptake. The main objective of this chapter is to introduce the basic concepts describing the mechanisms of oil absorption and those required to predict this phenomenon in fried foods.

THE PHENOMENON OF OIL ABSORPTION

As shown in Chapter 5, several methods were used to describe oil distribution in fried products. However, these techniques are all postfrying data acquisition, i.e., they are based on analysis after removing the product from the fryer. Therefore, they do not give conclusive evidence as to when and how the oil is actually transferred to the products.

An initial interpretation of the mechanism of oil absorption in fried products proposed that most of the oil enters the product from the adhering oil being pulled into the product when it is removed from the fryer, due to the condensation of steam in the product pores, which produces a vacuum (Gamble et al., 1987). It suggested that oil absorption is dependent not only on the amount of moisture lost but also on how the moisture is lost and that products with large quantities of small "moisture loss sites" will absorb less oil. However, no experimental results were presented to prove this theory.

Oil absorption is expected to be product dependent. A study on tortilla chips suggests that most of the oil does not penetrate the product during frying, but rather during the cooling period, when the product is removed from the fryer (Sun and Moreira, 1994). This was determined experimentally by dipping the chips immediately after frying into a beaker containing petroleum ether. The oil that is dropped in the beaker is collected by evaporating the petroleum ether, and this oil is defined as *surface oil content* (SOC) (% w.b.) = $(W_1/(W_1+W_2))\times100$; where W_1 is the weight of oil collected in the beaker (g) and W_2 is the weight of 10 fried chips after the surface oil has been extracted (g). The remaining oil content in the chips (% w.b.) is defined as *core oil content* and is determined using the Soxhlet extraction method. Some of the interesting findings are described below.

During frying (60 seconds), the amount of oil absorbed by the chips was around 4.3% (w.b.) for the overbaked chips (initial moisture content = 27.2% w.b.), 5.8% (w.b.) for optimally baked chips (44.1% w.b.), and 7.8% (w.b.) for the unbaked chips (56.7% w.b.) (Figure 7–1). These values remained constant throughout the entire frying period, with exception of the unbaked chips, which showed a slight increase in oil content during the first 20 seconds of frying. These values represented about 20–24% of the final oil content in all cases. During cooling, when the chips were removed from the fryer, the oil content increased sharply, reaching a maximum value at about 160 seconds after frying, i.e., when the chips' temperature was reduced to the ambient value. Almost 64% of the total oil content was absorbed by tortilla chips during the cooling (postfrying) process, indicating that the mechanism of oil absorption may be related to capillary pressure differences and, thus, interfacial tension between the oil and gas within the pore spaces. The effect of cooling time on the oil uptake of raw cubic shaped masa (2.54 × 2.54 × 2.54 cm) fried for 5 minutes at 190°C in fresh soybean oil can be observed in Color Plate 7. As the cooling time increases, the samples become darker, indicating that the oil content increases and concentrates mainly in the crust region.

Oil uptake has been primarily described as a surface phenomena involving an equilibrium between adhesion and drainage of oil during the postfrying process (cooling period) (Ufheil and Escher, 1996). Oil did not penetrate the potato slices during frying but did when the slices were removed from the fryer.

Oil absorption and deterioration increase nonlinearly with frying time. A comparison between optimally baked tortilla chips fried in fresh and degraded (1.2% FFA; 66.0% TPM) soybean oil showed that more oil is absorbed during cooling by the chips fried in fresh oil (20.5% w.b.) than by the chips fried in the degraded oil (15.0% w.b.) (Figure 7–2). The total final oil content, however, was roughly the same for the chips fried in the degraded oil (24.0% w.b.) and those fried in fresh oil (25.5% w.b.) (Tseng et al., 1996; Moreira et al., 1997). Most of the oil absorbed by the chips fried in degraded oil was concentrated at the surface. The

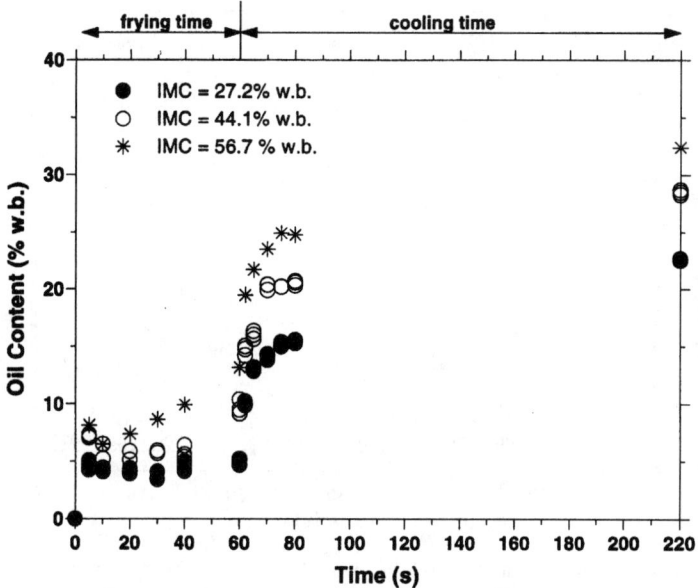

Figure 7–1 Oil Absorption in Tortilla Chips during Frying and Cooling Time. IMC, initial moisture content.

higher viscosity of the degraded oil could cause the oil to adhere to the product's surfaces, making it more difficult for the oil to be drained off from the chip's surface during cooling.

Reducing the residence time of the product in the takeout conveyor of a continuous fryer (Chapter 9) results in less oil allowed to drain off from the product surface and, therefore, in higher oil content during cooling. In the takeout conveyor, the product is not in the oil but is still under the fryer hood, thus maintaining its temperature closer to the oil temperature. As the temperature of the product is reduced during cooling (after the product is removed from the fryer), the capillary pressure inside the pore spaces increases, allowing for the surface oil to flow into the product. The decrease in temperature of the oil also causes the viscosity at the surface of the product to increase, making it more difficult for the oil to drip off from the product's surface. Therefore, an increase in the takeout conveyor speed results in more oil in the final product (Brescia and Moreira, 1997).

Experimental evidence suggests that oil absorption in fried foods can be described by the theory of fluid flow in a porous media. What follows is an introduction to these concepts as they are applied to frying.

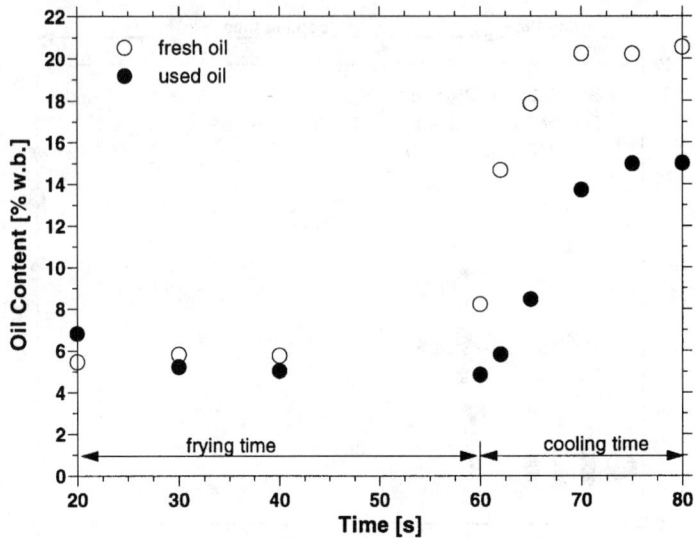

Figure 7–2 Effect of Fresh and Used Soybean Oil in the Oil Absorption of Tortilla Chips (Fried for 60 seconds at 140°C)

THEORETICAL TREATMENT OF POROUS MEDIA

Any material that contains pores or void spaces within the solid material is considered a porous media. Among the commonly analyzed porous media such as petroleum reservoir cores, soils, concrete, and other granular materials, solid material can be considered a packed structure of spherical grains of various sizes. The volume fraction of voids within the media (or porosities) range from 0.03 to 0.4 for spherical-type granular materials. Batches (deep-beds) of agricultural products such as cereal grains also represent porous media composed of either spherical or cylindrical solid matter. Bulk grain porosities can vary from 0.37 for sorghum to 0.51 for oats, both at a moisture content of 9.4% (w.b.). Porosities in foods can change from 0.01 to 0.6 during frying, usually at the crust region of the product.

The composition, or pore structure, of food affects such properties as porosity and density, as well as the transport of fluids through the media. For example, the permeability of fluid flow through porous food is dependent upon interconnectedness of the pores. At the pore level (microscopic scale) the properties are highly dependent upon the location within the media. For example, porosity in the medium can vary from 1.0 at the pore to 0.0 at the solid phase. However, there is great variability of porosity over different regions of consideration in a porous medium, making it difficult to predict the flow behavior or properties using a microscopic approach.

Whitaker (1977) developed the volume-averaging approach to treat the transport phenomena in porous media as continuum. Typically, any porous media are considered a continuum if any bulk behavior can be described by an average value. The elementary volume described by Whitaker (1977) is shown in Figure 7–3. This volume is large enough to mute individual, microscopic effects, so any properties averaged or any phenomena described over this volume are considered to be fairly constant, macroscopic properties or effects.

The mass transport theory in foods is not well developed because of their complex physical and chemical composition that may change during processing. Most foods are complex systems composed of materials in various forms, making mass transfer analysis more complicated than simple homogeneous systems. In the case of solid foods, mass transfer operations are based partially on empirical knowledge (Saravacos, 1995).

During frying, most fried products develop a porous structure, due to the removal of water as vapor. A *porous food product* can be defined as a solid body that contains holes or pores. These pores are filled with substances different from the solid structure, i.e., gases or liquids. A more mathematical definition is given by Scheidegger (1954), who describes the porous medium as a set A of space occupied by solid matter and its complementary set A^{o}, which is the pore space. The two sets are defined by the function:

$$f(x)=1 \text{ for } x \in A \text{ and } f(x)=0 \text{ for } x \in A^{o} \qquad [1]$$

These spaces should be distributed more or less frequently through the material if

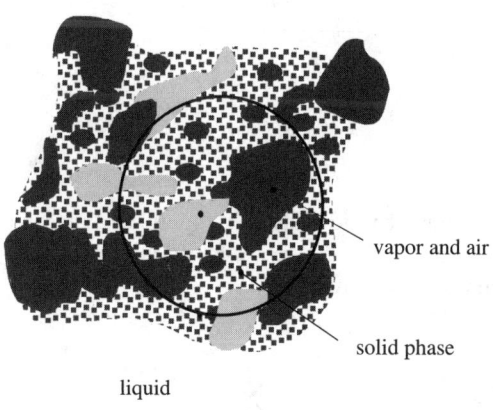

vapor and air

solid phase

liquid

averaging volume

Figure 7–3 Schematic of a Porous Medium.

it is to be called *porous*. Extremely small voids in a solid are called *molecular interstices*, and very large ones are called *caverns*. Pores are void spaces intermediate in size between caverns and molecular interstices; the limitation of their size is therefore intuitive and rather indefinite (Chatzis, 1980).

The pores in a fried product may be *interconnected* or *noninterconnected*. The interconnected part of the pore system is called the *effective* pore space of the porous medium. This connectivity is what matters in studying oil absorption in fried foods. The oil should be able to be transported throughout the entire porous structure, just as in a network of pipelines. An example of a material that is not considered suitable to conduct fluids is swiss cheese because, in spite of having holes distributed throughout the entire structure most of the holes are blind or isolated. The effective porosity in this system is zero or negligible.

PROPERTIES OF POROUS FOOD MATERIALS

Only the properties that may affect oil transfer in frying processes will be discussed in this section.

Porosity

A porous structure can be classified by the bulk porosity *(ϕ)* of the material. This property plays an important role in determining oil absorption in fried products. In that case, the pore volume may be filled with both gas and liquid. The *true* porosity is then defined as:

$$\phi = \frac{\sum_{k=1}^{p'} V_k}{V_b} \qquad [2]$$

where p' is the number of fluid phases, V the volume, and k can be oil, gas, or water. Most of the time, the measured porosity is the gas porosity (ε), which is related to the total porosity as follows:

$$\phi = \frac{\sum_{k=1}^{p'-1} V_k}{V_b} + \varepsilon \qquad [3]$$

If the calculation of porosity is based upon the interconnected pore space rather than on the total pore space, the resulting quantity is termed the *effective* porosity.

The pore system of a porous body forms a very complicated surface, one that is difficult to describe geometrically. Figure 7–4 shows an environmental electron scanning micrograph of a tortilla chip. For the purpose of analysis, one would like to talk about the "size" of pores, such as the "diameter," which is related to the cross-sectional area available for flow. However, the term *diameter* makes sense geometrically only if all the pores were spherical—unless some further specifications are made. In general, the pores will not be circular, and they will not even possess a normal cross-section as the walls may be irregularly diverging and converging. For simplification, pores must be visualized as rather tube-shaped objects and the pore throat diameter (or bottle neck) must be associated with the smallest diameter of the cross-sectional area available for flow (Nicholson and Petropoulus, 1973).

Pore Size Distribution

Pore size distribution, as the statistical description of a porous medium, is simply defined by determining what fraction of the total pore space has a "pore diameter" between α and $\alpha + d\alpha$. It is easily seen that the pore size distribution thus defined is normalized as follows:

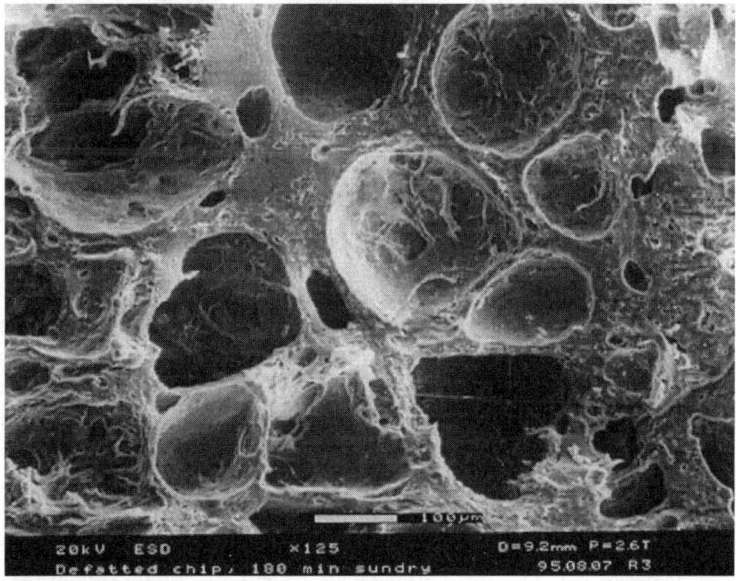

Figure 7–4 Cross-Sectional Area from ESEM of a Tortilla Chip Fried for 60 s at 190°C

$$\int_{a_{min}}^{a_{max}} f(a)da = 1 \qquad\qquad [4]$$

Instead of dealing with individual pore sizes, it is often more convenient to deal with the cumulative pore size distribution. This is denoted as $F(\alpha_i)$, and this quantity is defined as the fraction of the pore space, which has a pore diameter smaller than α_i.

Once the "pore size distribution" is defined, it can be characterized by fitting various standard distribution curves known from mathematical statistics, such as the Gauss curves. A mean value and a variance characterize the normal distribution. A wider shape implies a more heterogeneous medium. The parameters of the standard distribution curves will then serve as parameters for the pore size distribution curves (Tiab and Donaldson, 1996).

Tortuosity

Another geometric property of importance is the *tortuosity* (τ). It represents how much the pores deviate from being straight capillary conducts, and it is defined as the ratio of the length of a true flow path for a fluid to the straight-line distance between inflow and outflow. It is thus a dimensionless quantity whose physical meaning is quite clear, but for computation, it is often treated as an adjustable parameter (Collins, 1976).

STATISTICAL DESCRIPTION OF POROUS MEDIA

Predicting the macroscopic properties of porous materials is an important area of research. Large-scale computational methods exist today that allow calculation of the properties of porous materials based on the digital representation of their microstructure (Roberts, 1997). Simulating the porous media using first principle is complicated since it requires a detailed knowledge of the physics and chemistry of the system, which in most cases is not available. Therefore, an alternative is to reconstruct the microstructure of these materials using statistical methods. The system is considered an ensemble of porous media equivalent to each other. The system becomes a random porous medium whose geometrical properties can be related to some easily measurable statistical properties of the medium.

A stochastic description of fried tortilla chips used Monte Carlo techniques and a cluster counting code, which indicated the number of clusters and size of particles of a given type (Moreira and Barrufet, 1996). The spatial distribution of the product was described in terms of oil, water, gas, and solid sites randomly distributed in a cubic lattice. Color Plate 8A shows a two-dimensional (2-D) realization of this media before and after frying, respectively. For the purpose of a graphic representation, these sites are represented as geometrical spheres; however, these are randomly allocated throughout the structure. The oil was uniformly distributed in the lattice, not following a preferential orientation. Cluster counting for water, oil, solids, pores, and oil-coated solids after frying showed that most of the pores are interconnected in a single cluster (i.e., uninterrupted paths connecting all boundaries of the tortilla chip) and that most of the oil and water are located in finite isolated clusters. Even though there is a continuum in the pore space available for oil uptake, this space remains oil free at the end of the process. This cluster analysis represents a volume balance in which the tortilla chip structure is discreted in 10,000 cells, each occupied with oil, water, air, or dry matter. For example, the volume occupied by water is 60/10,000—this is translated into a 1.1% water content. The volumetric mass balance has been derived out of the material balance as:

$$V_{TC} = \frac{m_w}{\rho_w} + \frac{m_o}{\rho_o} + \frac{m_s}{\rho_s} + \frac{m_g}{\rho_g} \qquad [5]$$

where V_{TC} is the total volume of tortilla chips and m and ρ are the mass fraction and density of water (w), oil (o), solid (s), and gas (g).

This technique helps visualize the spatial oil and pore size distribution in a piece of tortilla chip. This code can be used as 3-D image analysis software so that the structure formation of the product can be visualized in a dynamic way. Because this is a stochastic model, the generated structures (Color Plate 8B) will not be alike. However, they do mimic what is observed by MRI techniques (Color Plate 9). In general, most of the oil concentrates inside of the puffed areas of the chips.

In conclusion, statistical modeling of porous media has great interest in the prediction of effective properties of the media. To optimize the properties of porous food systems, for example, it is necessary to understand how morphology influences effective properties such as permeability, conductivity, diffusivity, elastic moduli, etc. This understanding is crucial in the simulation of foods during drying, frying, freezing, and many other processes encountered in the food industry.

FLUID TRANSPORT MECHANISMS

During frying of foods, the transport phenomena of fluids such as water (liquid and vapor), oil, and air must be analyzed. These fluids each contribute to the overall transport of mass and heat in the food during the frying process.

Food products are considered hygroscopic materials in which some water is tightly bound to the solid matrix. Geankopolis (1993) defined *bound water* as the equilibrium moisture content of a material that is continued to its intersection with the 100% humidity line. This water in the solid exerts a vapor pressure less than that of liquid water at the same temperature. The excess water above the 100% humidity line is called *unbound water*, and it is held primarily in the pores of the solid. Bound water may exist under different conditions: moisture in the cell walls (may have solids dissolved in it) and liquid water in capillaries of very small diameter (concave curvature at the surface).

Water vapor and air can be lumped together as a single gas phase, and oil will flow as liquid to fill the pore space during the process. Therefore, the convective contribution to the frying process due to bulk fluid flow motion can be ideally characterized as a multiphase flow in porous media problem.

The most common macroscopic momentum equation used to characterize fluid velocities for flow in porous media is Darcy's law. For a laminar flow (Re<1), homogeneous porous material, Newtonian fluid, no slip, and isotropic media, the generalized form of Darcy's law for a single phase flow in any direction is:

$$\vec{u} = -\frac{K}{\mu}(\nabla P - \rho\vec{g}) \qquad [6]$$

where u is the velocity, K the absolute permeability, μ the Newtonian viscosity, P the pressure, and g the acceleration due to gravity.

The macroscopic equations for multiphase flow in a porous medium are typically written as generalizations of Darcy's law for single phase laminar flow in a porous medium that apply to both steady and unsteady state flow (Marle, 1981):

$$\vec{u} = -\frac{K_{ri}K}{\mu_i}(\nabla P_i - \rho_i\vec{g}) \qquad [7]$$

where $K_{ri}K$ is defined as the *effective permeability*, K_i, of the medium to each fluid i, and K_{ri} the *relative permeability* for the fluid i. The pressure differential between two immiscible fluids (wetting and nonwetting) at any point in the porous media is defined as the capillary pressure, P_c. Therefore, the pressure gradients of the two phases can be related by their capillary pressure difference (Leverett, 1941):

$$\nabla P_c = \nabla P_{nw} - \nabla P_w \qquad [8]$$

In the case of frying, the two immiscible fluid combinations present in the food are oil-gas, water-gas, and water-oil. For the water-gas and oil-gas interfaces, water and oil are considered the wetting fluids and the gas the nonwetting fluid, since the contact angle at a liquid/gas and liquid/solid interface is between 0° and 90° when measured from the liquid side. The contact angle defines the *wettability*, i.e., the ability at which the liquid will spread over a solid surface. The water-oil interface mainly occurs during the first seconds of frying and can be considered negligible. The concepts of wettability and capillary pressure are discussed in the next sections.

Eq.(7) is generally valid for a porous medium in which the contact angle between the two fluids is very small (Dullien, 1992). In this situation, the two fluids are separated by a stable interface, and capillary forces at the liquid/gas interface control the two-phase flow. Preferentially, the larger pores will be filled with the nonwetting fluid, the smaller pores with the wetting fluid. However, some wetting fluid will remain in the large pores as thick film along the pore wall forming a continuous network, while the nonwetting fluid will flow through the center of the pores. In this case, the fluids are hydrodynamically independent, and Darcy's law can be applied to the individual fluids.

Permeability

A porous medium is said to be permeable if a fluid can be passed through one surface to another under the influence of an external force (e.g., gravity, pressure). If two or more phases coexist in a porous medium under conditions where each phase is mobile, then Darcy's law may be applied to each phase independently. However, the permeability will be different for each phase and different from that for the material when saturated with only a single phase.

If the porous medium is 100% saturated, the permeability of the medium is a property of the material and not of the fluid that flows through it. *Absolute permeability*, K, of the material is permeability at 100% single fluid saturation. When the material is not 100% saturated by a single fluid, the permeability of the material to a particular fluid is called *effective permeability*, K_i. It is defined as the permeability to a particular phase in the presence of specific saturations of one or more other phases. The sum of the specific permeabilities is always less than the absolute permeability. Another useful term is the *relative permeability*, K_{ri}. It is defined by the ratio of effective permeability of one phase to the absolute permeability. The relative permeability for a multiphase flow is a function of the amount of fluid that is in the pore volume. Relative permeability is dimensionless, ranging from 0 to 1.

The absolute permeability, K, as described in Eq.(6) defines the conductivity of a porous medium with respect to permeation by a Newtonian fluid (Dullien, 1992). It depends on the pore size distribution of the media and is proportional to the cross-sectional area of pores open to the flow and inversely related to the average flow length. Absolute permeability represents the permeability of a liquid or gas at a fully saturated state and corresponds to its maximum value. The standard unit for permeability is the *darcy, d*. A porous medium has a permeability of one darcy if a pressure differential of 1 atmosphere produces for a fluid with 1 centipoise (*cP*) viscosity a flow rate of *1 cm³/s* through a *1 cm³* cube. Therefore, the conversion factor becomes *1 d = 0.987x10⁻¹² m²*.

When two or more phases are present in a porous material, the absolute permeability of each phase is less than the effective permeability (except at the irreducible saturation of the liquid phase in a wet-porous media). With two or more phases simultaneously mobile, the relative permeability of each phase drops rapidly from its value when all the other phases are at their irreducible saturation (i.e., when they are not mobile). Typical relative permeability curves are shown in Figure 7–5. It is shown that $K_g + K_w$ is less than unity for all saturation except possibly 0 and 1.

Permeability data for food materials are not available except for the work on meat by Contreras (1987) and on bread by Goedeken and Tong (1993). It is difficult to obtain experimental data for food materials because permeability values are quite low and the materials deform (Ni, 1997). Absolute permeability data for some porous media are listed in Table 7–1. No study exists for relative permeability values for food materials in the literature.

Permeability Measurements

Absolute permeability is measured by flowing a gas or liquid through a porous sample. To determine permeability to air, the pressure difference across the sample and the rate of flow through the sample are measured. Corrections (due to slippage effect) need to be made when working with gas permeability, specifically for low permeability values. For liquid permeability measurements, data should be collected at three or more flow rates (Collins, 1976).

Although permeability cannot usually be estimated from porosity alone, there is often a high correlation observed between porosity and the logarithm of permeability (Collins, 1976), particularly in sandstones and granular limestones. No attempts have been made to model the absolute permeability in terms of porosity for porous food materials.

Relative permeability measurements are performed using either a steady-state method or unsteady-state method. Each fluid phase is injected at a constant rate in the steady-state method until pressure and saturation equilibrium are obtained. Equilibrium points at several different flow rates are required to define the relative permeability curves. The unsteady-state test is performed by injecting one phase

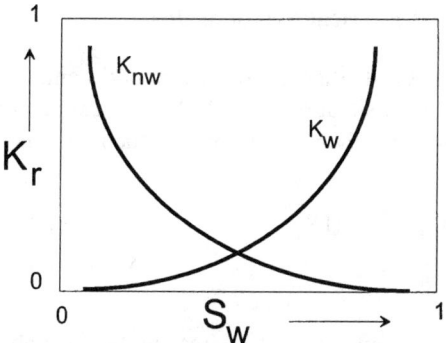

Figure 7–5 Typical Relative Permeability Curves

until the other phase has been completely produced from the sample (Collins, 1976).

Nuclear magnetic resonance (NMR) techniques have been used to estimate the permeability of porous materials. NMR can be used to determine porosity, water saturation, pore size distribution, pore surface to volume ratio, wetting phases, and diffusion, as well as to image the distribution and flow of fluids through porous media (Gubelin and Boyd, 1997). A method has been developed (Borgia et al., 1992; Norgaad et al., 1995) in which a measured parameter, NMR relaxation time, T_1, and the electrical formation factor, F, and/or porosity, ϕ, have been used successfully to predict permeability.

Table 7–1 Absolute Permeability Data in Porous Media

Material	*Permeability [$\times 10^{-14} m^2$]*	*References*
Silica powder	5.0	Collins (1976)
Sandstone	1.8	Wei et al. (1985)
Brick	0.2	Collins (1976)
Light concrete	2.5	Ilic and Turner (1989)
Softwood	0.01	Nasrallah and Perre (1988)
Wood	0.1	Stanish et al. (1986)
Leather	9.4	Collins (1976)
Flour dough	0.8 ($\phi = 0.1$)	Goedeken and Tong (1993)
(50% w.b.)	2270 ($\phi = 0.6$)	
Meat	14–46	Contreras (1987)

Relative Permeability Empirical Relationships

Many expressions exist that represent experimental data of relative permeabilities for different situations (Ni, 1997). An expression for gas relative permeability that is used widely in the porous media models is:

$$K_{gr} = 1 - 1.1 S_w \qquad S_w < 1/1.1$$
$$K_{gr} = 0 \qquad S_w > 1/1.1 \qquad [9]$$

showing that K_{gr} increases linearly with decreasing liquid content.
For liquid relative permeability, K_{wr}, the most used expression is:

$$K_{wr} = \left(\frac{S_w - S_{ir}}{1 - S_{ir}} \right)^3 \qquad S_w > S_{ir}$$
$$K_{wr} = 0 \qquad S_w < S_{ir} \qquad [10]$$

where S_{ir} is the irreducible liquid saturation or percolation threshold assumed to be 0.09 (Nasrallah and Perre, 1988). Eq.(10) shows that the value of K_{wr} decreases very rapidly with decreasing liquid content and approaches zero as the irreducible liquid saturation is reached.

THE INTERACTION OF FLUIDS WITH SURFACES

Adsorption

The walls of the pores may adsorb the molecules of a fluid if this fluid is a vapor near its condensation point. This signifies that within a few molecular distances from the wall there is a strong attractive potential between it and the molecules of the fluid. Thus, a much greater concentration of molecules of the fluid may be found at the walls of the pores (container) than in its interior (Collins, 1976).

During the process of adsorption, energy is released at the walls of the pores. This energy is called the *heat of adsorption*. Thus, if any pressure measurements are made during the process of adsorption, the temperature of the system has to be carefully monitored. In general, conditions are arranged in such a manner that the liberated heat is conducted away so that the process is isothermal.

Because of the large potential present in the adsorbed layers, the rheologic state of the fluid inside the pores may be different from that on the walls. In particular, if the fluid is a vapor near its condensation point, the adsorbed molecules may well form a film of *liquid* along the walls. This phenomenon will affect the relative easiness with which one fluid may move within the porous material. For example, if gas and liquid are to be displaced within the porous system, the resistance to gas flow will be larger than if liquid were not condensing at the walls.

Boundary Tension

When two immiscible fluids (i.e., air-water or oil-water) are in contact, these fluids are separated by a well-defined interface. This interface is a thin layer, on the order of few molecular diameters, with properties that are different from both bulk phases. Within each fluid and away from the interface and the walls of the pores, molecules in the liquid are uniformly attracted in all directions to each other. However, at the boundary between these two immiscible fluids, there are no similar molecules beyond the interface. Because molecules at the surface are attracted more strongly from below, there is an inward-directed force that attempts to minimize the surface by pulling it into the shape of a sphere. This surface activity creates a film-like layer of molecules that are in tension, which is a function of the specific free energy of the interface (Miller and Neogi, 1985).

The creation of this surface requires work. The work in "ergs" required to create 1 cm² of surface is termed *boundary energy*. Conventionally, when this boundary tension is between a gas and a liquid it is known as *surface tension* and termed *interfacial tension* when between a solid and a fluid or between two immiscible liquids.

Boundary tension has the dimensions of force per unit length. Tables 7–2 and 7–3 show some typical values of liquid-gas surface tensions and of liquid (1)-liquid (2) interfacial tensions reported in the literature, respectively.

Wettability

When an interface is in intimate contact with the walls of a container—for example, a capillary tube, the interface intersects the solid surface at an angle, Θ, which depends on the relative boundary tension of the liquids to the solid.

Consider a drop of water immersed in oil in contact with a solid surface, as indicated in Figure 7–6. There will be different boundary tensions among these systems that will dictate the shape and the contact area of the droplet with the solid. The adhesion tension (A_T) is defined as the difference between the two interfacial tensions of the fluids and the solid:

$$A_T = \gamma_{ws} - \gamma_{os} = \gamma_{ow} \cos\Theta \qquad [11]$$

Table 7–2 Liquid-Vapor Surface Tension of Some Liquids

Liquid	σ_{lv} (Dyne/cm)
Water[a]	72.7
Glycerol[a]	63.1
Mercury[a]	435.5
Whole milk[a]	46.7
Olive oil[a]	33.0
Coconut oil[a]	33.4
Cotton seed oil[a]	35.4
Soybean oil[b]	30.1
Oleic acid[a]	32.5

Source: Data from (a) M.J. Lewis, *Physical Properties of Foods and Food Processing Systems*, © 1996, Woodhead Publishing Limited; and (b) Y.-C. Tseng, R.G. Moreira, and X. Sun, Total Frying-Use Time Effects on Soybean-Oil Deterioration and On Tortilla Chip Quality, *International Journal of Food Science and Technology*, Vol. 31, pp. 287–294, © 1996, Blackwell Science Ltd.

The contact angle, Θ, is described by the Young's equation as:

$$\cos\Theta = \frac{\gamma_{ws} - \gamma_{os}}{\gamma_{ow}} \qquad [12]$$

where γ_{ws} is the interfacial tension between the solid and water, γ_{os} the interfacial tension between the solid and the oil, and γ_{ow} the interfacial tension between oil and water.

If the solid is water-wet then $\gamma_{ws} > \gamma_{os}$ the adhesion tension A_T is >0, and the contact angle is between $0°$ and $90°$. Conversely, an oil-wet solid will exhibit negative adhesion tensions with $\gamma_{ws} < \gamma_{os}$ and the contact angle between $90°$ and $180°$ (Miller and Neogi, 1985). Methods used to measure surface tension, interfacial tension, and contact angles are discussed in Chapter 3.

Consider a water-wet system such as a sponge in which the pore space is partially filled with oil/gas and water. If this water-wet system is immersed in water, some oil will be spontaneously expelled from the core as water is imbibed along the walls and into the smaller pores until a state of equilibrium is attained between the solid-fluid specific surface energies (interfacial tensions). The entering wetting fluid will accumulate in the pores that create the greatest fluid-fluid interfacial curvature. Thus, the wetting phase accumulates in the smallest pores.

Table 7–3 Liquid (1)-Liquid (2) Interfacial Tension of Different Materials with Air

Liquid (2)	Temperature (°C)	Interfacial Tension (Dyne/cm)
Butter oil	40	19.2
n-hexane	20	51.1
Benzene	20	35.0
Olive oil	25	17.6
Coconut oil	25	12.8
Cotton seed oil	25	14.9
Peanut oil	25	18.1

Note: Water is Liquid (1).

Source: Adapted from M.J. Lewis, *Physical Properties of Foods and Food Processing Systems*, p. 181, © 1996, Woodhead Publishing Limited.

CAPILLARY PRESSURE

Capillary pressure is the difference in pressure between two immiscible fluids across a curved interface at equilibrium. Curvature of the interface is the consequence of preferential wetting of the capillary walls by one of the phases (Anderson, 1987). In Figure 7–7, two immiscible fluids are shown in contact with a capillary. The fluid labeled *w* wets the walls of the capillary, and the fluid labeled *o* is nonwetting and is resting on a thin film of the wetting fluid. The pressure within the nonwetting fluid is greater than the pressure in the wetting fluid and, consequently, the interface between the fluids is curved convex with respect to the nonwetting fluid. By definition, the capillary pressure is the pressure difference between the nonwetting and wetting phases at this interface.

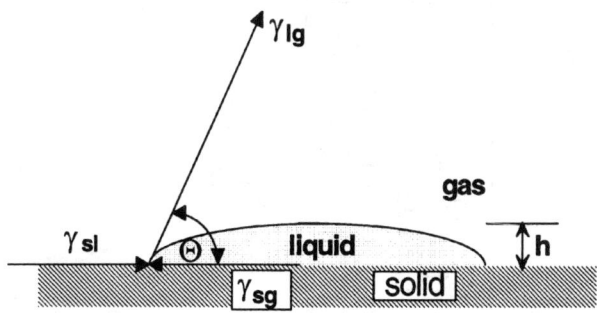

Figure 7–6 Forces Involved in Wettability of Liquid Dropped on a Solid Surface

Derivation of the Plateau Equation

Capillary pressure is related to the curvature of the interface by the expression developed by Plateau (Lake, 1989). Consider an arbitrary portion of interfacial surface separating two fluids with a pressure difference across the interface, producing a curvilinear rectangle as illustrated in Figure 7–8. As indicated previously, the concave side faces the nonwetting fluid, and the pressure will be greater from this side. The convex side faces the wetting fluid, and the pressure is lower from this side. In this arbitrary surface, the two centers of curvature are on the same side. Therefore, r_1 and r_2 are both positive.

The expansion of the surface from its initial state $ABCD$ to its final state $A'B'C'D'$ represents an expansion work by increasing the pressure on the convex side. The areas are:

$$ABCD = l_1 \times l_2 \tag{13}$$

$$A'B'C'D' = \left[l_1 + \left(\frac{l_1}{r_1} \right) dz \right] \times \left[l_2 + \left(\frac{l_2}{r_2} \right) dz \right] \tag{14}$$

$$A'B'C'D' = l_1 l_2 \times \left[1 + \frac{dz}{r_1} + \frac{dz}{r_2} + \frac{dz^2}{r_1 r_2} \right] \tag{15}$$

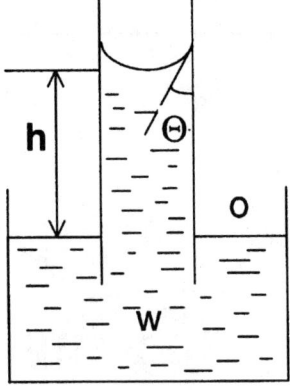

Figure 7–7 Capillary Pressure between Air and Water

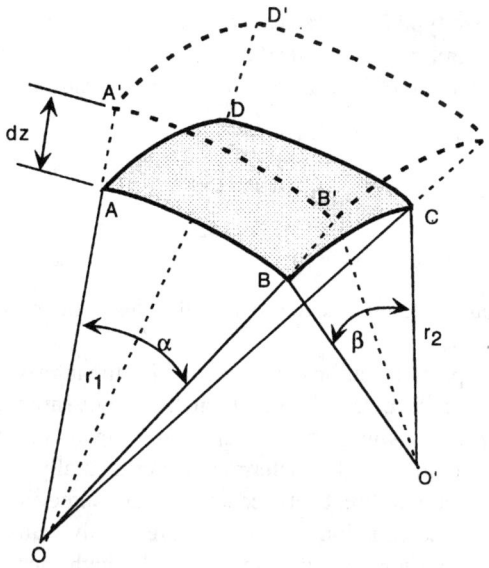

Figure 7–8 Portion of Interfacial Surface Separating Two Fluids with a Pressure Difference across the Interface

The net increase in the area neglecting the second order differential is:

$$ABCD - A'B'C'D' = l_1 l_2 \times \left[\frac{dz}{r_1} + \frac{dz}{r_2} \right]$$ **[16]**

The isothermal work required to expand this area is against the surface tension force:

$$\text{Surface Tension Work} = \gamma \times l_1 l_2 \times \left[\frac{dz}{r_1} + \frac{dz}{r_2} \right]$$ **[17]**

Notice the dimensions [(dyne/cm) × (cm²) = dyne/cm]. The isothermal work done to advance the surface a distance *dz* is equal to:

$$\text{Displacement Work} = (P_o - P_w) \times l_1 l_2 \times dz$$ **[18]**

At equilibrium, the two types of works cancel out. Equating the two work quantities and canceling common terms yields the Plateau, or Laplace, equation for capillary pressure as a function of interfacial tension and the radii of curvature:

$$P_c = \gamma \left(\frac{1}{r_1} + \frac{1}{r_2} \right)$$

[19]

In the case of a porous material, due to the irregularities of the pores, the r_2 curvature radii may be negative.

In principle, the Laplace equation can be solved for the interface configuration in any situation in which the external forces (usually gravity) are transmitted to the interface as pressure and in which the pressures can be evaluated as a function of position. Nearly all methods used to determine experimentally the boundary tension require either direct or indirect detailed knowledge of interface configuration.

In general, the Laplace equation cannot be solved analytically because of the nonlinearity in the expressions for the curvature, although there is a number of important special cases for which analytical solutions can be obtained, such as in cylindrical surfaces.

If the radii of curvature are equal, as in a capillary tube, Eq.(19) reduces to:

$$P_c = \frac{2\gamma}{r_c}$$

[20]

This r_c is not the radius of the capillary tube, but it is related to the radius of the capillary tube by the interfacial geometry of a wetting fluid in a capillary (radius of curvature). Figure 7–7 shows that the radius of the spherical interface is much larger than the radius of the capillary, and the two radii are related by the cosine of the contact angle as follows:

$$\cos \Theta = \frac{r}{r_c}$$

[21]

where r is the radius of the capillary tube and r_c the radius of the spherical interface.

Substituting r_c in terms of r in Eq.(20) yields the expression for capillary pressure in terms of the interfacial tension, contact angle, and radius of the capillary tube:

$$P_c = \frac{2\gamma \cos\Theta}{r} \qquad [22]$$

Microscopic observations of immiscible fluids using glass beads and sand grains have established the complex geometric aspects of the liquid-liquid and liquid-solid contacts. The curvature of the fluid interfacial boundary is a function of grain size, interstitial volume, chemical composition, and surface tensions. The angle of contact (Θ) between the liquid interface and the solid, measured through the denser phase, is a function of the relative wetting characteristics of the two fluids with respect to the solid. When the radii of curvature have their centers of rotation on the same side of the interface, they are positive, but when the radii are on opposite sides, r_1 is positive and r_2 is negative. This would be a more realistic scenario in irregular porous food materials.

Capillary Rise

When a capillary tube is immersed in water vertically with one end touching the water, the water rises in the tube to a height above the surface (Figure 7–7); the narrower the tube, the greater is the height, h, to which the water rises. The phenomenon is known as *capillarity*; it is commonly observed in most porous materials, e.g., ink soaking into blotting paper or water soaking into porous solids, as in the reconstitution of freeze-dried foods.

A capillary pressure equation can be derived from a balance of forces acting in these systems. Water will rise in the capillary tube until the forces in the upward and downward directions equilibrate. The force responsible for lifting the water is the adhesion force that acts along the perimeter of the water-oil-solid interface:

$$F_{up} = 2\pi r A_T \qquad [23]$$

where A_T is described by Eq.(11). The total force down equals the weight of the column of water (gravity force) minus the weight of the column of oil (buoyancy force). That is:

$$F_{down} = \pi r^2 gh(\rho_w - \rho_o) \qquad [24]$$

The liquid will rise in the tube until the vertical component of the surface tension equals the total force down. Thus, the capillary rise is:

$$h = \frac{2\gamma_{ow}\cos\Theta}{rg(\rho_w - \rho_\sigma)} \qquad [25]$$

The pressure below the interface within the capillary tube (P_w) can be expressed as the pressure at the bulk interface (P^*) minus the hydrostatic column of water (h) within the capillary:

$$P_w = P^* - \rho_w gh \qquad [26]$$

Similarly, the pressure above the interface (oil phase) can also be expressed as the difference between the pressure at the bulk level minus a hydrostatic column of oil:

$$P_o = P^* - \rho_o gh \qquad [27]$$

The difference between Eqs.(26) and (27) is the capillary pressure:

$$P_c = (\rho_w - \rho_\sigma)gh \qquad [28]$$

Many techniques use the capillary rise phenomenon to measure capillary pressures. The shape and height of the meniscus depends on the relative magnitudes of the molecular cohesive forces and the molecular adhesive forces between the liquids and the walls of the capillary.

Semipermeable Disk Measurement of Capillary Pressure

A capillary pressure (moisture potential) curve requires the measurement of the equilibrium pressures at various saturations of a porous medium during a displacement process. This is usually achieved by enclosing the porous medium in a cell and displacing one fluid with another slowly by changing the pressure in the displacing phase. Unless the displaced phase is a gas, an outlet has to be provided for it; this can be done by applying a semipermeable medium that is impermeable to the displacing phase. Various applications of the above principle of measurement have been used in the oil industry and in soil physics.

The *capillary pressure* is then measured directly by a gauge or by an open-end manometer; the latter procedure is customary in soil physics as one obtains directly the length of the moisture potential. The *saturation* at any pressure can be determined by measuring the amount of displacing fluid that entered the porous medium, by electrical measurements, or by MRI techniques.

Capillary pressure curves have been obtained in large numbers by workers in the oil industry, with oil as displaced and water or gas as displacing fluid.

The derivation of capillary pressure equations thus far has been based on a single uniform capillary tube. Porous food, ceramic, or geologic materials, however, are composed of interconnected pores of various sizes. In addition, the wettability of the pore surfaces may vary from point to point within the system, due to the variation in the mixture of chemicals in contact with the fluids. This leads to variation of the capillary pressure as a function of fluid saturation and an overall mean description of the system wettability.

This process can be repeated in the laboratory by displacing water from a core with a gas or oil. The pressure required for the equilibrium displacement of the wetting phase (water) with the nonwetting gas or oil is the water drainage capillary pressure, which is recorded as a function of the water saturation.

A core is saturated with water containing salts (NaCl, $CaCl_2$, or KCl) to stabilize the clay minerals, which tend to swell and dislodge when in contact with fresh water. The saturated core is then placed on a porous disk, which also is saturated with water. The porous disk has finer pores than does the porous sample and should have a permeability at least one order of magnitude lower. The pore sizes of the porous disk should be small enough to prevent penetration of the displacing fluid until the water saturation in the core has reached its irreducible value.

The pressure of the displacing fluid is increased in small increments. After each increase of pressure, the amount of water displaced is monitored until it reaches static equilibrium. The capillary pressure is plotted as a function of water saturation. If the pore surfaces are preferentially wet by water, a finite pressure (the threshold pressure) will be required before any of the water is displaced from the core. If the core is preferentially oil-wet and oil is the displacing fluid, oil will imbibe into the core, displacing water at zero capillary pressure.

Placing the core on another porous disk that is saturated with oil may reverse the displacement, and the core is covered with water. If the core is preferentially wet by water, water will imbibe into the core and displace the oil spontaneously.

Drainage and Imbibition

Depending on the wetting properties of the fluids, there are essentially two different types of displacement in two-phase flow in porous media. Drainage displacements are where a nonwetting invading fluid displaces a wetting fluid. The opposite case, imbibition, occurs when a wetting fluid displaces a nonwetting fluid. The mechanisms of the displacements in drainage and imbibition are quite different, and the two cases should not be confused. Typically, slow drainage is characterized by pistonlike motion inside the pores, where the invading nonwetting fluid enters a pore only if the capillary pressure is equal to or greater than the threshold pressure of that pore. The threshold pressure corresponds to the capillary pressure in the narrowest part of the pore. However, in imbibition at low injection rate, the invading fluid will enter the narrowest pores before any other is considered. Imbibition is a spontaneous process, whereas drainage is forced.

OIL ABSORPTION IN POROUS FOOD DURING COOLING

The mechanism involved in oil adhesion and absorption during deep fat frying is complex, and several simplifications must be made to attempt modeling the oil absorption during the postfrying period (cooling). During cooling, the surface tension between oil and gas increases as temperature decreases, resulting in an increasing capillary pressure that causes the surface oil to flow into the porous product, thus increasing its internal oil content. The following is a description of the oil absorption mechanism in tortilla chips during cooling, i.e., when most of the oil (80% of the total final oil content) is absorbed by the product, as described in the work of Moreira and Barrufet (1998).

The frying process of tortilla chips can be divided into three distinct stages:

1. *Prefrying*. This stage is characterized by the raw material conditions before frying. Oil content is zero at this stage. The product's physical properties of importance are the initial moisture content, initial porosity, and density of each component of the product (solid, water, and air).

2. *Frying*. The product's properties change during this stage. Moisture content is reduced; the physical properties such as thermal conductivity, specific heat, and bulk density decrease; porosity increases as a result of water evaporation; and the product's thickness increases as a result of puffing. Only 20% of the total oil is absorbed during this stage.

3. *Postfrying (Cooling)*. This period starts immediately after frying, when the product is removed from the fryer. The temperature inside the chip decreases until it reaches the ambient (or cooling) temperature. The water content is constant, and the physical properties—with exception of the bulk density—remain essentially the same. Most of the oil (80%) is absorbed during this period.

Based on these observations, the following assumptions can be made to develop a mathematic model to describe tortilla chip oil absorption during the postfrying period (cooling):

1. A tortilla chip is an infinite slab.

2. The thickness of a tortilla chip increases during frying as a result of an expansion caused by an increase in porosity and remains constant throughout cooling.
3. A "microscopically uniform" porous medium is formed and the surface of the chip is covered with a uniform layer of oil after frying.
4. Oil is absorbed during frying, but most of the oil is absorbed during cooling.
5. A fried product is composed of inert gas (air), solid (dry matter), water, and oil.
6. The moisture remaining inside the chip after frying is negligible, and all the capillary pores are occupied by gas and oil.
7. Capillary tubes are independent of each other and run across the product's thickness.
8. The ideal gas law is used to calculate the air pressure within the pore.
9. The effective surface tension follows a linear relationship with temperature.

The second assumption is based on observations by Moreira et al. (1995) using MRI measurements of the cross-sectional area of tortilla chips.

Assumption 3 is used for simplification, assuming an average pore size distribution. Assumption 4 is based on the results and observations as presented by Sun and Moreira (1994) and Moreira et al. (1997). During frying, only 20% of the total oil content (31.8% w.b.) was absorbed by the tortilla chips, and the rest (80%) remained at the chip's surface. During cooling, about 64% of the total oil content was absorbed by the chips, and only 36% remained at the surface of the chips.

Assumption 6 is justified by the fact that at the end of the frying period, the product contains only about 1–2% water, i.e., in a volumetric basis, this water may be considered negligible compared with the volume of oil and air in the pores. In addition, it has been shown experimentally (Moreira et al., 1997) that most of the oil is absorbed by the tortilla chip during the first 20 seconds of cooling, when its temperature is well above the condensation temperature (100°C), suggesting that oil absorption during the postfrying period may be controlled by capillary phenomena and that the effect of water vapor condensation is negligible. Figure 7–9 illustrates this phenomenon.

Assumptions 7 through 9 are valid for analysis of fluid flow in porous media. For modeling purposes, assumption 7 implies a simplifying media that will provide results equivalent to those from a random structure. Assumption 8 is justified as gas tends to behave as ideal at low pressure. Assumption 9 has been confirmed by Moreira and Barrufet (1998) (see Table 7–4 and Chapter 3).

In summary, the oil absorption model during the postfrying period represents oil uptake in a microscopically uniform porous media. Convection and conduction

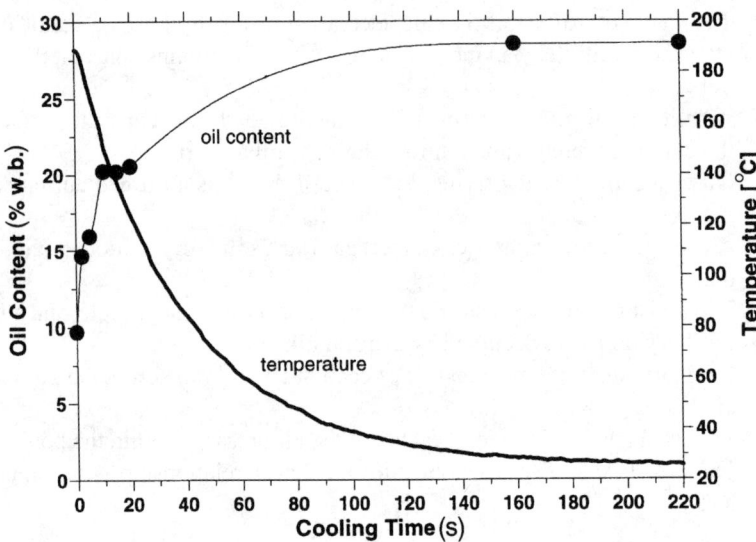

Figure 7–9 Oil Content and Temperature Changes of Tortilla Chips during Cooling

control the transfer of heat. Mass transfer (oil and gas) from the surface oil adhered to the product's surface and the surrounding inert gas (air) is controlled by capillary pressure. The average mass of gas within the pore increases during frying and cooling. The pore size increases during frying but remains constant during cooling. During cooling, both gas and oil compete for filling the fried product void space. The bulk volume of the tortilla chip and its water content are constant during cooling.

Governing Equations

Figure 7–10 presents the condition that exists when a tortilla chip pore (as a capillary tube) of radius r_i is immersed in an extensive pool of a wetting liquid (oil). The pressure in the liquid phase beneath the gas-liquid interface is less than the pressure in the gaseous phase above the interface. This difference in pressure across the interface is referred to as *capillary pressure* of the system. Using Eq.(22), the pressure relationship for a liquid slug in a small vertical tube can be expressed as:

$$P_1 - P_2 = \frac{2(\gamma \cos \Theta)}{r}$$

[29]

Table 7–4 Surface Tension Values of Ethyl Alcohol and Water in the Temperature Range 0–30°C

Liquid	Temperature (°C)	σ_{lv} (Dyne/cm)
Ethyl alcohol	0	24.05
	10	23.61
	20	22.75
	30	21.89
Water	0	75.60
	10	74.22
	20	72.75
	30	71.18

Source: Adapted with permission from M.J. Lewis, *Physical Properties of Foods and Food Processing Systems*, p. 173, © 1996, Woodhead Publishing Limited.

$$P_2 - P_3 = -\rho_l gh \qquad [30]$$

where ρ_l is the density of liquid, g the acceleration due to gravity, and h the capillary height. In the case of fried products, the capillary pressure increases as it is cooled down to an ambient air temperature. Because the product's surface is covered with a thin layer of oil after frying, this pressure difference becomes the driving force for fluid absorption during cooling.

The pressure within the pore can be determined by adding Eqs.(29) and (30):

$$P_1 = \frac{2(\gamma \cos \Theta)}{r} - \rho_f gh_f + P_3 \qquad [31]$$

where $P_1 = P_g$ = gas pressure inside the capillary tube, $P_3 = P_{atm}$ = atmospheric pressure, ρ_f the density of fluid mixture underneath the meniscus of the capillary (gas and oil), and h_f is the capillary height of fluid mixture (oil and/or gas) absorbed during the cooling process. Therefore, at any time during cooling, the gas pressure within the pores of a tortilla chip can be expressed as:

$$P_g(t) = P_{atm} + \frac{2\left[\gamma(t)\cos \Theta\right]}{r} - P_f(t)gh_f(t) \qquad [32]$$

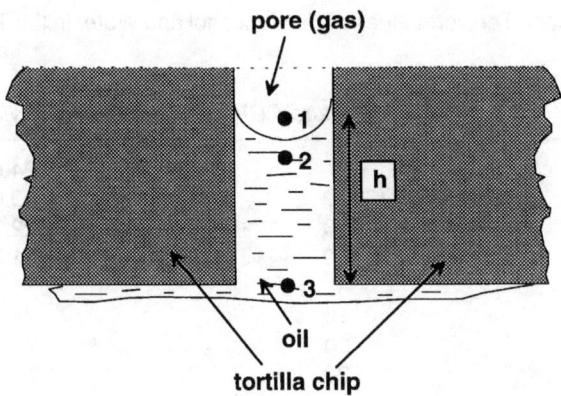

pore (gas)

oil

tortilla chip

Figure 7–10 Pressure Relationship for a Tortilla Chip Undergoing Cooling. *Source:* Reprinted from *Journal of Food Engineering*, Vol. 35, R.G. Moreira and M.A. Barrufet, A New Approach to Describe Oil Absorption in Fried Foods: A Simulation Study, Pages No. 1–22, Copyright 1998, with permission from Elsevier Science.

where t is the time required to cool the tortilla chips. Using the ideal gas law to describe $P_g(t)$ and assuming that the fluid volume can be described as the volume of a bundle of capillary tube of uniform radius and height, Eq.(32) can be expressed in terms of average capillary radius and height of tortilla chips, as follows:

$$P_g(t) = \frac{P_{atm}\theta(t)y}{\theta(o)}\left\{\frac{h(o)-h_w(o)}{h(i)-h_o(t)}\right\} \qquad [33]$$

where $y = [r(o)/r(i)]^2[w_g(i)/w_g(o)]$, w_g is the number of moles, $h_w(o)$ the height of water column in the capillary (a function of the tortilla initial moisture content) at the initial condition (before frying), $h_o(t)$ the height of oil column in the capillary at any time t, and $h(o)$ and $h(i)$ the capillary height (equal to the thickness of tortilla chips) at the initial and intermediate (at time zero during cooling) conditions, respectively.

The product temperature (θ) and the product oil content (M_o) change during cooling. Energy and mass balances are written on a differential volume. The basic one-dimensional heat transfer and capillary pressure equations are employed:

- The governing differential equation describing the temperature change in the product during cooling is:

$$\frac{\partial^2 \theta}{\partial x^2} = a_{eff} \frac{\partial \theta}{\partial t}$$

[34]

- The oil content (during cooling) changes with time as:

$$M_o(t) = \frac{\rho_o h_o(t)}{\rho_b(t) h(e)} \phi(e)$$

[35]

where a_{eff} is the thermal difusivity, x the space coordinate, and $\phi(e)$, the porosity at equilibrium (final), is related to the final measured gas porosity, $\varepsilon(e)$, as:

$$\phi(e) = \varepsilon(t)\left(\frac{h(e)}{h(e) - h_o(t)}\right) = \varepsilon(e)\left(\frac{h(e)}{h(e) - h_o(e)}\right)$$

[36]

and

$$\rho_b(t) = \phi(e)\left\{\rho_g\left(\frac{h(e) - h_o(t)}{h(e)}\right) + \frac{h_o(t)}{h(e)}\rho_o - \rho_s\right\} + \rho_s$$

[37]

where ϕ is the true porosity (see Eq.2), ε is the gas porosity, ρ_b is the bulk density, r_o is the oil density, and r_s is the solid density.

Eqs.(34) to (36) describe the temperature and oil content changes with time in the tortilla chips. Two boundary conditions and one initial condition are needed to solve Eq.(34), at $x = 0$ (center line) for any time:

$$\frac{\partial \theta}{\partial x} = 0$$

[38]

at $x = \pm L/2$ (surface) for any time:

$$h'_{air}(\theta_{surf} - T) = -k_{eff}\frac{\partial \theta}{\partial x}$$

[39]

where h'_{air} is the convective heat transfer coefficient and k_{eff} the effective thermal conductivity of the product. Two conditions are required to solve Eq.(34) for $h_o(t)$, the height of the oil column in the capillary at different times:

1. *at t = 0; θ = 190°C; h$_o$(o) = h$_o$(i)*
2. *at t = ∞ or θ (e) = T (cooling temperature):*

$$h_o(e) = \frac{\rho_f(e)h(e)}{\rho_o} \tag{40}$$

The second constraint is related to the final condition, which depends on the amount of oil available for absorption; i.e., tortilla chips can be covered by a thin layer of oil after being removed from the fryer (limited oil available) or can be left in the fryer until all tortilla chips are cooled down (oil available is unlimited).

For the case of limited oil available, the effective fluid density is evaluated as:

$$\rho_f(t) = \left(\frac{V_o}{V_o + V_g}\right)\rho_o + \left(\frac{V_g}{V_o + V_g}\right)\rho_g = f_o(t)\rho_o + (1 - f_o(t))\rho_g \tag{41}$$

i.e., this effective fluid density changes during cooling. Initially, the effective density will be closer to the oil density. However, as the surface oil enters the pore space, the effective density becomes less dense because the oil volume is being depleted. The gas volume will be depleted as well, however; gas from the surroundings makes up for that loss. The variable V in Eq.(41) refers to the volume of oil (subscript o) or gas (subscript g).

The relationship proposed for $\gamma \cos \Theta$ is the following (Macleod, 1976):

$$\gamma \cos \Theta = \Gamma(t)\theta(t) = A - B\theta(t) \tag{42}$$

where Γ is the effective interfacial tension, and the coefficients A and B are obtained from two extreme values, Γ_{max} and Γ_{min}, at known temperatures and left as parameters for sensitivity analysis. That is, when $\theta = \theta_{min}$, $\Gamma = \Gamma_{max}$, and, when $\theta = \theta_{max}$, $\Gamma = \Gamma_{min}$,

$$A = \Gamma_{min} + \theta_{max}\left(\frac{\Gamma_{max} - \Gamma_{min}}{\theta_{max} - \theta_{min}}\right) \tag{43}$$

and

$$B = \frac{\Gamma_{max} - \Gamma_{min}}{\theta_{max} - \theta_{min}}$$ [44]

The values of Γ_{max} and Γ_{min} are within the range of γ_{max} and γ_{min} (as $\cos\Theta \le 1$).

The average value of the chip temperature through its thickness is calculated, and for every value of $\theta(t)$, Eq.(32) is solved for $h_f(t)$, replacing $h_o(t)$ from Eq.(40).

The developed cooling model is a function of the initial and final structure of the product. It takes also into account the initial moisture content of the product (prior to frying) and the initial oil content (precooling), which corresponds to the total oil uptake during frying.

The initial structure is characterized by an average pore size r and an initial gas porosity $\varepsilon(o)$. The final structure is assumed to be completed at the end of the frying period and is characterized by an intermediate gas porosity $\varepsilon(i)$, and a puffing coefficient. The puffing coefficient p indicates the expansion undergone by the product in the axial direction (the thickness). The oil uptake during frying can be determined experimentally by removing the surface oil when dipping the fried product into ether (Sun and Moreira, 1994).

The cooling model calculates the amount of oil driven into the product by capillary action by determining the transient capillary pressure as a function of temperature. The pressure within a pore can be evaluated using the ideal gas law and a bundle of capillaries of height h equal to the product thickness to model the pore space medium (Eq.33) and the structure parameter y. The product temperature $\theta(t)$ as a function of time is obtained from solving the differential heat transfer equation (Eq.24).

Another expression for the pore pressure is obtained from the definition of capillary pressure (Eq.31). This includes the average pore radius r, the effective interfacial tension $\gamma\cos\Theta$, fluid density of the imbibing fluid—which will change with time—and instantaneous capillary height of the imbibing fluid h_f.

A quadratic equation for the capillary height h_f is obtained by equating the two pore pressure equations (Eq.32 and Eq.33). This capillary height is then converted to fat content using Eq.(35).

Analysis of the Cooling Process

Model Validation

Figure 7–11A presents the rate of oil absorption in tortilla chips during cooling as a function of initial moisture content, as measured and predicted by the model. The predicted and experimental values show good agreement. Oil content in torti-

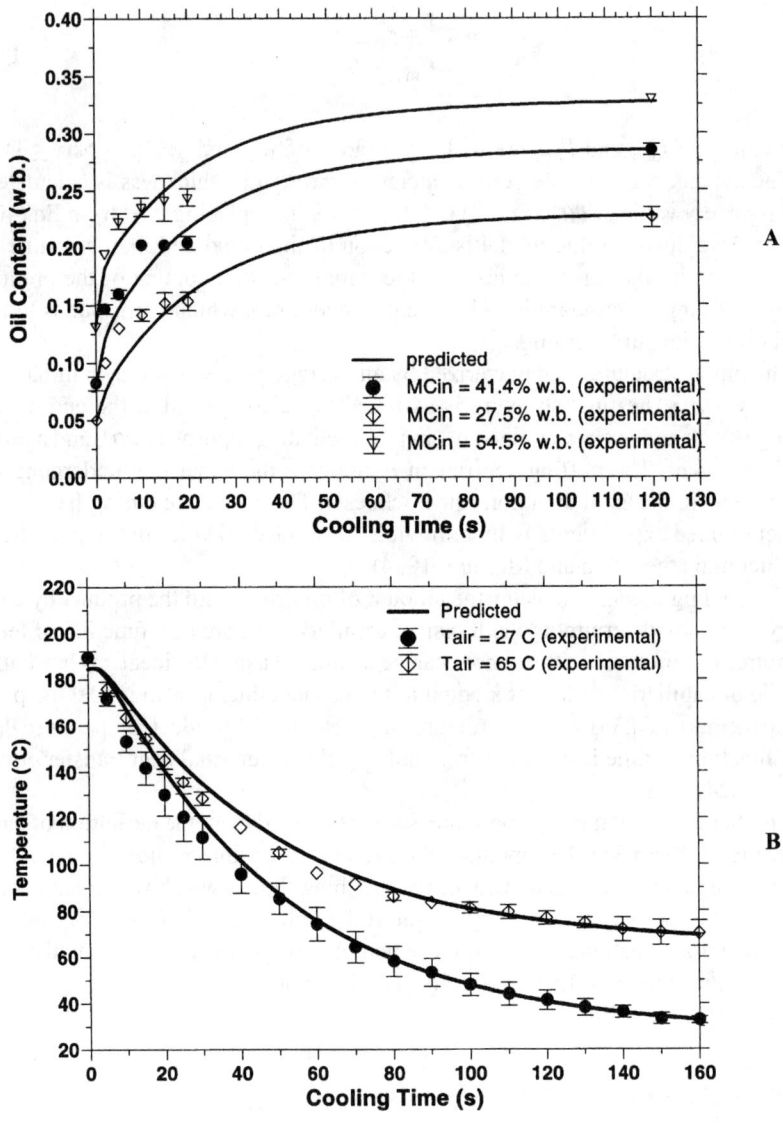

Figure 7–11 A, Observed and Predicted Oil Absorption During Cooling; **B,** Observed and Predicted Temperature History During Cooling of Tortilla Chips. Thickness, 2.6 mm. *Source:* Reprinted from *Journal of Food Engineering*, Vol. 35, R.G. Moreira and M.A. Barrufet, A New Approach to Describe Oil Absorption in Fried Foods: A Simulation Study, Pages No. 1–22, Copyright 1998, with permission from Elsevier Science.

lla chips increased exponentially, reaching an equilibrium value at about 120 seconds of cooling.

The initial moisture content of the product affects dramatically the final oil uptake. The more water the product has initially, the more that oil will be absorbed during frying and, in large extent, during cooling.

Clearly shown is the fact that the tortilla chip with the lowest initial moisture content (27.5% w.b.) absorbs less oil (Figure 7–11A). Increasing the initial moisture content to 41.4% w.b. and to 54.5% w.b. results in 26% and 46% more oil uptake, respectively. These are considerable differences.

The temperature history at the center of the tortilla chip (thickness of 2.6 mm) during cooling is presented in Figure 7–11B. Temperature was measured by inserting a thin thermocouple (type E, 0.25 mm) in the center of the chips. The temperature of the frying oil was measured with a thermocouple (type E, 0.81 mm) (Chen and Moreira, 1997). A good agreement between the predicted and the observed values is shown. It took about 160 seconds for the temperature at the center of the tortilla chips to reach the cooling temperature. It is also observed that it takes about 40 seconds for the center temperature of tortilla chips to reach 100°C when most of the oil has already been absorbed (Figure 7–11A).

Based on the good agreement between the experimental and theoretic values, a parametric sensitivity analysis was conducted to evaluate oil absorption during cooling of the tortilla chip. The parameters investigated were: tortilla's cooling air temperature, pore's radius, and interfacial tension.

Sensitivity Analysis

The fat content depends on the following variables: cooling temperature; temperature of the product; effective interfacial tension; initial moisture content [through $h_w(o)$]; product thickness h and puffing p; average pore radius; fluid density of the imbibing fluid (oil and gas); and structure coefficient y.

The structure coefficient y is defined as:

$$y = \left(\frac{r(o)}{r(i)} \right)^2 \frac{w_g(i)}{w_g(o)} \qquad [45]$$

The derivation of this factor takes into account that the pore size (r) of the product before (o) and after frying (i) may be different, as well as may the mass of gas (w_g) within the pore.

Structural Change

The pore space characteristics and fluid contents change during frying. The only variables that can be measured are the initial moisture content [$M_w(o)$], initial oil content [$M_o(o)$], and part of the structural changes that take place during frying [indicated by the puffing coefficient (p)]. When no structural (radial) changes are assumed (i.e., the structure is forced to remain constant) during frying, a limited set of combinations in $M_w(o)$, $\varepsilon(o)$, and $\varepsilon(i)$ are possible that may or may not yield physically sound results. However, using the mechanistic model other structural changes may be determined from simulation, as for example the structure changes that take place when the product is constrained to have a certain value for initial oil content [$M_o(o)$]. The following is an analysis on the effect of structural changes during frying on final oil content during cooling.

Let us consider, for simplicity, that the amount of gas in the pore space before and after frying is essentially the same. During frying, it is assumed that most of the water leaves the pore space while some oil is driven in. However, this does not substitute completely the vacated water volume. A large fraction of the pore volume is occupied by inert gas, which expands substantially during frying, due to the increase in temperature. This is reflected in the value of intermediate gas porosity (developed during frying), which is substantially larger than the initial porosity. The relative rates of gas expansion and of water loss may be responsible for the development of the structure during frying. The rate at which water leaves the pore space may not be sufficiently high as to allow the gas to expand in an essentially isobaric process. Therefore, it is proposed that the gas expansion takes place with a pressure increase (due to temperature increase) that hinders the complete substitution of water by oil.

This higher gas pressure will introduce stresses in the pore walls that will deform in a rather irregular fashion. To evaluate these structural changes quantitatively, this deformation can be represented in terms of two parameters: (a) the measured puffing coefficient (p) and (b) the intermediate average pore size $r(i)$.

Equating the pore pressure from capillary theory and from thermodynamics at initial conditions results in two different case studies:

- The first one is to constrain the product to have a fixed $M_o(o)$ that is equivalent to a certain capillary height [$h_f(o)$]. In this case, the model will determine the changes in the product structure y and final oil content.
- The second case consists of imposing no deformation (at least in the radial direction) and letting the model determine the oil content that would correspond to that particular structure.

An extensive parameter sensitivity analysis involving the structural variables—initial and intermediate gas porosity, and fluid contents $M_w(o)$ and $M_o(o)$, which are intimately related to initial and intermediate gas porosity, respectively—has been developed.

The simulation involved a matrix with various levels of these dominant variables. The other variables remained fixed at the base case (see Table 7–5). The

Table 7–5 Parameters Used to Simulate Oil Absorption in Tortilla Chips during Cooling

Parameters	Values
$M_w(o)$	0.44 w.b.
r	20 μm
P_{atm}	101,300 Pa
T	25°C
L=h	1.3 mm
C_p	2.56 kJ/kg K
k_{eff}	0.091 W/m K
g	9.8 m/s²
p	1.4
θ	190°C
ρ_s	1,400 kg/m³
ρ_w	1,000 kg/m³
ρ_g	1.2 kg/m³
ρ_o	900 kg/m³
$\varepsilon(o)$	0.32
$\varepsilon(e)$	0.58
h′	51 W/m² K
Γ_{min}	0.032 N-rad/m
Γ_{max}	0.02 N-rad/m

Source: Reprinted from *Journal of Food Engineering*, Vol. 35, R.G. Moreira and M.A. Barrufet, A New Approach to Describe Oil Absorption in Fried Foods: A Simulation Study, Pages No. 1–22, Copyright 1998, with permission from Elsevier Science.

ranges taken for each variable were the following: $M_w(o) = 0.2 – 0.6$; $M_o(o) = 0.0 – 0.3$; $\varepsilon(o) = 0.2 – 0.5$; and $\varepsilon(i) = 0.6 – 0.9$.

The number of different cases run is then equal to the product of all levels, i.e., 1260. This allowed a systematic analysis of the effect of each variable on the final oil content and the structure coefficient (y).

The maximum and minimum values of the structural parameter were about 2.4—corresponding to shrinkage in the radial direction—and 0.8—corresponding to a radial expansion. For an average radius of 40 μm, this corresponds to a final radius of 26 μm and 44.7 μm, respectively.

The highest and lowest fat contents were 0.6 and 0.1, respectively. As observed previously, the highest and lowest final porosities correspond to the maximum and minimum fat contents, respectively; and the highest and lowest final bulk densities correspond to the lowest and highest fat contents, respectively.

The following section summarizes the results observed from this simulation.

Adjusting structure to satisfy $M_o(o)$. The final oil content will increase with increasing $M_o(o)$ and increasing $\varepsilon(i)$ (Figure 7–12). For a fixed intermediate porosity $\varepsilon(i)$ and $M_o(o)$ the oil uptake during cooling is *independent* of $M_w(o)$ and of initial porosity $\varepsilon(o)$. Figure 7–12 shows a practical graph that can be used to determine the fat content for different values of these four variables. This is specific for the frying conditions and the raw product properties of tortilla chips. However, it can be easily used for other fried material, as well, once the characteristic parameters are provided.

Because $M_o(o)$ has been forced to have a certain value, this imposes a structural change reflected in y. The structural parameter y *does change* substantially with $M_w(o)$ and with the initial porosity (Figure 7–13). As $\varepsilon(o)$ increases, y decreases, whereas an increase in $M_w(o)$ causes an increase in y. This is reasonable, as a product with a higher $M_w(o)$ will have a lower $\varepsilon(o)$. This also indicates that we cannot have arbitrary combinations of these variables. From a material balance computation and using the model of capillary tubes, the "water porosity" can be determined and the suitable combinations of $M_w(o)$ and $\varepsilon(o)$ can then be found. A product with higher $M_w(o)$ will be more distorted [pores shrunk, i.e., $r(i) < r(o)$] than a product with higher initial porosity.

The structural parameter y also changes substantially with the values of $\varepsilon(i)$ and $M_o(o)$. As the $M_o(o)$ increases, y decreases, whereas an increase in $\varepsilon(i)$ causes an increase in y. Additionally, the final fat content increases for both cases (Figure 7–12). Again, this opposite trend in the structural parameter is logical as an increase in $M_o(o)$ represents a lower $\varepsilon(i)$. As indicated earlier with the pair [$M_o(o)$, $\varepsilon(i)$ and $\varepsilon(o)$], there is an allowable range of combinations for $M_o(o)$ and $\varepsilon(i)$. A product with higher $M_o(o)$ is less distorted [$r(i) > r(o)$], whereas a product with higher $\varepsilon(i)$ is more distorted [$r(i) < r(o)$].

The largest value of y (2.4) corresponds to the following combination of parameters: $\varepsilon(o) = 0.3$, $\varepsilon(i) = 0.6 – 0.9$, $M_o(o) = 0.0$, $M_w(o) = 0.6$ (Repeated for the range in $\varepsilon(i)$ because $M_o(o) = 0.0$). The lowest value of y (0.8) corresponds to the following combination of parameters: $\varepsilon(o) = 0.7$, $\varepsilon(i) = 0.6$, $M_o(o) =0.3$, $M_w(o) = 0.2$.

Figure 7–14 shows contours of the structure parameter versus initial oil content and the other three variables used.

Free structure. When the structure is not adjusted in the radial direction, the initial oil content is not imposed but determined from solving the capillary height directly by combining the two pore pressure equations (Eqs.32 and 33). In this case, all the three variables affect the final fat oil content. However, because the structure parameter is set equal to 1 there is a limited set of results that have physical meaning. Certain combinations of $M_o(o)$, $\varepsilon(i)$, and $\varepsilon(o)$ may give a capillary height larger than the thickness of the tortilla chip, which would give unrealistic results (fat content >100%).

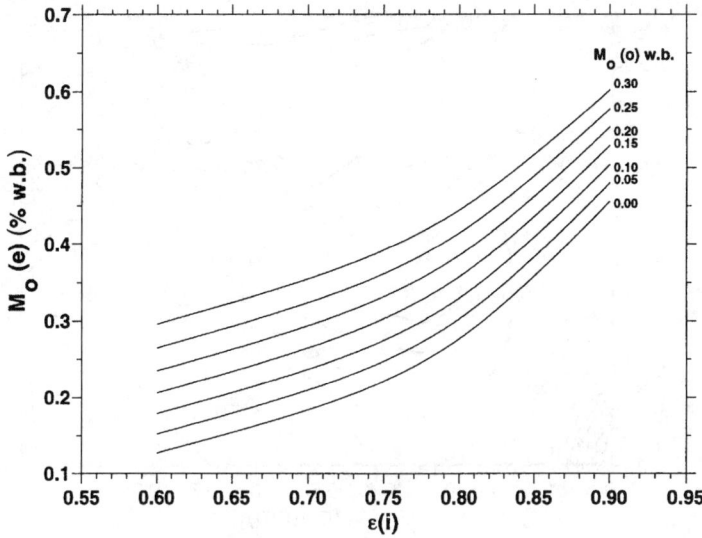

Figure 7–12 Effect of Initial Oil Content $M_o(o)$ and Intermediate Porosity $\varepsilon(o)$ on the Product Final Oil Content $M_o(e)$

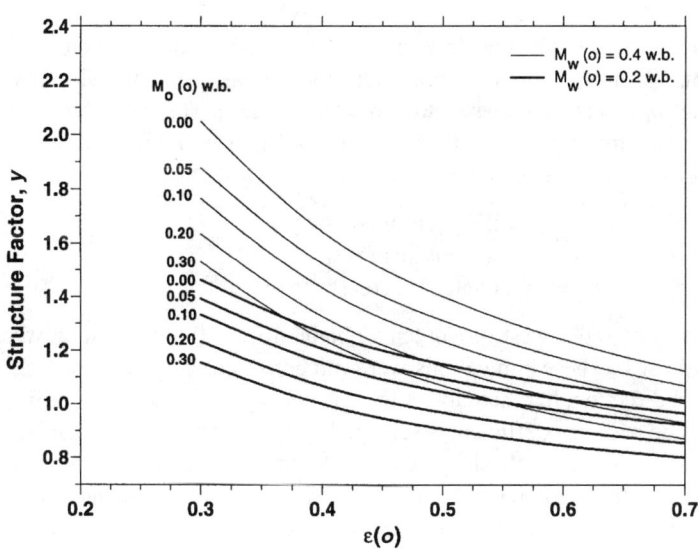

Figure 7–13 Effect of Initial Moisture Content $M_w(o)$, Initial Oil Content $M_o(o)$, and Initial Porosity $\varepsilon(o)$ on the Product Structure (y)

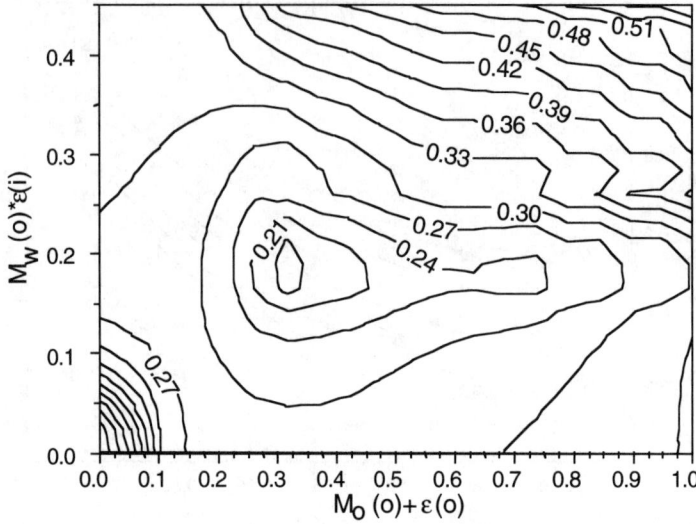

Figure 7–14 Contour Plot of the Product Structure *y* Versus Initial Moisture Content $M_w(o)$, Initial Oil Content $M_o(o)$, Initial Porosity $\varepsilon(o)$, and Intermediate Porosity $\varepsilon(i)$

Because $M_o(o)$ is not considered a variable, a total of 180 permutations (1260/7) was run. When the structure parameter is set equal to 1, only 71 cases provide meaningful results from these 180 permutations used. These combinations do not allow a value for $M_w(o)$ larger than the value used for $\varepsilon(o)$. From these 71 observations, it can be inferred that the final oil content varies as follows:

- as $M_w(o)$ increases, oil content increases;
- as $\varepsilon(i)$ increases, oil content increases;
- as $\varepsilon(o)$ increases, oil content decreases.

The largest and lowest oil contents obtained from this simulation were 0.968 and 0.134 w.b. These simulations were run considering no structural changes in the radial direction, but including a fixed *experimentally observed* puffing coefficient. If the puffing coefficient is set equal to 1, the results are similar but a larger number of results have physical meaning. Additionally, by not allowing any structural change, the final oil content is generally lower. The following is a discussion of the effect of pore radius, cooling temperature, and interfacial tension on the final oil content of tortilla chips simulated, assuming no structure changes in the radial directions.

Pore radius (r). The pore radius, r, of the tortilla chips varied from 2 µm and 40 µm based on the results obtained from scanning electron microscopy image analysis (Lujan-Acosta, 1996).

Figure 7–15 shows the chip average oil content after 120 seconds of cooling (equilibrium). Increasing the pore radius from 2 µm to 15 µm results in a 9% decrease in oil uptake. However, increasing the pore size from 15 µm to 20 µm leads to only 1% decrease in tortilla chip final oil content.

Cooling temperature (T). The larger the difference between θ and T during cooling, the larger will be the pressure gradient within the product's pore and, thus, the higher will be the oil absorption rate during cooling (Figure 7–16). Decreasing the cooling temperature by 33% (i.e., from 75°C to 50°C) and by 67% (i.e., 75°C to 25°C) results in 12% and 24% more oil uptake, respectively, during 120 seconds of cooling. These are considerable differences.

Interfacial tension (γ). The interfacial tension controls whether the oil will overcome the energy barrier to flow into the porous medium due to pressure gradients. The effect of $\Gamma = \gamma cos(\Theta)$ is shown in Figure 7–17. As expected, the higher

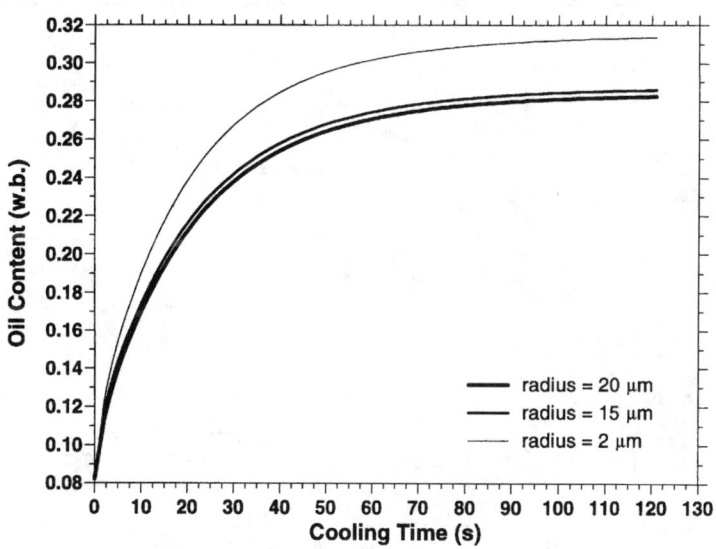

Figure 7–15 Effect of Tortilla Chip Pore Radius on Oil Absorption during Cooling. *Source:* Reprinted from *Journal of Food Engineering*, Vol. 35, R.G. Moreira and M.A. Barrufet, A New Approach to Describe Oil Absorption in Fried Foods: A Simulation Study, Pages No. 1–22, Copyright 1998, with permission from Elsevier Science.

the value of γ, the more oil is absorbed by the product, resulting in a less porous product (higher bulk density). Decreasing γ by 42% and by 55% results in 3% and 1% less oil absorption, respectively.

It is important to note that this mechanistic model for oil absorption during the postfrying period depends on the conditions developed during the frying process, i.e., porosity, initial oil content, and puffing coefficient. Current research combines the frying and cooling processes to predict oil absorption in fried products.

NOTATIONS

A_T adhesion tension [N/m]
C_p specific heat [kJ/kg K]
g   acceleration of gravity [m/s²]
h   capillary height [m]
h'_{air} convective heat transfer coefficient of air and tortilla surface [W/m² K]
k_{eff} effective thermal conductivity [W/m K]
K   absolute permeability [darcy]

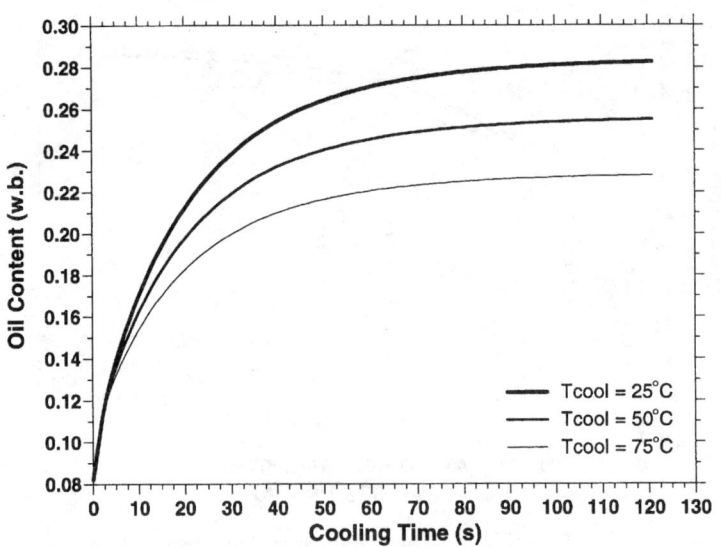

Figure 7–16 Effect of Cooling Air Temperature on Tortilla Chip Oil Absorption during Cooling. *Source:* Reprinted from *Journal of Food Engineering*, Vol. 35, R.G. Moreira and M.A. Barrufet, A New Approach to Describe Oil Absorption in Fried Foods: A Simulation Study, Pages No. 1–22, Copyright 1998, with permission from Elsevier Science.

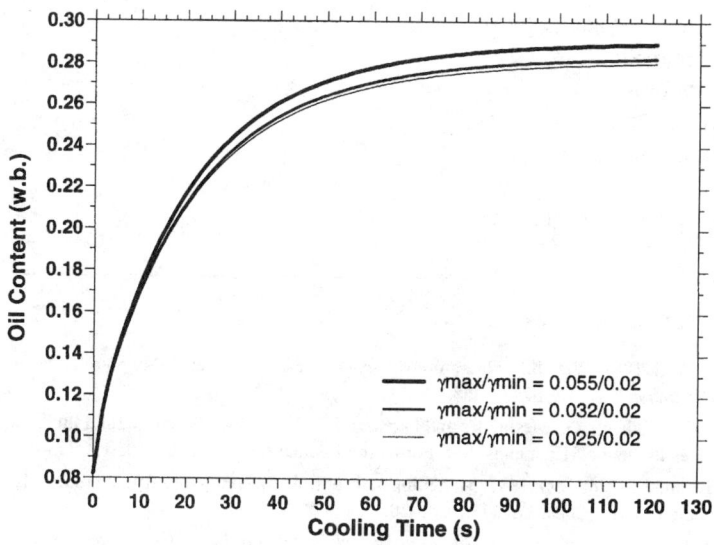

Figure 7–17 Effect of Interfacial Tension on Tortilla Chip Oil Absorption during Cooling. *Source:* Reprinted from *Journal of Food Engineering*, Vol. 35, R.G. Moreira and M.A. Barrufet, A New Approach to Describe Oil Absorption in Fried Foods: A Simulation Study, Pages No. 1–22, Copyright 1998, with permission from Elsevier Science.

K_i	effective permeability [darcy]
K_{ri}	relative permeability [ ]
L	tortilla chip thickness [m]
L'	length of plate [m]
m	mass fraction [kg/kg]
P	pressure [Pa]
P_c	capillary pressure [Pa]
P_{atm}	atmospheric pressure [Pa]
p'	number of components
r_c	radius of curvature [m]
r	capillary radius [m]
S_{ir}	irreducible liquid saturation
T	cooling temperature [°C]
t	time [s]
u	velocity [m/s]
V	volume [m³]
x	coordinate related to thickness [m]

α_{eff} thermal diffusivity [m²/s]
ε gas porosity
ϕ total porosity
θ temperature of the product [°C]
Θ contact angle [rad]
ρ density [kg/m³]
γ interfacial tension [N/m]

REFERENCES

Anderson, W.G. 1987. Wettability literature survey—the effects of wettability on relative permeability. *J Petroleum Tech*, 39:1453–1468.

Borgia, G.C.; Brighenti, G.; Mesini, E.; and Fantazzini, P. 1992. Specific surface and fluid transport in sandstones through NMR studies. SPE Formation Evaluation. Sept:206–210.

Brescia, L., and Moreira, R.G. 1997. Modelling and control of a continuous frying process: II. Control development. *Trans Chem Eng*, 75(c):12–20.

Chatzis, I. 1980. *A network approach to analyze and model capillarity and transport phenomena in porous media*. PhD diss., chemical engineering. Ontario, Canada: University of Waterloo.

Chen, Y., and Moreira, R.G. 1997. Modeling of a deep-fat frying process. *Trans Chem Engr*, 75(c):181–190.

Collins, R.E. 1976. *Flow of fluids through porous materials*. Tulsa, OK: The Petroleum Publishing Company.

Contreras, R. 1987. Hot air and microwave drying of deformable and rigid porous media. PhD diss. University of Minnesota.

Dullien, F.A. 1992. *Porous media: fluid transport and pore structure*. San Diego. CA: Academic Press.

Gamble, M.H.; Rice, P.; and Selman, J.D. 1987. Relationships between oil uptake and moisture loss during frying of potato slices from c.v. Record U.K. Tubers. *Int J Food Sci Technol*, 22:233–241.

Geankopolis, C.J. 1993. *Transport process and unit operations*. Englewood Cliffs, NJ: Prentice Hall.

Goedeken, D.L., and Tong, C.H. 1993. Permeability measurements of porous food materials. *J Food Sci*, 58(6):1329–1331.

Gubelin, G., and Boyd, A. 1997. Total Porosity and bound-fluid measurements from an NMR tool. Slumberger Wireline and Testing. *J Petrol Technol*, July:718–720.

Ilic, M., and Turner, I.W. 1989. Convective drying of a consolidated slab of wet porous material. *Int J Heat Mass Trans*, 23(1):2351–2362.

Lake, L.W. 1989. Enhanced oil recovery. In *Basic equations for fluid flow in permeable media*. Englewood Cliffs, NJ: Prentice Hall.

Leverett, M.C. 1941. Capillary behavior in porous solids. *Trans AIME*, 142:152–169.

Lewis, M.J. 1996. Physical properties of foods and food processing systems. Cambridge, England: Woodhead Publishing Limited.

Lujan-Acosta, F.J. 1996. Production of low-fat tortilla chips using alternative methods of drying before frying. Masters thesis. College Station, TX: Texas A&M University.

Macleod, D.B. 1976. On a relation between surface tension and density. *Trans Faraday Soc*, 19:38–42.

Marle, C.M. 1981. *Multiphase flow in porous media*. Houston, TX: Gulf Publishing Company.

Miller, C.A., and Neogi, P. 1985. Interfacial phenomena equilibrium and dynamic effects. New York: Marcel Dekker.

Moreira, R.G., and Barrufet, M.A. 1996. Spatial distribution of oil after deep-fat frying from a stochastic model. *J Food Eng*, 27(2):205–220.

Moreira, R.G., and Barrufet, M.A. 1998. A new approach to describe oil absorption in fried foods: a simulation study. *J Food Eng*, 35:1–22.

Moreira, R.G.; Palau, J.; and Sun, X. 1995. Deep-fat frying of tortilla chips: an engineering approach. *Food Technol*, 49(4):146–150.

Moreira, R.G.; Sun, X.; and Chen, Y. 1997. Factors affecting oil uptake in tortilla chips in deep fat frying. *J Food Eng*, 31:485–498.

Nasrallah, S.B., and Perre, P. 1988. Detailed study of a model of heat and mass transfer during convective drying of porous media. *Int J Heat Mass Trans*, 31(5):957–967.

Ni, H. 1997. Multiphase moisture transport in porous media under intensive microwave heating. PhD diss., Department of Agricultural and Biological Engineering. Ithaca, NY: Cornell University.

Nicholson, D., and Petropoulus, J.H. 1973. Capillary models for porous media: IV. flow properties of parallel and serial capillary models with various radius distributions. *J Phys D:Appl Phys*, 6:1737–1744.

Norgaad, J.V.; Olsen, D.; Springer, N.; and Reffstrup, J. 1995. Capillary pressure curves for low permeability chalck obtained by NMR imaging of core saturation profiles. Paper #SPE-20605. SPE Annual technical Conference and Exhibition, Dallas, Oct. 22–25.

Roberts, A.P. 1997. Statistical reconstruction of three-dimensional porous media from two-dimensional images. *Phys Rev*, 56:3203–3212.

Saravacos, G.D. 1995. Mass transfer properties of foods. In *Engineering properties of foods*, ed. Rao and Rizvi. New York: Marcel Dekker.

Scheidegger, A.E. 1954. Statistical hydrodynamics in porous media. *J Appl Phys*, 25:994–1001.

Stanish, M.A.; Schajer, G.S.; and Kayihan, F. 1986. A mathematical model of drying for hygroscopic porous media. *AIChE J*, 32(8):1301–1311.

Sun, X., and Moreira, R.G. 1994. *Oil distribution in tortilla chips during deep-fat frying*. St. Joseph, MI: American Society of Agricultural Engineers, Paper No. 94-6506.

Tiab, D., and Donaldson, E.C. 1996. *Petrophysics: theory and practice of measuring reservoir rock and fluid transport properties*. Gulf Publishing Company.

Tseng, Y.; Moreira, R.G.; and Sun, X. 1996. Total frying-use time effects on soybean oil deterioration and on tortilla chip quality. *Int J Food Sci Technol*, 31:287–294.

Ufheil, G., and Escher, F. 1996. Dynamics of oil uptake during deep-fat frying of potato slices. Food Sc. & Technol; *Lebensm—Wiss U-Technol*, 29(7):640–644.

Wei, C.K.; Davis, H.T.; Davis, E.A.; and Gordan, J. 1985. Heat and mass transfer in water-laden sandstone: convective heating. *AIChE J*, 31(8):1338–1348.

Whitaker, S. 1977. Simultaneous heat, mass and momentum transfer in porous media: a theory of drying. *Adv Heat Trans*, 13:119–203.

CHAPTER 8

Deep-Fat Frying Systems

The processes used to fry food products can be divided into two broad categories: (1) those that are static and smaller, classified as batch fryers used in the catering restaurants, and (2) those that fry large amounts of products in a moving bed, used in the food industry, classified as continuous fryers. These fryers can operate at atmospheric and low- or high-pressure conditions. In this chapter, the descriptions of different frying systems will be presented in detail.

BATCH FRYING SYSTEMS

Batch fryers should be of the appropriate size and installed in the proper number. In addition, other factors, such as fuel source, speed of temperature recovery, and safety, should be taken into consideration when selecting a frying apparatus.

A deep-fat fryer (batch or continuous) consists of a chamber where heated oil and a food product are placed. The frying oil is directly heated by means of electricity, gas, or fuel oil. The simplest heating system consists of gas flames placed directly against the bottom of a frying kettle. Some batch fryer designs use an infrared gas heater on a V-bottom (called open-pot fryers), as the one shown in Figure 8–1A. The open-pot fryer shows a deep cold zone that helps to trap food particles. The tube-type fryers (Figure 8–1B) use tube-fired burners to heat the frying oil. Generally, the tube-type fryers (Figure 8–1B) are less efficient than the open-pot fryers (Figure 8–1A). Tube-type fryers can accumulate food particles at the bottom of the vat, resulting in an insulating layer between the heat source and the oil, thus reducing heat transfer and accelerating fat breakdown, due to local overheating and contamination.

Figure 8–2 shows a schematic of a gas-fired batch fryer. The fryer body is made of stainless steel. The product is placed in a basket, then lifted down to the pan containing the frying oil. Some fryers have a basket-lifting system that automati-

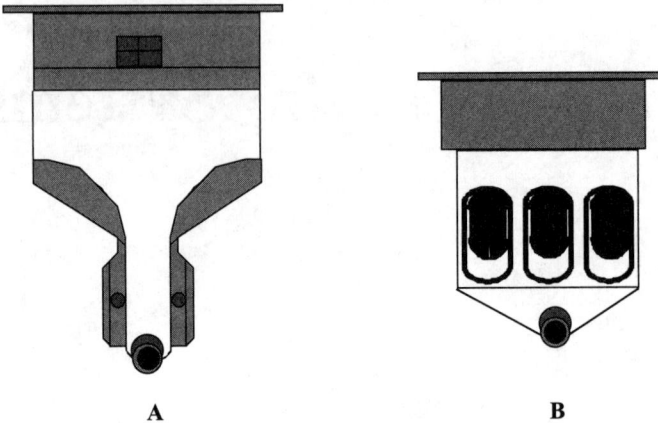

Figure 8–1 Schematic of **A**, Open-Pot Fryer; **B**, Tube-Type Fryer Gas Heating System. Courtesy of Frymaster, Shreveport, Louisiana.

cally raises the basket when the frying time is finished. A valve is used at the bottom of the fryer for easy draining of the oil. Some models also include built-in filtering systems with automatic washdown of the oil. The main switches in the fryer consist of the oil temperature set point, frying time, and the manual basket-raising button.

The frying oil temperature depends on the product to be fried and the food quality requirements. Typically, the temperature ranges from 160°C to 200°C. Thermostats or solid-state controllers are used to control the oil temperature in the fryer. An electric batch fryer is shown in Figure 8–3; the difference in the design is the heating source. In this case an electric lift-out heater element is used to heat the frying oil.

The characteristics of a commercial batch fryer (Frymaster, Shreveport, Louisiana) are shown in Table 8–1. Batch fryers can be of different types. Countertop types are small-capacity fryers (oil capacity 8–11 L), and economic and high-efficiency fryers can have a capacity ranging from 17 to 28 L of oil. The high-efficiency fryers are equipped with turbo-jet infrared burners using 30–40% less energy than the same capacity standard gas-fired fryers. The throughput can range from 12 kg/hour to 35 kg/hour of finished frozen french fries. Electric fryers are available with different energy inputs to produce the desired quantity of fried products.

Pressure fryers were developed primarily to be used in the food service industry for chicken frying. The time required to fry chicken parts is faster in a pressure

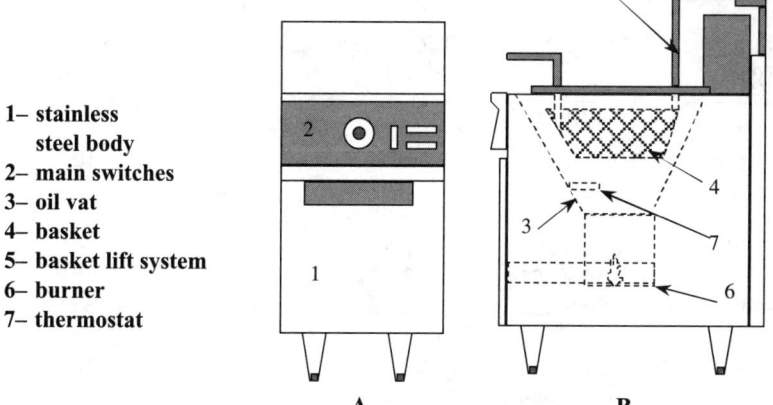

1– stainless
   steel body
2– main switches
3– oil vat
4– basket
5– basket lift system
6– burner
7– thermostat

Figure 8–2 Schematic of a Gas-Fired Batch Fryer. **A**, Front view; **B**, Side view. Courtesy of Frymaster, Shreveport, Louisiana.

fryer than in atmospheric (open) fryers, and the food is more moist and uniform in color and appearance.

Pressure batch fryers can be electric or gas fired and are available in many sizes. Figure 8–4 presents a schematic of a commercial pressure fryer (Henny Penny, Eaton, Ohio). Oil capacity can range from 11 L to 25 L and food capacity from 5.6

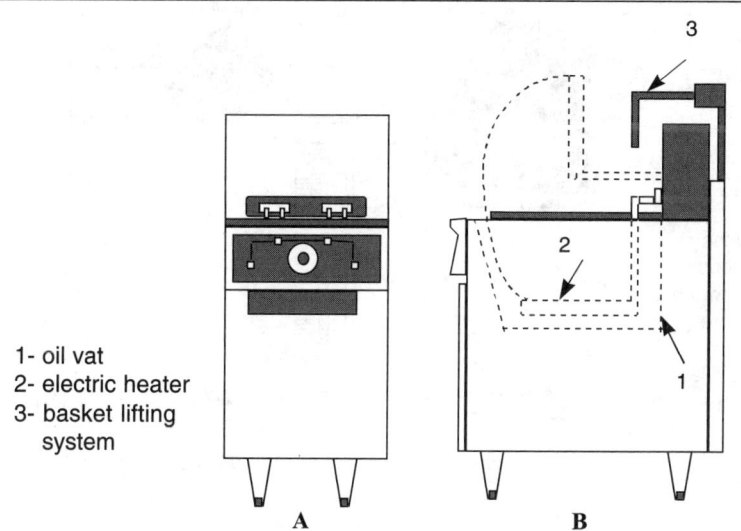

1- oil vat
2- electric heater
3- basket lifting
   system

Figure 8–3 Schematic of an Electric Batch Fryer. **A**, Front view; **B**, Side view. Courtesy of Frymaster, Shreveport, Louisiana.

Table 8–1 Characteristics of a Commercial Batch Fryer[a]

Type	Frying Area (mm)	Oil Capacity (L)	Energy Input (kW)	Throughput[b] (kg/h)
Gas-fired				
Countertop	279 × 305	8-11	13.2	12
Economic	305 × 381	17-22	29.3	27
	356 × 381	22-28	35.8	36
Standard	305 × 381	17-22	32.2	30
	356 × 381	22-28	35.8	36
High-efficiency	356 × 381	22-28	23.4	36
Electric fryers	356 × 394	22-28	14	32
	356 × 394	22-28	17	36
	356 × 394	22-28	22	41

[a] Frymaster; [b]french fries (frozen to done).

Source: Data from Frymaster.

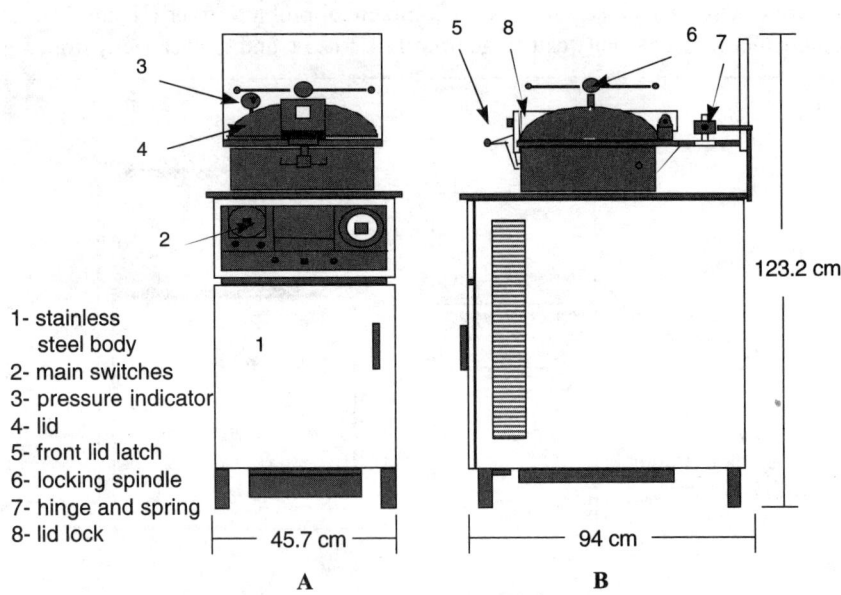

1- stainless steel body
2- main switches
3- pressure indicator
4- lid
5- front lid latch
6- locking spindle
7- hinge and spring
8- lid lock

45.7 cm

94 cm

123.2 cm

A

B

Figure 8–4 Schematic of a Pressure Batch Fryer. **A**, Front view; **B**, Side view. Adapted from Henny Penny, Eaton, Ohio.

kg to 10.9 kg. The operating temperature of pressure fryers can vary from 160°C to 177°C and the frying time from 7 to 10 minutes. Two pressure ranges are used. Some fryers employ 9–14 psi and others 28–32 psi.

Some disadvantages of pressure fryers are that the frying oil will degrade faster (steam from the food is retained within the fryers, thus increasing buildup of FFA), and the oil must be filtered more frequently (bread and crumbs can burn or oxidize, causing off flavors) (Lawson, 1985).

HIGH-CAPACITY FRYING SYSTEMS

The industrial fryer consists of several basic components, which are similar for both continuous and batch systems. In designing a continuous fryer, factors such as the amount of food, the conveyor system, the food characteristics, and the handling system after frying are important to effectively produce high-quality products. An oversized fryer can be very inefficient, causing severe oil degradation, creating cleanup problems, and resulting in poor-quality final product. Multipurpose frying systems tend to be inefficient, too. It is better to design a fryer for maximum efficiency in producing one product type than to produce multiple products inefficiently (Stier, 1996a).

Many determining factors come into play when deciding what type of fryer to select, including the type of product being produced, cost, space, manufacturer, and safety. According to Belshaw (1976), an effective automated frying system consists of four requirements: accurate temperature control, efficient heat transfer, minimum oil contamination, and rapid oil turnover.

A continuous fryer system consists of at least five independent sets of equipment: (1) the kettle or tank containing the frying oil; (2) a heating unit with a control system for generating thermal energy; (3) a conveying system for moving the product into, through, and out of the frying process; (4) a fat system, which pumps and filters the frying oil; and (5) an exhaust system for removing the hot vapors emerging from the product.

Heating Systems

In practice, three major design systems exist for heating and controling cooking oil: direct, indirect, and external.

Direct heating systems (Figure 8–5) use natural gas, fuel oil, or electric heaters to increase the frying oil temperature. The heating elements (or heating tubes) are immersed in the cooking oil directly below the product zone. Heating is achieved by firing these heating elements.

Heating is transferred directly into the cooking oil by conduction through the heating elements. Oil temperature control is achieved by modulating the fuel input

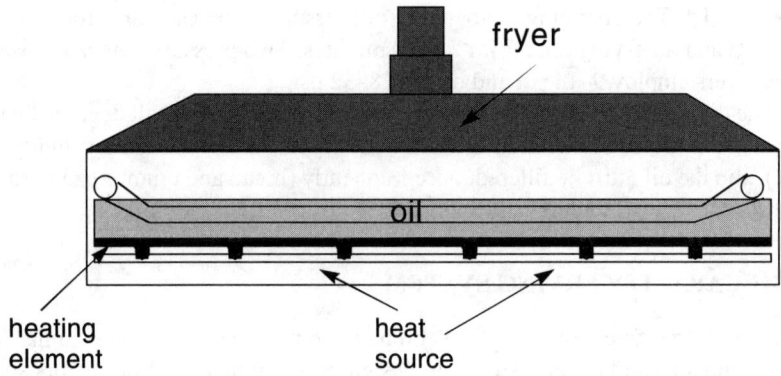

Figure 8–5 Schematic of a Direct Heating System

to the burners or by switching the electrical elements. Immersion tubes can be arranged widthwise, lengthwise (multiple burners), or through S-shaped single burner tubes. Examples of food products fried in direct heated systems include kettle-style potato chips, coated products (fish, meat, poultry, and vegetables), oil-roasted nuts, doughnuts, and other snack products.

Some advantages and disadvantages of the direct heating systems are (Stier, 1996a):

Advantages
- compact packages
- lower initial cost
- some temperature zoning
- flexibility (multipurpose)

Disadvantages
- no control of overall temperature
- relatively high oil volume
- limited heat capacity
- lower thermal efficiency
- mechanical removal of fines
- settlement of fines on heating pipes
- lack of temperature uniformity
- limited product pack depth

Indirect heating systems (Figure 8–6) use external sources, such as a heater fired with gas, oil, or electricity to heat a thermal fluid (chlorinated hydrocarbons). The temperature of the thermal fluid is maintained above the target temperature of the frying oil. The same heating-tube arrangement described in the direct heating system is used to heat the frying oil. Oil temperature is controlled by regulating,

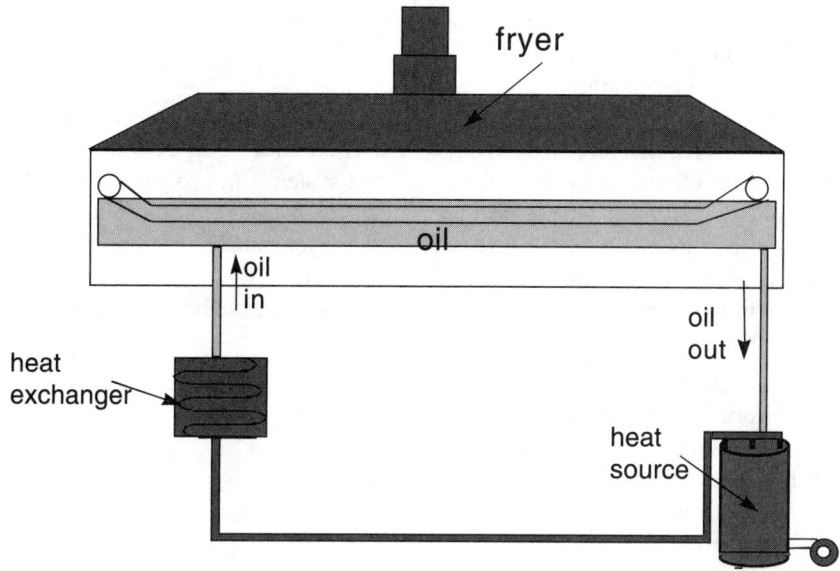

Figure 8–6 Schematic of an Indirect Heating System

with a three-way valve, the heated thermal fluid entering the frying system. The thermal fluid flow direction is controlled by a feedback controller that tells the valve to release the thermal fluid to the fryer or back to the heater, based on the oil temperature.

Some of the products processed in this system are batter and breaded foods and oil-roasted nuts. This system is more popular in Europe than in the U.S. because of the erroneous belief that the frying oil can be contaminated by the thermal fluid (Stier, 1996a).

Some of the advantages and disadvantages of the indirect heating systems are (Stier, 1996a):

Advantages
- more uniform temperature control
- no gas equipment in cooking area
- little noise
- high thermal efficiency
- some temperature zoning

Disadvantages
- no control of overall temperature
- limited heat capacity
- removal of fines by mechanical means

- settlement of fines on heating pipes
- limited product pack depth
- difficulty in cleaning

External heating systems (Figure 8–7) use external heat exchangers to maintain the frying oil temperature. The frying oil is passed through an external heat exchanger and returned directly to the fryer. The system can use gas, fuel oil, or electricity to heat the oil in the heat exchanger. Temperature is controlled by inserting a thermocouple on the inlet side of the fryer, thus controlling the fluid (fuel, steam, thermal) to the heat exchanger.

Some of the advantages and disadvantages of the external heating systems are (Stier, 1996a):

Advantages
- uniformity of temperature control
- no gas equipment in cooking area
- high heat capacity
- low oil volumes
- high thermal efficiency
- temperature zoning possible
- automatic cleaning-in-place system included
- ability to match oil/product flow

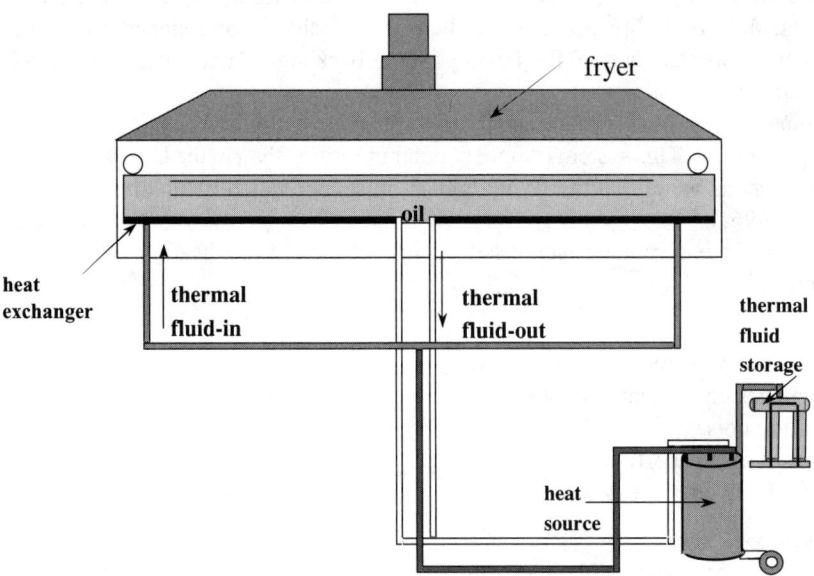

Figure 8–7 Schematic of an External Heating System

Disadvantages
- high initial cost
- requires more floor space
- pump/motor noise

Temperature Control

Generally, two methods are used to control the oil temperature in the fryer, the on/off method and the modulating system. The on/off is the simplest system and consists of turning the heat source on or off by means of a thermostat switch connected to a thermocoupler placed in the frying oil. This system is used when accurate temperature control is not required. A temperature control range of $\pm\,5°C$ is possible (Stier, 1996a).

Modulating systems are more common in high-production operations. Temperature probes are used to accurately measure the oil temperature and digital controllers are used to efficiently maintain the oil temperature closely to the set point.

Conveyor Systems

According to Stier (1996b), food products can be classified according to the way they behave during frying as:

- surface-fried, such as doughnuts
- immersion fried, for example, pellets
- partially floating, including chips and tempura
- nonfloating, referring to meat products and french fries

Therefore, different types of conveyor systems are required to push the product through the fryer. Doughlike products, such as doughnuts, corn chips, and fabricated potato chips, are extruded directly into the frying oil. Semisolid or solidlike products, such as potato chips, french fries, fish sticks, meat patties, pellets, and most coated products, require feed conveyors that drop the products into the fryer.

In continuous operation, the design of the conveyor system through the fryer depends on the food type. For surface-type products such as doughnuts, the product moves in a screen-type conveyor that automatically turns the doughnut pieces over at the right time. The same processing applies to potato chips. During the process, the chips lose water and float to the oil surface. In this case, turning the chips is needed to fry them uniformly.

Some food products, such as batter and breaded products, have the tendency to float to the surface during frying. Conveyor systems, as the one shown in Figure 8–8A, are designed to keep the products under the oil so that they are cooked as they pass through the fryer.

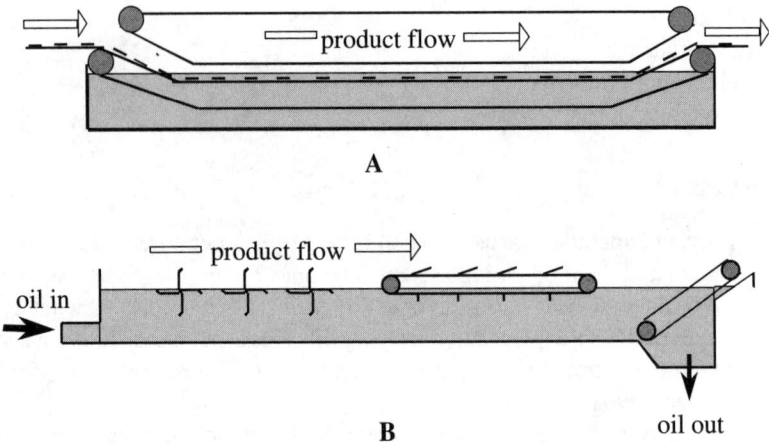

Figure 8–8 Conveyor Systems. **A,** Belt System; **B,** Rotating Paddles and Submerger Conveyor.

Potato chips, tortilla chips, tempura, and extrudates sink and then float to the surface during frying. The conveyor systems have to be designed to allow the product to rise and then keep it below the surface part way during the process. Figure 8–8B presents a fryer with two conveyor systems: one used at the beginning of frying, consisting of a series of rotating paddles that dunk, separate, agitate, and control the advance of the product as it is cooked; and another that will transfer the product through the final cooking zone in the oil by means of a flighted submerger-type conveyor that holds the product under the surface while controlling its advance through the fryer.

Typical Continuous Fryers

A number of manufacturers have worked continually over the past years to improve the design of continuous fryers to produce efficiently high-quality final products. Some examples of these companies include Heat and Control and Stein in the U.S. and Florigo in Europe (the Netherlands).

Figure 8–9 shows schematics of a continuous fryer in open and closed positions. This is a "split-apart" type of fryer. The hood and the product conveyor system can be raised for cleaning and inspection. When in open position, the heat exchanger remains in a fixed position between the hood and the kettle, making it easier to clean.

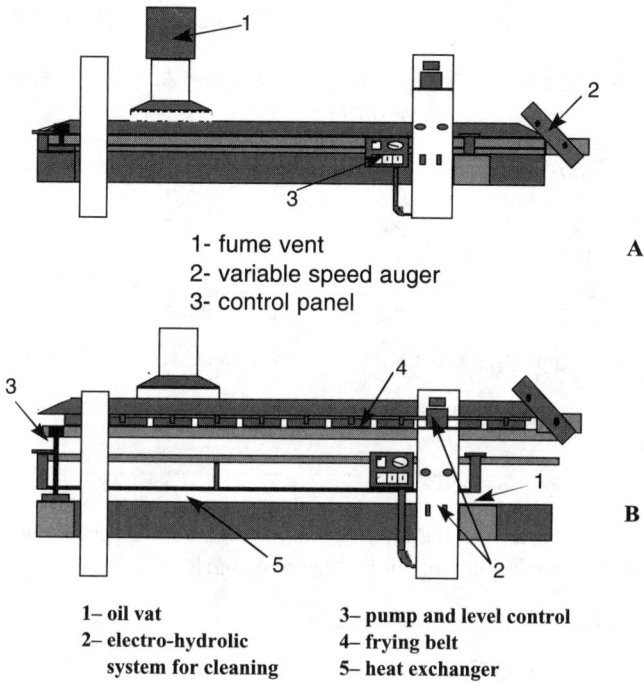

1- fume vent   **A**
2- variable speed auger
3- control panel

1- oil vat   3- pump and level control
2- electro-hydrolic   4- frying belt
   system for cleaning   5- heat exchanger

B

Figure 8–9 A Schematic of a Continuous Fryer. **A**, closed position; **B**, open position

---

The output capacity of continuous fryers varies from 250 to 25,000 kg/hour for french fries, 100 to 2,000 kg/hour for tortilla chips, and 100 to 2,500 kg/hour for potato chips. A continuous frying system used for frying nuts can have a capacity from 500 to 3,000 kg/hour. Table 8–2 shows an example of a commercial continuous fryer specification for frying nuts.

Continuous vacuum frying is a concept developed by Florigo (H&H Industry Systems B.V., the Netherlands) in the early 1970s to produce high-quality french fries. Due to the improvement in quality of the raw materials and blanching techniques, the use of vacuum fryers almost died out, with exception of one or two production companies that still produce a nonblanching product. Today, the Florigo automatic continuous vacuum fryers (Figure 8–10) are used mainly to produce fruit chips and very delicate snack products.

French fries processed in a vacuum fryer can achieve the necessary degree of dehydration without excessive darkening or scorching of the product (the frying oil temperature is much lower than in an atmospheric fryer). Given the current

Table 8–2 Technical Description of a Commercial Continuous Nut Fryer[a]

Capacity[b]	Belt Width (mm)	Frying Surface Area (m²)	Cooling Surface Area (m²)	Energy Input (kW)
500	600	1.5	1.8	7.0
750	750	2.0	2.7	8.0
1,000	750	2.6	3.2	8.0
1,500	940	4.0	5.0	11.4
2,000	940	5.0	6.5	12.0
3,000	940	7.0	9.0	17.5

[a] Florigo; [b]based on frying time of 5 min and product layer of 50–60 mm.

Source: Courtesy of Florigo, Woerden, The Netherlands.

preoccupation with lowering the fat contents of diets, vacuum frying would be an efficient way of reducing the oil content in fried snacks.

Conventional Potato Chip Frying Process

The most common frying conditions for conventional continuous potato chip processing utilizes external heating and continuous oil circulation. The chips are immersed initially in hot oil at temperatures of about 182°C to 199°C and conveyed through the fryer. The product and the frying oil flow concurrently along the frying path. The fried chips are withdrawn from the oil at a temperature of about 160°C to 177°C. There is usually a 16–25°C drop in temperature during the course of continuous frying. In some cases, multizone fryers (Figure 8–11) are used to provide a different set of frying parameters at points along the fryer to improve product quality. Whenever the temperature drops along the frying path in one zone, it rises as the next zone is entered, resulting in a "sawtooth" temperature profile along the frying path.

Hard-Bite Potato Chips

Also called *home style chips*, hard-bite chips are usually fried in a batch process but may also be cooked in a continuous process and usually have a harder bite than does a conventional chip. These chips are usually sliced from 1.27 to 1.78 mm thickness. The potatoes may be washed to remove surface starch or cooked unwashed. The slices are fried at temperatures lower than the conventional chips, at about 121–177°C for a longer period of time. The fried chips are usually lighter in color than are the conventional chips and have an average of 32–40% w.b. oil content (Benson et al., 1992).

Figure 8–10 Vacuum Fryers (Florigo). *Source:* Courtesy of Florigo, Woerden, The Netherlands.

Another type of hard-bite chip is the so-called Maui-style or open kettle chips. These chips have normal to heavier thickness and are darker (nonuniform) in color than are the conventional chips. These chips are not washed prior to frying, and the oil content ranges from 32% to 40% w.b. or higher. The temperature of the oil during the process of making Maui-style chips initially may range from 143°C to 166°C.

Figure 8–12 shows a block diagram of a multizone continuous fryer designed to produce snack products with low oil content. The following description is based on the invention by Benson et al. (1992) to produce hard-bite type potato chips. The frying oil is heated by an external heat exchanger. Recirculated oil from the heat exchanger is introduced to the kettle in several points along the fryer. At the end of the fryer kettle, the oil is withdrawn from the container for heating and circulation.

The raw material is dispensed from a conveyor belt (not shown) dropped into the hot oil, and conveyed along the cooking zone by means of a series of rotating

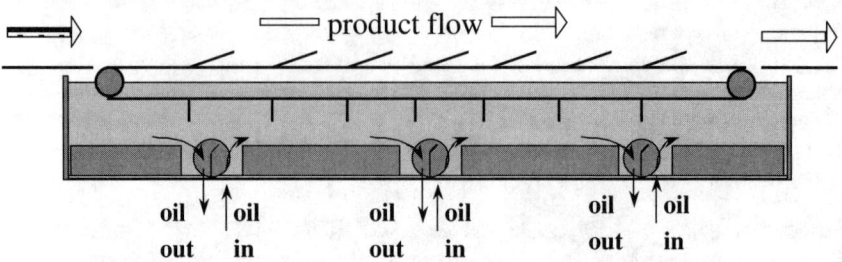

Figure 8–11 Multizone Fryers

paddles that dunk, separate, agitate, and control the advance of the chips as they are fried. In zone A, the temperature set is in the range of 151 ± 3°C and drops slowly along the frying path to 118–124°C. In zone B, the oil temperature decreases from around 121 ± 3°C to about 115 ± 3°C. Zones C and D are characterized by a time-temperature frying profile along the frying path, with a 5–20°C temperature difference between the beginning and the end of each zone. The final temperature at the end of zone D is in the range of 146 ± 3°C.

Oil flows from the frying zones through exit pipes while fresh oil and/or recycled oil is introduced through inlet pipes. After the products pass zone C, they will contact a series of flighted submerger conveyors (zones E and F) that hold the chips below the surface of the oil while controlling their advance through the

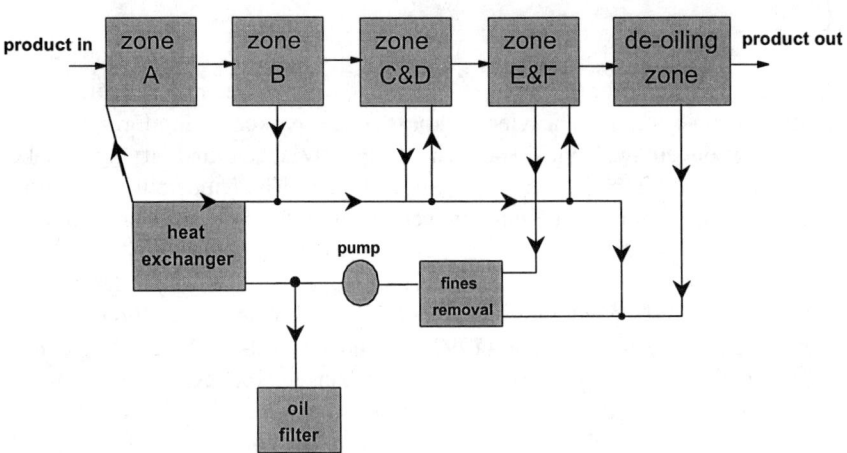

Figure 8–12 Continuous Fryer Diagram for Hard-Bite and Reduced Oil Content Potato Chips

fryer. The fried products are then removed from the fryer by means of a takeout conveyor, which deposits the products onto a de-oiling conveyor. The de-oiling conveyor is of open-weave mesh construction and is located within a chamber that contains an opening to an exhaust fan. Drainage of oil from the chips along the takeout and de-oiling conveyors is enhanced by heat, which may be provided by flow of fryer exhaust gases and/or overhead heaters. The fried, de-oiled chips are then deposited into a rotary discharge airlock drum and exit the fryer.

Meatballs/Pizza Toppings/Bacon Bits Frying System

An innovative frying system, ContinuTherm, designed and manufactured by Blentech (Rohnert Park, California), uses a screw system to push the product through the fryer. Pizza toppings, bacon bits, and meatballs can be cooked in hot oil at temperatures up to 129°C. The oil is heated indirectly with steam in the fryer jacket and in a heat exchanger. Figure 8–13 shows a schematic of the system. The system capacity varies from 45 to 45,000 kg of product/hour.

Safety

Safety is a vital aspect of good design when it comes to frying systems. Some of the elements of the frying system that have potential for severe injury include the hot frying oil contained in the fryer kettle, the flames or heat exchangers, the flammable heating media, belts and conveyors, and any moving parts. A list of safety features that are included by manufacturers to reduce injury risks are (Stier, 1996c):

1. *Systems to remove water*: A number of serious injuries may result from water being discharged into the hot oil, causing explosions that throw oil in all directions and, sometimes, blow off the end of the fryer. All fryers must be designed with a system that drains water from cleanups to allow spot cleaning with a drain valve.
2. *Valves*: Use of ball valves in circulation piping must be avoided as they can accumulate water that can cause steam explosion in the hot oil during operation. Use of butterfly and plug valves is recommended.
3. *Addition of soda caustic*: Maximization of system design safety concerns is achieved by avoiding addition of soda caustic to hot oil.
4. *Oil level*: One cause of fires in fryers has been low oil levels in the kettle, thus exposing the heating tubes to the air. Oil residue build-up in the tubes can ignite from the heat and burst into flames. Controls should be used to ensure that the oil remains at the proper level. Air pressure valves are safer than ball or float valves.

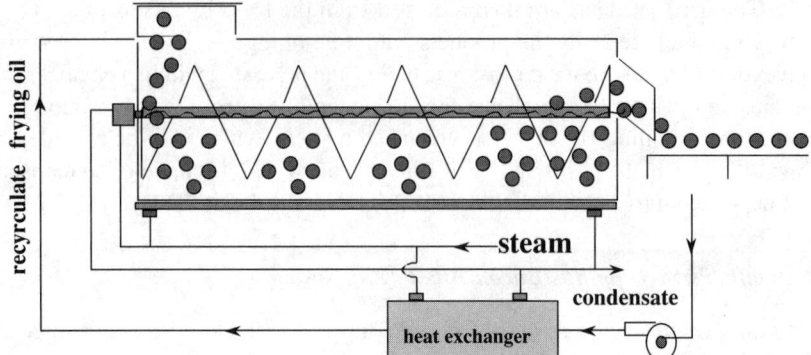

Figure 8–13 Screw Fryer System (Blentech, Colorado). Adapted from of Blentech, Rohnert Park, California.

---

5. *Conveyors*: Mechanical conveyors are generally easier to maintain than hydraulic units. Hydraulic conveyors must be monitored for damage of the fluid lines, and the fluids are under pressure and are flammable.
6. *Burners*: Manufacturers today produce fryers that fire within closed systems because open-flame burners are hazardous and can create problems when operators try to light them manually.

Other important aspects of safety when operating a frying system include operational procedure highlighted in the equipment, installation of CO_2-based extinguishing system above all fryers, and operation of the system with well-trained operators and a well-educated cleanup crew.

Future of Frying System Design

Stier (1996c) listed important areas that will improve frying system designs and, thus, product quality in the future. These are the following:

- *Combination Systems*: Oven and fryer combination systems that will improve product quality by reducing oil content.
- *Enhanced System Efficiency*: Understanding how oil quality can affect heat capacity and heat transfer ability will improve fryer design, thus making systems more efficient.
- *Reduced Fat Content*: Understanding the interaction between oil chemistry and heat, mass, and momentum transfer during frying will enhance frying system design and improve product quality by reducing oil content. In addition, with the approval by FDA of Olestra, a zero-calorie fat replacer, new

development will be required to understand the frying characteristics of this new product.

- *Odor Control*: Means of reducing emissions from frying systems will be of great concern in the future. In Europe, for example, frying systems today are allowed to emit only water and CO_2. Some examples of emissions control systems include catalytic converters, scrubbers, incinerators, and direct-fired heating of frying oil with incineration.
- *Process Control*: Automatic frying systems run with less supervision and greater productivity than systems with manual control. Chapter 9 describes an example of a continuous frying process control based on advanced control techniques.
- *On-line Sensors*: It is very important today to have sensors capable of measuring on-line in real time the quality of the frying oil (viscosity, TPM, FFA, etc.) and frying products (oil content, crispness, color, etc.) to better control frying systems efficiently. Through feedback loops to computer-based controllers, operation conditions can be altered continuously on-line to ensure high-quality product.

REFERENCES

Belshaw, T.E. 1976. Cutting and frying equipment. *Proc Am Soc Bakery Eng*, 112–120.

Benson, C.K.; Caridis, A.A.; and Klein, L.F. 1992. *Continuous food processing method*. Patent No. 5,137,740. United States Patent Office.

Lawson, H.W. 1985. Standards for fats and oils. In *The L.J. Minor foodservice standards series*. Vol. 5. New York: Van Nostrand Reinhold/AVI.

Stier, R.F. 1996a. Understanding high-volume frying, Part 1. *Baking and Snack*, February/1996:70–76.

Stier, R.F. 1996b. Understanding high-volume frying, Part 2. *Baking and Snack*, April/1996:51–54.

Stier, R.F. 1996c. Understanding high-volume frying, Part 3. *Baking and Snack*, May/1996:41–44.

CHAPTER 9

Continuous Fryer Control Systems

The Institute of Food Technologists (IFT) has identified the need to improve process design and operation efficiencies through closed-loop control strategies (IFT Research Committee, 1993). Process control, which maximizes throughput while optimizing product quality and extending the processing time between shut down and cleanup would obviously increase efficiency.

Automatic control of continuous fryers helps to improve final product quality, increase process efficiency, and reduce waste of raw materials. Continuous frying processes are multiple-input, multiple-output systems involving complex interactions that are generally characterized by strong relationships among mass, energy, and momentum transfer, including complex physicochemical transformations such as gelatinization of starch, denaturization of proteins, browning reactions, etc. Such changes are influenced by the chemical composition and physical state of the materials and the frying oil, and by the process conditions.

This chapter focuses on the development of control strategies for continuous-frying systems.

CONTINUOUS-FRYING SYSTEMS

In the continuous industrial frying processes, quality control is maintained by monitoring oil storage conditions, the frying process, and product testing at regular time intervals. Attention is also given to the design aspects of the continuous fryer to minimize the effects of heat degradation/oxidation in the frying oil while maintaining high throughputs of products of consistent quality.

Raw materials used in frying processes are mainly of biological origin, and their compositional and physical nature can vary considerably. Such variations can introduce significant, unmeasurable disturbances to the process that makes manual control unreliable. Besides the complexity caused by raw material variability, continuous fryers also exhibit larger dead times and are typically nonminimum phase systems.

In an industrial operation, it is not unusual to encounter raw material variability on the order of 5–10% in lots of products received from different growers or suppliers. Yet all the product has to be fried to the same safe-storage quality (moisture content, oil content, color, flavor, etc.). The challenge to the operator of the continuous fryer is to properly vary the residence time of the product in the fryer and/or the frying oil temperature.

Continuous fryers are usually manually controlled by changing the submerger speed and, thereby, the residence time of the product in the fryer. Depending on the fryer design, oil temperature and/or takeout conveyor speed are also altered during the process. These procedures often lead to overcooking of the product and take about 25% of the operator time.

Automatic controllers for continuous fryers should perform well with respect to the following characteristics:

1. *Accuracy:* The output (oil content, color, moisture content) must remain close to the set point.
2. *Stability:* The output values should not fluctuate very much.
3. *Response time:* A sudden change in the input(s) or output(s) should be quickly offset by the controller.
4. *Robustness:* The control system should be able to operate adequately under a wide range of disturbances (ambient condition, raw material composition, etc.)

Oil/moisture-content- (color)-activated controllers are used on continuous fryers; they should be judged according to these four criteria.

PROCESS CONTROL SYSTEMS

Assuming that the reader has not received formal training in process control, the basic concepts of control theory are briefly considered. In addition, a list of terminologies necessary to describe control system is presented at the end of this chapter.

A process control system is an automatic regulating system in which the output is a variable, such as temperature, moisture content, pressure, flow, etc.

The frying of food in a continuous fryer is called the *process*. The fryer-exit product moisture content, for example, is the output signal or the controlled variable. The submerger speed (or sometimes the frying-oil temperature) is the input signal or the manipulated variable. Those signals or variables, which tend to adversely affect the value of the controlled variable, are defined as *disturbances*; in a continuous fryer, they include the ambient conditions, the raw material composition, raw material variety, frying oil properties, etc.

The relationship between the controlled, manipulated, and disturbance variables constitutes the process control problem of the continuous fryer. Figure 9–1 shows a block diagram of a multi-input, multi-output (MIMO) continuous frying process. The manipulated variables consist of the oil temperature (OT), the submerger speed (SS), and the takeout conveyor speed. The controlled variables are, in this case, the oil content (OC), moisture content (MC), and other product quality attributes (PQA), such as color, crispness, porosity, etc. Random changes in the raw material composition (moisture content, particle size distribution, sugar content, etc.) in the ambient conditions, in the frying oil quality, and in the frying characteristics of the product affect the final product quality. The control problem consists of finding the value of the manipulated variables that minimizes the effect of the disturbances on the controlled variable. This can be done by using feedback control, feedforward control, or a combination of both.

In feedback control systems, the output signal has a direct effect on the control action. The actuating error signal, which is the difference between the input signal (i.e., the set point) and the feedback signal (i.e., the measured controlled variable), is fed to the controller in order to reduce the error and bring the output of the system to the desired value (set point). Figure 9–2 shows the input-output relationship of a feedback control system for a single-input, single-output (SISO) continuous fryer.

In feedforward control systems, the output has no effect on the control action, i.e., the output is neither measured nor fed back for a comparison with the input (see Figure 9–3). Feedforward control is the control of undesirable effects of measurable disturbances (e.g., inlet moisture content) by compensating for them before they occur. It can minimize the transient error, but as feedforward is an open-

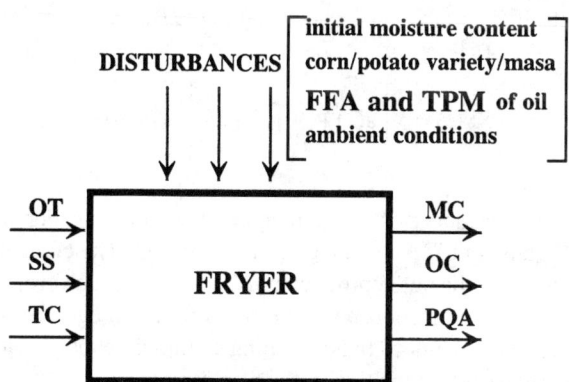

Figure 9–1 Block Diagram of a Multi-Input Multi-Output Continuous Frying Process. OT, oil temperature; SS, submerger speed; TC, takeout conveyer speed; MC, moisture content; OC, oil content; PQA, product quality attributes; FFA, free fatty acids; TPM, total polar materials.

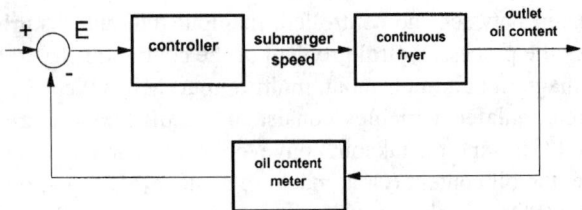

Figure 9–2 Feedback Control System. E, control error (set point-measured moisture content).

loop control system, there are limitations to its functional accuracy. It does not cancel the effects of unmeasurable disturbances (e.g., wear) under normal operating conditions. It is, therefore, necessary that a feedforward control system also include a feedback loop, as shown in Figure 9–4.

Classic Feedback Controller

A *classic control method* is generally defined as one in which a dynamic transfer function model is expressed in terms of Laplace transform variables. It is restricted to single-input/single-output systems. The control algorithms are typically analog digital versions of the proportional-integral-derivative (PID) algorithm.

The block diagram in Figure 9–5 represents the proportional and integral (PI) feedback control of a continuous fryer. The submerger speed (manipulated variable) at time t, $SS(t)$, is based on the error between the measured output and the set point of the controlled variable, $e(t)$, according to the standard continuous PI control algorithm:

$$SS(t) = K\left[e(t) + \frac{1}{T_i}\int E(t)dt\right] + SS(o) \qquad \textbf{[1]}$$

where $SS(o)$ is the initial speed of the submerger when the controller is started, K is the proportional gain, and T_i is the integral time constant. The controlled variable can be the oil content or the outlet product moisture content. The proportional term in Eq.(1) provides a rapid response to an error, and the integral term prevents a steady-state error. The Ziegler-Nichols tuning methods or parameter optimizations are often used to determine the best control parameters.

The presence of time delays in the process limits the performance of a conventional feedback control system. A time delay adds phase lag to the feedback loop,

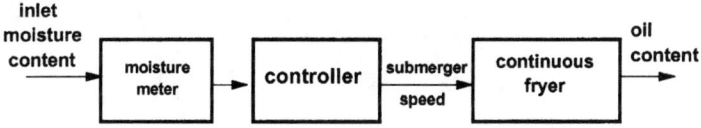

Figure 9–3 Feedforward Control System

---

thus adversely affecting the closed-loop stability. Consequently, the controller gain must be reduced below the value that could be used if no time delay were present, and the response of the closed-loop system will be slow, compared with that of the control loop with no time delay (Seborg et al., 1989).

For systems with significant time delays, such as continuous fryers, improved control performance over PID controllers has been achieved with the Smith predictor (Seborg et al., 1989) and the minimum variance (adaptive) controller scheme (Isermann, 1991) provided that the time delay is estimated accurately. Incorrect estimates of delays can cause poor performance for these control systems (Clarke et al., 1987a).

The last decade has seen the development of a class of controllers referred to as model predictive controllers (MPC). MPC are especially well suited for food processes because of their ability to deal with delays. One form of MPC, the general-

---

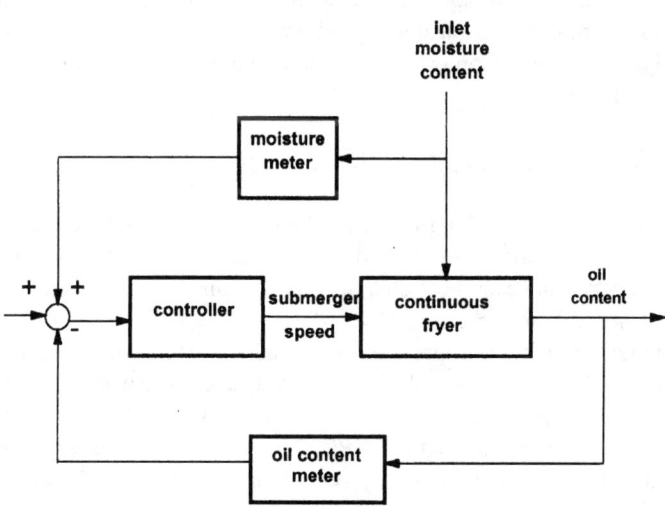

Figure 9–4 Feedforward Control System with Feedback Loop

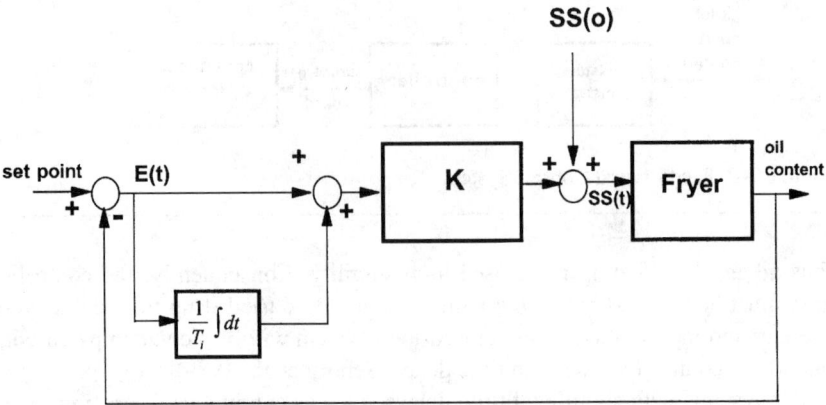

Figure 9–5 Block Diagram of Proportional and Integral (PI) Control of a Continuous Fryer

ized predictive controller (GPC), has been found to overcome the limitations associated with the minimum variance and the Smith predictor control schemes (Clarke et al., 1987a, 1987b). A self-tuning GPC is capable of controlling processes with variable parameters, variable dead time, and variable model orders. With regard to continuous fryers, the most important characteristics of the GPC are its robustness, the ease of tuning, and low process overshoot. Schonauer (1995) successfully employed a GPC with an ARX (autoregressive with an exogenous input) model to control a twin-screw food extruder. Haarshma (1994) applied dynamic matrix control (DMC) to a continuous frying process, achieving good results.

Model Predictive Controller

Control design using long-range prediction based on a dynamic model of the plant (as does the MPC family of controllers) has become an important contender for high-performance applications. MPCs make direct use of an explicit and separately identifiable model.

Although significant progress has been made in theoretical understanding and application of adaptive control, no single adaptive control scheme has been suitable as a general-purpose algorithm for the majority of real processes. Clarke et al. (1987a, 1987b) proposed that the GPC algorithm is effective with a plant that is simultaneously nonminimum phase, open-loop unstable and whose model is over- or underparameterized by the estimation scheme without special precautions being taken. Essentially, GPC tends to overcome these issues and creates a robust control scenario.

One form of MPC, the DMC, has been used extensively in the petrochemical industry for controlling complex processes that are multivariable and have operating constraints (Morari et al., 1988; Garcia and Morshedi, 1986). The method calculates moves on manipulated variables that minimize future projections of controlled variable errors and constraint violations in the least squares sense.

A primary difference between DMC and GPC is the underlying plant model. DMC is based on exogenous input (X) models or coefficients of the step response; GPC is based on ARX (autoregressive with exogenous input) or ARMAX (autoregressive moving average with exogenous input) models. The use of ARX and ARMAX models in GPC means less coefficients to adapt than the X-model-based algorithms (Schonauer, 1995). By using auto regressive integral moving average with an exogenous input (ARIMAX) models, which represent the disturbance as random steplike or Brownian motion, the controller will automatically have an integrator in it that will prevent offset. Although step/pulse models are easily obtained and make few assumptions about the system, the purpose of a model is to emulate the dynamic plant behavior so that accurate predictions can be made. The ratio of polynomials (ARX) will give a better fit than a simple polynomial (X) when approximating functions (Clarke and Mohtadi, 1988).

Figure 9–6 illustrates the moving horizon approach of GPC. The basic concept of GPC is that, given an appropriately parameterized model and vectors of past plant outputs, future set points, previous control inputs, and potential future controls, the predicted output over a range up to the specified prediction horizon, N, can be computed. Increments of control are considered rather than full-valued

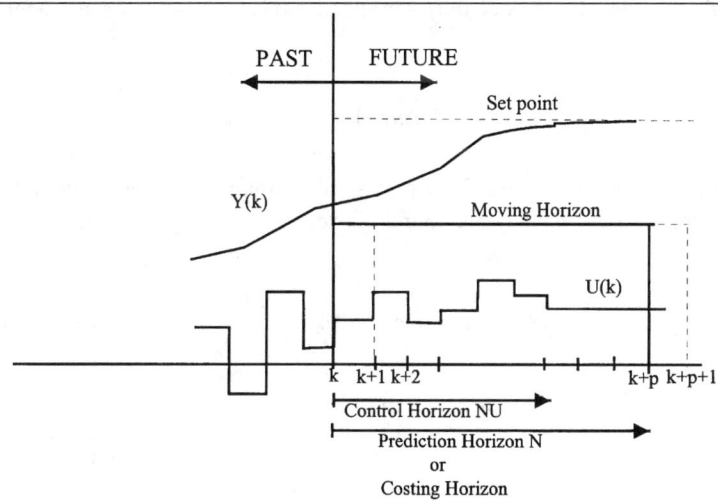

Figure 9–6 Moving Horizon Approach of GPC

controls, and beyond a specified control horizon, *NU*, the control increments are zero. The vector of control inputs is calculated based on optimization of a cost function, usually of quadratic form. The GPC involves a receding horizon philosophy where, at each sample time, the free response of the plant is computed, the control increment vector is calculated using the optimization routine with future set points (with or without constraints), the first control increment is implemented, all vectors are shifted in time, and the procedure is repeated at each sample time.

The GPC typically uses a time domain stochastic model of the process to calculate future changes in manipulated variables that will minimize a cost function. The GPC adopts an integrator as a natural consequence of its assumption about the basic plant model unlike a majority of designs where integrators are added in an ad hoc way. The ARIMAX model form is assumed, and successive recursion of the diophantine equation is used to develop the control equations. The ARIMAX model is of the form:

$$A(q^{-1})y(t) = B(q^{-1})u(t-nk) + C(q^{-1})e(t)/\Delta \qquad [2]$$

where nk is the time delay samples before the control signal $u(t)$ affects the output $y(t)$, $e(t)$ is a white noise random sequence, Δ is the operator $(1-q^{-1})$ so that $\Delta x(t) = x(t)-x(t-1)$, and A, B, and C are polynomials in the backward shift operator q^{-1}: $A(q^{-1}) = 1 + a_1 q^{-1}+...+a_{na}q^{-na}$; $B(q^{-1}) = b_1 + b_2 q^{-1}+...+b_{nb}q^{-nb+1}$, $C(q^{-1}) = 1 + c_1 q^{-1}+...+c_{nc}q^{-nc}$. The noise term is assumed to be white. If A and C polynomials in Eq.(2) are equal to 1, the model is called X, and ARX when $C=1$. For the case of $C=1$, resulting in an ARX model, the j step ahead that is a predictor for such a model is of the form:

$$\hat{y}(t+j/t) = G_j \Delta u(t+j-1) + F_j y(t) \qquad [3]$$

where $G_j(q^{-1}) = E_j B$, and E_j and F_j are polynomials uniquely defined, given A and the prediction interval j. They are found via recursion of the Diophantine equations:

$$1 = E_j \tilde{A} + q^{-j} F_j \qquad [4]$$

$$1 = E_{j+1} \tilde{A} + q^{-(j+1)} F_{j+1} \qquad [5]$$

where $\tilde{A} = A\Delta$.

The vector f can be defined as the component of the future plant outputs composed of known terms at time t, so that for example: $f(t+1) = [G_1(q^{-1})-g_{10}]\,\Delta(t) + F_1y(t)$, and $f(t+2) = q[G_2(q^{-1}) - q^{-1}g_{21}-g_{20}]\,\Delta u(t) + F_2y(t)$, where $G_iq^{-1}=G_{i0} + g_{i1}q^{-1} + \cdot\ \cdot\ \cdot$ Then the predictor becomes:

$$\hat{y} = G\tilde{u} + f \qquad [6]$$

where N is the maximum costing horizon, $\hat{y} = [\,\hat{y}(t+1), y(t+2),...,\hat{y}(t+N)]^T$, and $\tilde{u} = [\Delta u(t), \Delta u(t+1),..., \Delta u(t+N-1)]^T$, $f = [(t+1), f(t+1), f(t+2),..., f(t+N)]^T$.

The quadratic cost function is minimized:

$$J(N_1,N_2) = E\left\{ \sum_{j=N_1}^{N_2} \left[y(t+j)-w(t+j) \right]^2 + \sum_{j=1}^{N_2} \lambda(j)[\Delta u(t+j-1)]^2 \right\} \qquad [7]$$

where N_1 is the minimum costing horizon, N_2 is the maximum costing horizon, $\lambda(j)$ is a control-weighting sequence, $y(t+j)$ are the future plant outputs, $w(t+j)$ are the future set points for the plant outputs, and $u(t+j-1)$ are the controls. By manipulating the cost function expectation, the control increment is:

$$\Delta u(t) = g^{-T}(w - f) \qquad [8]$$

where g^{-T} in the first row of $(G^TG + \lambda I)^{-1}G^T$ is calculated.

A CASE STUDY

In this section, an example of application of GPC to a continuous fryer will be presented. The results are based on the work of Haarshma (1994) and Brescia and Moreira (1997a,b). The experimental data used for these studies came from a real continuous fryer and were collected at 5-second intervals (0.2 Hz) by two data acquisition systems, Genesis (Iconics, California) and In-Touch (WonderWare, California) through a 486-MHz personal computer. The steady-state data were analyzed on a Pentium personal computer using Matlab (The Mathworks, Natick, Massachusetts) software.

Process Description

A simplified schematic drawing of a continuous fryer is shown in Figure 9–7. The frying medium is vegetable oil. The amount of oil is about 327.5 kg. When this oil is at high temperatures, its density is 808.9 kg/m³. The oil is being "turned over" periodically (approximately every 10 hours) to keep a constant oil quality.

The food product (wheat-based snack extruded product) is dropped into the fryer by a conveyor. It travels freely through the oil a distance of 5.54 m. In some designs, this stage contains several rotating paddles (see Chapter 8). A submerger conveyor helps keep the product immersed in the oil and pushes it forward through the fryer. The total length of the submerger section is 4.27 m. The submerger conveyor speed can be changed to adjust the residence time of the product in the fryer.

A takeout conveyor positioned at an angle of 30° moves the product out of the frying medium through a series of paddles. The objective of this section is to re-move the product from the fryer and to de-oil the product. The product is main-tained at a temperature similar to that of the frying oil by having part of the con-veyor under the hood. This is done because most of the oil in the product is absorbed after frying when it comes in contact with the ambient temperature. Some design incorporates, in addition to the takeout conveyor, a de-oiling con-veyor, where drainage of the oil from the product may be enhanced by heat, which may be provided by flow of fryer exhaust air and/or by overhead heaters (see Chapter 8). The takeout conveyor is 2.5 m long.

The product is then transported by a conveyor to a cooling section that is open to ambient conditions. After this section, a weightbelt is in place, where the sen-

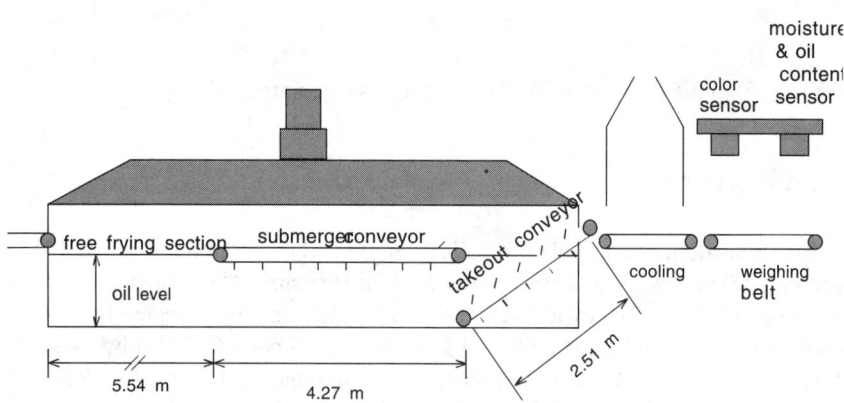

Figure 9–7 Schematic Representation of a Continuous Fryer. Not to scale.

sors for PQA measurements are located. The quality attributes measured on-line in this study were "color b" (CB), MC, and OC.

Color was measured with a Colorex sensor (Infrared Engineering Inc., Waltham, Massachusetts). The sensor measured color CIELAB units in L^*a^*b space. The "b" value of the CIELAB scale is positive when the product is yellow and negative when the product is blue. In other words, color b is a measure of yellowness/blueness in the product. In this study, the b value was used over "a" and "L" for modeling and control because it was affected the most by process variables.

Moisture content was measured on a wet basis as the quantity of water per unit mass of product. The OC measurements were also performed in terms of water content. Moisture content and OC were measured with a MM55 sensor (Infrared Engineering; Waltham, Massachusetts) located after the cooling section.

For control purposes, the OT, SS, and takeout conveyor (TC) speed were considered as manipulative variables. The chosen variables in the process are shown in Figure 9–1. A study was done to determine the effect of these variables on PQA. The PQA (CB, MC, and OT) were taken as controlled variables.

The OT is expressed in °C and typically ranges from 188°C to 193°C, with 190°C being the optimum temperature. The speed of the conveyors is expressed as a percentage (%) of the maximum attainable speed. Table 9–1 lists typical operating conditions, residence times, and lengths of some sections of the continuous fryer. The optimum operating conditions of the submerger and takeout conveyors are listed in Table 9–2. Optimum operating conditions result in a residence time of 105.27 seconds up to the color sensor and 115.37 seconds up to the moisture-color sensor.

Dynamic Analysis

Designing an effective control system for any process requires a thorough understanding of the process dynamics. In a control scenario, the number of manipulated (independent) variables must be at least as large as the number of controlled (dependent) variables. The controlled variables should measure product quality directly and should strongly affect it. The manipulated variables should have a large effect on the controlled variables (large gain), should rapidly affect the controlled variables (minimum delay, small time constant), and should affect the controlled variables directly rather than indirectly (Seborg et al., 1989).

The dynamic analysis can be done on the responses obtained from step tests. The first part of the analysis is the visual inspection of the effect of the different process variables on the product quality attributes. Calculations of the correlation between the process variables and the product quality attributes are also per-

Table 9–1 Typical Operating Conditions, Residence Times, and Lengths of Some Sections of the Continuous Fryer

Section	Length (m)	Op. Speed min-max (%)	Op. Speed min-max (m/s)	Minimum Res. Time (s)	Maximum Res. Time (s)
Free frying	5.54	constant	constant	13	13
Submerger Conveyor	4.27	50-75	0.12-0.20	20.88	34.63
Takeout Conveyor	2.51	30-90	0.07-0.17	14.79	33.39
Cooling Conveyor		constant	constant	27.4	27.4
Weightbelt/Color Sensor		constant	constant	18.6	18.6
Weightbelt/Moist-oil Sensor		constant	constant	10.1	10.1
Total				104.77	137.12

Source: Adapted with permission from L. Brescia and R.G. Moreira, Modelling and Control of a Continuous Frying Process: I. Dynamic Analysis and System Identification, *Transactions of the Institute of Chemical Engineers*, Vol. 75(C), pp. 3–11, © 1997, Institute of Chemical Engineering.

Table 9–2 Optimum Operating Conditions of the Submerger and Takeout Conveyors

Conveyor	Operating Speed (%)	Operating Speed (m/s)
Submerger	67	0.17
Takeout	70	0.12

Residence time (color sensor): 105.27 s. Residence time (moist-color sensor): 115.37 s.

Source: Adapted with permission from L. Brescia and R.G. Moreira, Modelling and Control of a Continuous Frying Process: I. Dynamic Analysis and System Identification, *Transactions of the Institute of Chemical Engineers*, Vol. 75(C), pp. 3–11, © 1997, Institute of Chemical Engineering.

formed. Gains are calculated and system linearity determined from step up and down tests. The Relative Gain Array is calculated to evaluate the degree of interactions in the system.

Step Test

The step test can be designed as follows. With the process operating at steady state, the input variables are perturbed by applying a positive/negative step. After the system reaches equilibrium, a step back to the operating point is applied until the system reaches steady state. These step changes are applied to the input variables, one at a time. Table 9–3 lists the set points for the inputs and outputs used in the system.

The responses of product quality attributes to the input changes are then analyzed. For convenience, the step responses are modeled as first-order functions.

Table 9–3 Input and Output of the System Operating at Optimum Conditions

Input/Output	Set point
Oil temperature [°C]	190.0
Submerger speed [%]	67.0
Takeout conveyor [%]	70.0
Moisture content [%]	2.2
Oil content [%]	22.0
Color b [ ]	57.0

Source: Adapted with permission from L. Brescia and R.G. Moreira, Modelling and Control of a Continuous Frying Process: I. Dynamic Analysis and System Identification, *Transactions of the Institute of Chemical Engineers*, Vol. 75(C), pp. 3–11, © 1997, Institute of Chemical Engineering.

Table 9–4 Effect of Input Positive Steps on Color b, Moisture Content, and Oil Content of Fried Food Product

Input/Output	Color b	Moisture Content	Oil Content
Oil Temperature			
Delay (s)	90	95	95
Time constant (s)	70	75	200
Settling time (s)	205	190	90
Submerger Speed			
Delay (s)	55	70	60
Time constant (s)	60	50	35
Settling time (s)	160	150	90
Takeout Conveyor Speed			
Delay (s)	55	70	60
Time constant (s)	80	55	70
Settling time (s)	190	110	150

Source: Adapted with permission from L. Brescia and R.G. Moreira, Modelling and Control of a Continuous Frying Process: I. Dynamic Analysis and System Identification, *Transactions of the Institute of Chemical Engineers*, Vol. 75(C), pp. 3–11, © 1997, Institute of Chemical Engineering.

Delays, time constants, and settling times are extracted from the step tests. The *settling time* is defined as the time required for the process output to reach and remain inside a band whose width is equal to ± 5% of the total change in the response. *Time constant* is defined as the time value at which the response is 63.2% complete. Delay is the time it takes for the signal to have a response. The effect of each input variable on the system output is then evaluated in detail.

Oil temperature. Table 9–4 shows the delays, time constants, and settling times of the PQA for a positive step change in the OT. Figure 9–8 shows the output responses to the OT changes. The step change was introduced at approximately 50 seconds after the test started for all conditions. The CB of the product decreased by 2.81 when an increase of 2.8°C in the temperature of the oil was applied. The response took place after approximately 90 seconds. The color turned to a darker yellow (more blue) as a result of a browning reaction because the product cooked more.

The MC of the product also responded inversely to OT positive change. Increasing the temperature caused more water to evaporate from the product during frying. Its response time was similar to the CB, about 95 seconds.

The OC increased 0.9% w.b. due to an increase in OT. The oil occupied the spaces left by the water during frying. Therefore, increasing water loss during frying resulted in increased final OC of the product.

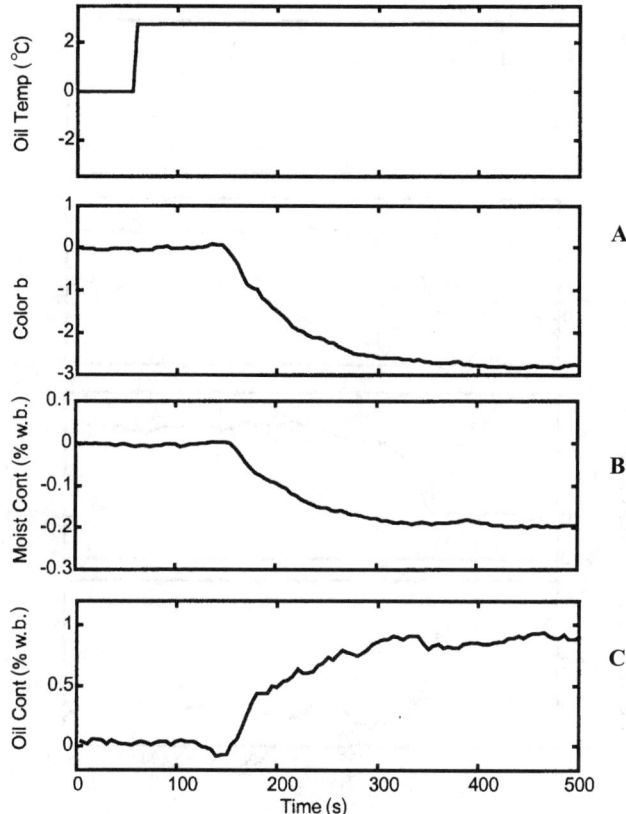

Figure 9–8 Effect of Oil Temperature Step Change (at 50 s) on: **A**, Color b; **B**, Moisture Content; **C**, Oil Content. *Source:* Adapted with permission from L. Brescia and R.G. Moreira, Modelling and Control of a Continuous Frying Process: I. Dynamic Analysis and System Identification, *Transactions of the Institute of Chemical Engineers*, Vol. 75(C), pp. 3–11, © 1997, Institute of Chemical Engineering.

Submerger speed. Table 9–4 shows the delays, time constants, and settling times of the PQA for a positive step change in the SS. Figure 9–9 shows output responses to SS changes. All PQA responded within 55 and 70 seconds to SS changes. The CB increased 1.93 and the MC 0.11% w.b., while the OC decreased 0.45% w.b. as the SS increased 10%.

By increasing the submerger conveyor speed, the residence time of the product in the fryer decreased. The product was exposed to the hot oil for a shorter period, and, therefore, the color turned lighter (less blue) as less water evaporated during frying. As a result of less cooking, less oil was absorbed by the product.

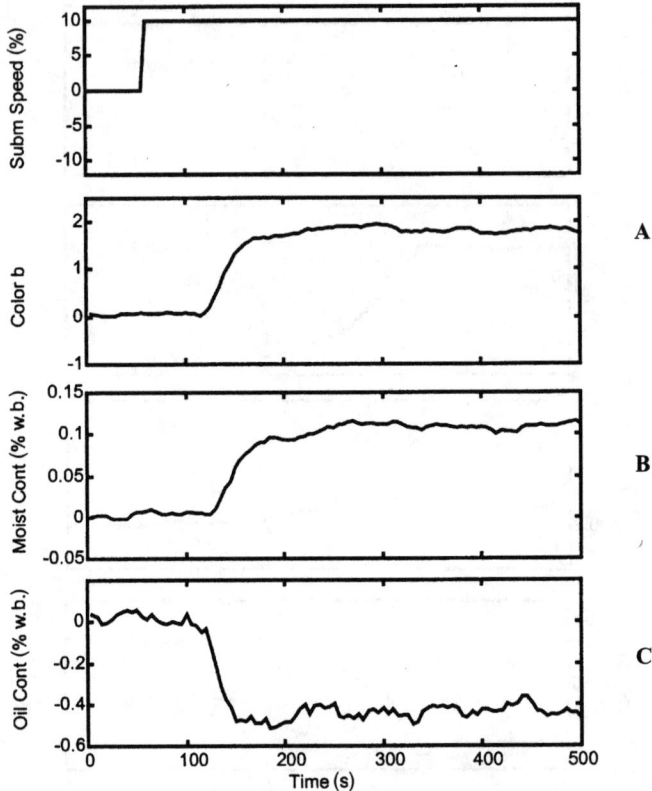

Figure 9–9 Effect of Submerger Speed Step Change (at 50 s) on: **A**, Color b; **B**, Moisture Content; **C**, Oil Content. *Source:* Adapted with permission from L. Brescia and R.G. Moreira, Modelling and Control of a Continuous Frying Process: I. Dynamic Analysis and System Identification, *Transactions of the Institute of Chemical Engineers*, Vol. 75(C), pp. 3–11, © 1997, Institute of Chemical Engineering.

Takeout conveyor speed. Table 9–4 shows the delays, time constants, and settling times of the PQA for a positive step change in TC. Figure 9–10 shows the effect of 15% step changes in the TC on the system outputs. The CB value increased 0.74 and delayed 55 seconds to respond to the change in the input. The MC was almost constant, increasing only 0.03% w.b. The OC increased, with a 0.8% w.b. and a delay of 60 seconds.

Clearly, the effect of the TC on final OC is observed in this experiment. By reducing the residence time of the product in the takeout conveyor, less oil was allowed to drip from the product surface, resulting in higher OC during cooling. In

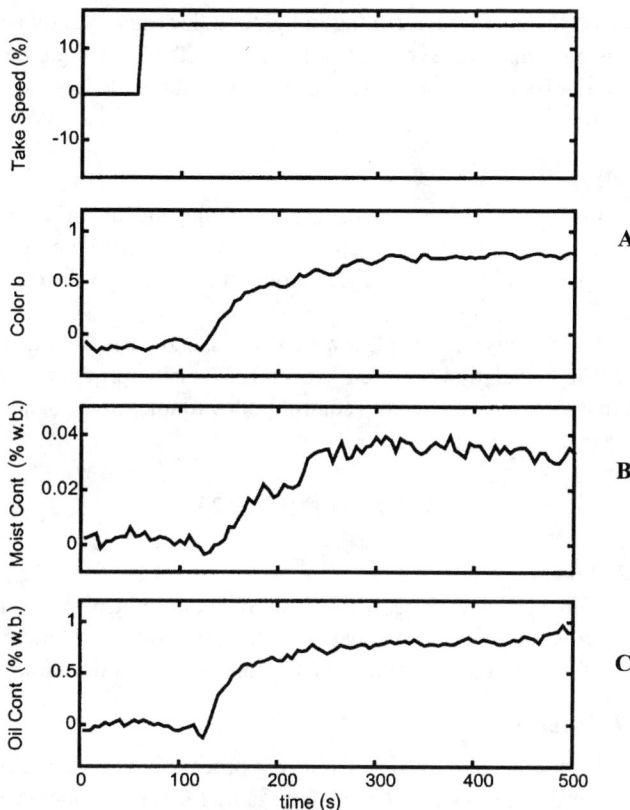

Figure 9–10 Effect of Takeout Conveyor Speed Step Change (at 50 s) on: **A**, Color b; **B**, Moisture Content; **C**, Oil Content. *Source:* Adapted with permission from L. Brescia and R.G. Moreira, Modelling and Control of a Continuous Frying Process: I. Dynamic Analysis and System Identification, *Transactions of the Institute of Chemical Engineers*, Vol. 75(C), pp. 3–11, © 1997, Institute of Chemical Engineering.

the takeout conveyor the product is not in the oil but is still under the fryer hood, thus maintaining its temperature higher than the ambient temperature. Moreira and Barrufet (1996) discussed that as the temperature of the product is reduced during cooling (after the product is removed from the fryer), the capillary pressure increases, allowing for the surface oil in the product to diffuse through the product. The decrease in temperature of the oil also causes the viscosity at the surface of the product to increase, thus making it more difficult for the oil to drip off the product's surface. Therefore, an increase in the TC resulted in more oil in the final product. Color b changes, in this case, could be the result in density reduction of the product as OC increased, resulting in lighter color (more yellowish).

In conclusion, step tests revealed that the CB and the OC should be used as the controlled variables in the control scenario, as they show the largest and quickest response to input changes.

Correlations

The relationships between the dependent and independent variables are studied using the steady-state values obtained by the step tests. There was a high correlation between the studied variables. The correlation values were on the order of 0.89–0.94.

The correlations between the PQA were studied, and the results are presented in Table 9–5. The correlation between MC and CB of the product was 0.94. This suggested that there was no need to control both variables at the same time. Their relationship can be presented as:

$$B = 14.31 \, (MC) + 29.75 \qquad [9]$$

with an R^2 of 0.88.

In conclusion, all the analyzed variables turned out to be highly correlated. An important correlation was the one between the CB and the MC indicating that it would be inefficient to try to control the CB and the MC at the same time.

Gains and Linearity

In general, for control purposes, the frying process is assumed to be linear. To evaluate linearity, the system steady-state gains (SSG) are calculated based on step up and down responses in addition to the effect of step size on the system outputs.

The SSG is the direction and the magnitude of the change in an output variable when a change is made in an input variable. It is calculated as:

Table 9–5 Correlation between Product Quality Attributes

Attribute	Color b	Moisture Content	Oil Content
Color b	1	**0.94**	−0.35
Moisture Content	**0.94**	1	−0.46
Oil Content	−0.35	−0.46	1

$$SSG = \frac{\Delta response}{\Delta input} \tag{10}$$

The larger the difference found between the gains, the more nonlinearity there is in the process. In this study, this was true. Gains from larger steps were higher than gains from small steps.

Product MC gains were all minuscule. It would require a large change in the forcing function to cause a statistically significant change in the measured value of MC. For instance, to obtain a 0.1% change in MC by the SS, a 10% change would be required.

Oil content exhibited larger changes than did MC when driving forces were applied. Takeout conveyor speed provoked the largest changes in OC whereas the SS affected it very little.

Color b appears to have a practical range of values for the tested steps. All driving forces, with exception of TC, will induce a significant change (1–2.81 units) on CB. Schonauer and Moreira (1995) demonstrated that CB had a strong correlation with other PQA, such as bulk density in an extrusion process.

In conclusion, the fastest response was obtained with the SS, and the slowest one with the OT. Oil content reacted very quickly to changes in SS and TCS but slowly to OT changes.

Relative Gain Array

Bristol (1966) developed the first systematic approach for the analysis of multi-variable process control problems (Seborg et al., 1989). It is called the *relative gain array* (RGA). It measures process interactions and provides recommendations for effective pairing of controlled and manipulated variables. The relative gain between a controlled variable C and a manipulated variable M is defined as:

$$v_{ij} = \text{open - loop gain} / \text{closed - loop gain} \tag{11}$$

for $i = 1,2,...,n_c$ and $j = 1,2,...,n_m$; where n_c and n_m are the number of controlled and manipulated variables, respectively. These relative gains can be arranged in an RGA.

The RGA can be calculated as follows. If

$$C' = KM \tag{12}$$

where C' is the matrix of controlled variables, M is the matrix of manipulated variables, and K is the matrix of steady state gains,

$$v_{ij} = K_{ij}H_{ij}, \qquad\qquad [13]$$

where

$$H = (K^{-1})^T \qquad\qquad [14]$$

Table 9–6 shows the RGAs for a positive step test conducted in this study. Bristol (1966) suggested the pairing of the controlled and manipulated variables so that the corresponding relative gains are positive and as close to 1 as possible. For the case of the RGA on Table 9–6, the recommendation is to pair CB-SS, MC-OT, and OT-TC. The RGA for the MC-TC, 0.21, is closer to 1 than to 5.45. This pair would have been a better choice following Bristol's guidelines, but it would have meant pairing OC-OT, which have a negative gain. Similar results were obtained for the other step tests (negative and larger size).

In conclusion, the RGA study showed the high interaction among control loops, suggesting the use of a multivariable control scheme. The best combinations were SS and CB, OT and MC, and TC and OC.

Identification of the Frying Process

To thoroughly understand the process dynamics, a process model is needed before a control system can be designed. Deep-fat frying modeling has been explored using analytical approaches (see Chapters 6 and 7). However, all the models existent in the literature are steady-state models. The literature does not give any information in fundamental modeling of deep-fat frying process dynamics. Stochastic identification techniques would provide the greatest benefit for modeling and control in the snack industry (Schonauer and Moreira, 1995). Recently, Haarshma (1994) and Bullock (1995) modeled a continuous frying process using a black box stochastic modeling approach. This method connects directly the observed characteristics of the product with the process variables by equations obtained from statistical analysis.

Table 9–6 Relative Gain Array from Positive Step on First Step Test

Attribute	Oil Temperature	Submerger Speed	Takeout Speed
Color b	−4.14	**5.24**	−0.1
Moisture Content	**5.45**	−4.65	0.21
Oil Content	−0.31	0.42	0.89

Source: Adapted with permission from L. Brescia and R.G. Moreira, Modelling and Control of a Continuous Frying Process: I. Dynamic Analysis and System Identification, *Transactions of the Institute of Chemical Engineers*, Vol. 75(C), pp. 3–11, © 1997, Institute of Chemical Engineering.

Stochastic modeling seems appropriate for food systems because it deals better with uncertainties and allows multiobjective optimization. Schonauer and Moreira (1995) employed stochastic modeling to describe the dynamics of a twin-screw extruder. An ARX model having eight parameters described the process dynamics very well.

Therefore, stochastic modeling seems to be a good approach to model the continuous frying process. The process was taken as a black box from which we know nothing but data from its inputs and outputs. Statistical analysis of the data was performed and the parameters describing the process were found. This method is called *system identification*.

Identification Tests

Before modeling, it is necessary to develop an excitation signal to produce the input-output data required for system identification. The process was taken as an MIMO process. The OT (°C), the SS (%), and the TC (%) were taken as inputs. The oil content [% w.b.] and CB were taken as outputs.

Excitation Signals

Obtaining models that adequately describe the process requires excitation of the process across all important frequencies. Pseudorandom binary sequence (PRBS) input signals are often used to identify stochastic processes (Ljung, 1987), and were used as inputs for this process. A plot of the PRBS signals is shown in Figure 9–11.

A PRBS signal is defined by its amplitude, period, and bit interval. The period equals the total number of positive and negative states. The bit interval is the shortest time between signal state change. The parameters of the PRBS signals used in this case study are presented in Table 9–7.

Model Selection

The continuous frying process was modeled as an ARX model. The data used for modeling were obtained by running the process with the PRBS described above. A total of 1200 data points were available. A subset of the data was used for estimation. Another subset was used for validation.

The ARMAX model has this form:

$$CB(t) + \sum_{i=2}^{ny_{CB}} (a_{CB_i} * CB(t-i+1)) + OC(t) + \sum_{i=2}^{ny_{OC}} (a_{OC_i} * OC(t-i+1)) =$$
$$\sum_{i=1}^{nu_{OT}} \left(b_{or_i} * OT(t-1) \right) + \sum_{i=1}^{nu_{SS}} (b_{SS_i} * SS(t-1)) + \sum_{i=1}^{nu_{TC}} (b_{TC_i} * TC(t-1)) + e(t) \quad [15]$$

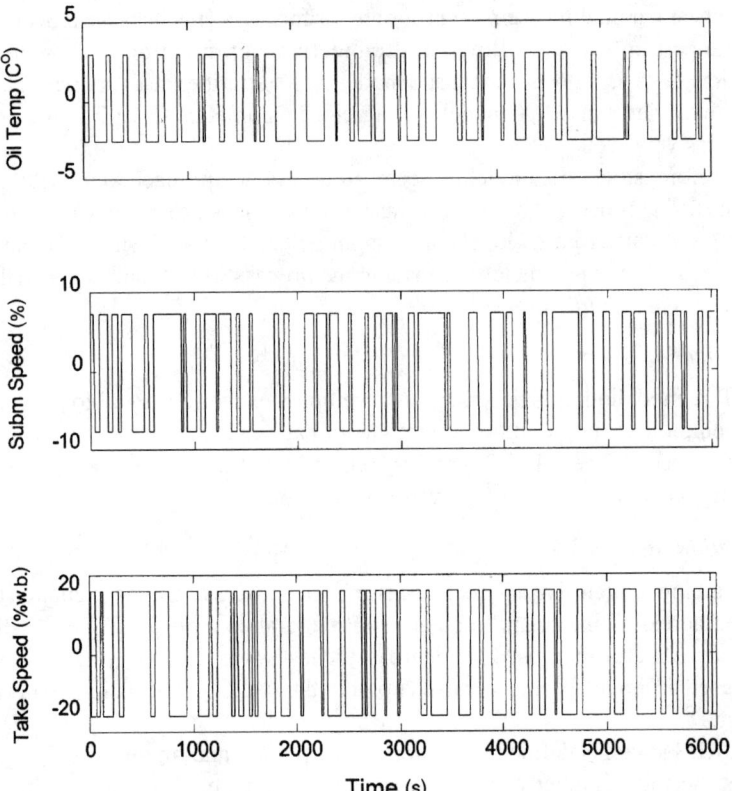

Figure 9–11 Pseudorandom Binary Sequence Signal. *Source:* Adapted with permission from L. Brescia and R.G. Moreira, Modelling and Control of a Continuous Frying Process: I. Dynamic Analysis and System Identification, *Transactions of the Institute of Chemical Engineers*, Vol. 75(C), pp. 3–11, © 1997, Institute of Chemical Engineering.

where a_i, b_i, and c_i are the coefficients starting at $a = 1$, $ny_{CB,OC}$ are the number of outputs, and $nu_{OT,SS,TC}$ are the number of past inputs. *CB* and *OC* are the outputs minus their steady-state means, and *SS, OT,* and *TC* are the inputs minus the steady-state means. The time delay nk (see Eq.2) is absorbed in the B polynomials so that its leading elements are zero.

Ljung (1991) presents two techniques for model order selection—Akaike's final prediction error (FPE) criteria and its closely related Akaike's information criteria (AIC). Both simulate the cross-validation situation, where the model is tested on another data set (Soderström and Stoica, 1989).

Table 9–7 Parameters for PRBS Used in this Study

Variables	Amplitude	No. States	Bit Interval (Samples)
Oil Temperature (°C)	5.61	141	8.5
Submerger Speed (%)	15	162	7.4
Takeout Speed (%)	40	176	6.8

Source: Adapted with permission from L. Brescia and R.G. Moreira, Modelling and Control of a Continuous Frying Process: I. Dynamic Analysis and System Identification, *Transactions of the Institute of Chemical Engineers*, Vol. 75(C), pp. 3–11, © 1997, Institute of Chemical Engineering.

The FPE is defined as:

$$FPE = \frac{1 + \dfrac{n}{N}}{1 - \dfrac{n}{N}} * V \qquad [16]$$

where n is the total number of estimated parameters, N is the total number of recorded data samples, and V is the loss function (quadratic fit) for the model structure in question.

The AIC is defined as:

$$AIC = \log[(1 + 2n / N) * V] \qquad [17]$$

According to Akaike's theory, in a collection of different models, the one with the smallest FPE or AIC should be selected. The AIC was used in this study.

Another tool for selecting a model is to perform a residual analysis. The residuals are the difference between the observed data and the data predicted by the model. The residuals associated with the data and a given model should be ideally white and independent of the input for the model to correctly describe the system. Another approach for model selection is to check whether a model is capable of reproducing the observed output when driven by the actual input (Ljung, 1991).

Table 9–8 summarizes the best MIMO model for the continuous frying process.

Model Validation

Validation for the models in Table 9–8 is shown in this section. One-step ahead and infinite-step ahead predictions were made on all the validation data subsets

Table 9–8 Number of *a* Coefficients, *b* Coefficients, Delays, and Total Number of Parameters for the ARX-MIMO Model

	a_1	a_2	b_1	b_2	b_3	nk_1	nk_2	nk_3	No. of
	CB	OC	OT	SS	TC	OT	SS	TC	parameters
CB	3	0	3	4	4	18	12	11	27
OC	0	1	3	5	4	19	13	12	

Source: Adapted with permission from L. Brescia and R.G. Moreira, Modelling and Control of a Continuous Frying Process: I. Dynamic Analysis and System Identification, *Transactions of the Institute of Chemical Engineers*, Vol. 75(C), pp. 3–11, © 1997, Institute of Chemical Engineering.

using the calculated coefficients. Residuals were computed and whiteness and independence analyses were performed. The autocorrelation function of the residuals and the cross-correlation function between the residuals and the process inputs were calculated.

AIC values, loss function values, and numbers of parameters for the ARX MIMO model were –642.74, 0.01, and 27, respectively. Figure 9–12 shows plots of one-step ahead and infinite-step ahead predictions, respectively, for the ARX model on top of the validation data set. The predictions do not fit the data exactly but follow the dynamics very well. The one-step ahead prediction fits the observed data better than does the infinite-step ahead prediction, as expected.

Figure 9–13 shows the autocorrelations and cross-correlations for the model residuals. Residuals were white and independent of the inputs.

Control Development and Implementation

The GPC was used in this study. The controller was adapted for the continuous frying process into Matlab (The Mathworks Inc., Natick, Massachusetts) code. The process was simulated in closed-loop with a 10% random noise (normal distribution with mean 0.0 and variance 1.0) on the outputs, and its performance was tested. A tracking test was used to test the controller qualitatively. Also, standard deviations were calculated.

Choice of Controller Parameters

The choice of output horizons, control horizon, and weighting sequence is critical in controller performance. Clarke et al. (1987a) developed suggestions for selecting output and control horizons. If the plant's dead time (nk) is exactly known, there is no point in setting the minimum output horizon (N_1) to be less than nk. If

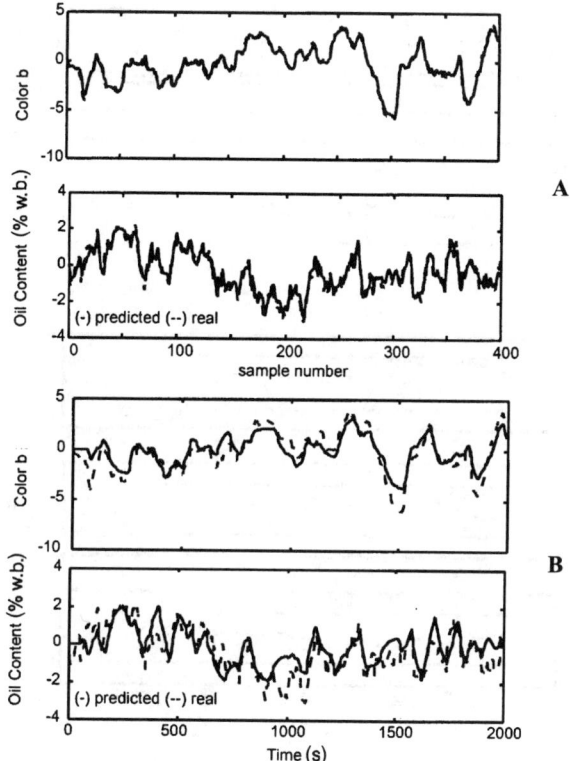

Figure 9–12 A, One-Step-Ahead Prediction; **B,** Infinite-Step-Ahead Prediction for the ARX-MIMO Model. *Source:* Adapted with permission from L. Brescia and R.G. Moreira, Modelling and Control of a Continuous Frying Process: I. Dynamic Analysis and System Identification, *Transactions of the Institute of Chemical Engineers*, Vol. 75(C), pp. 3–11, © 1997, Institute of Chemical Engineering.

nk is not known or is variable, N_1 can be set to 1 with no loss of stability and the degree of $B(q^{-1})$ increased to encompass all possible values of nk.

In practice, a large maximum output horizon (N_2) is suggested, corresponding to the rise time of the plant. For simple plants, a control horizon (NU) of 1 gives generally acceptable control. Increasing NU makes the control and the corresponding output response more active until a stage is reached where any further increase in NU makes little difference.

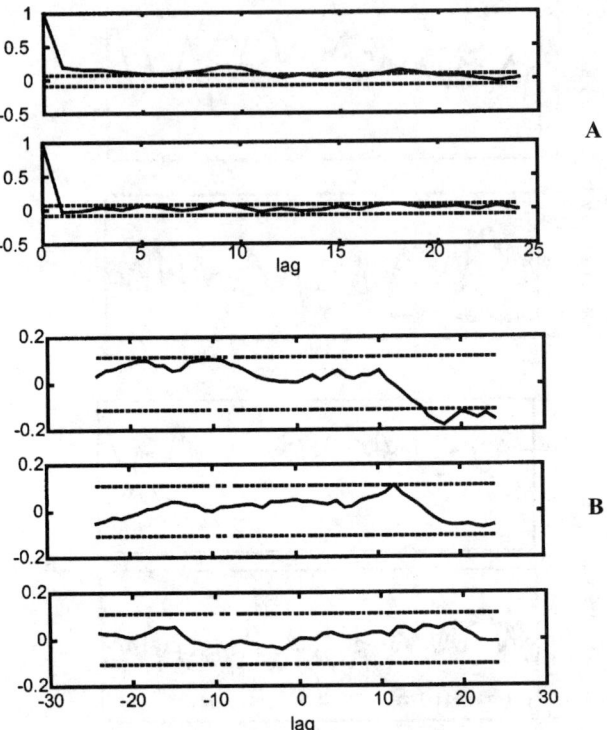

Figure 9–13 A, Correlation Function of Output Residuals; **B,** Cross-Correlation Functions between Inputs and Color b Residuals for the ARX-MIMO Model. *Source:* Adapted with permission from L. Brescia and R.G. Moreira, Modelling and Control of a Continuous Frying Process: I. Dynamic Analysis and System Identification, *Transactions of the Institute of Chemical Engineers,* Vol. 75(C), pp. 3–11, © 1997, Institute of Chemical Engineering.

Table 9–9 presents the N_1, N_2, and NU values chosen via simulations. The weights used in the tracking tests, also obtained through simulations, are shown in Table 9–10.

Controller Performance

A tracking test was designed to see how well the controller responded to set point changes. The performance of the MIMO-ARX controller was analyzed. Standard deviations from the set points were calculated for the simulations. Input variables were forced to be within the ranges specified in Table 9–11.

The product MC and OC set points were changed simultaneously in the tracking test utilizing the set points in Table 9–12. The set points were changed from set point 1 to set point 2, i.e., from 61.39 to 59.76 for CB, and from 23.72 to 25.53% w.b. for OC at 500 seconds. After the process reached equilibrium, set points were changed from set point 2 to set point 3, i.e., from 59.76 to 61.39 for CB, and from 25.53 to 23.77 % w.b. for OC at 3,000 seconds. Figure 9–14 presents the ARX controller simulation.

Figure 9–14A shows the simulated process and the set point changes. The controller tracked the set points well. Figure 9–14B shows the input changes that took place in order to reach the new set points. To track a +1.63 change in CB and a −1.76 change in OC set points, the ARX controller increased the OT about 3°C, reduced the SS about 7%, and increased the TCS by about 17%. Clearly, we can see that the takeout conveyor had the greatest effect in the oil absorption of the snack food. The same results were obtained for an increase and a decrease in CB and OC set points.

The standard deviation from the set points was calculated to compare controller performance quantitatively as:

$$S_m = \sqrt{\frac{\sum_{i=1}^{n}(y_i - y_{spt_i})^2}{n-1}}, \qquad [18]$$

where y_i is the measured output at sample time i, y_{spt_i} is the set point at sample time i, and n is the number of data points in the segment.

Table 9–13 shows the calculated standard deviations from the set points for the controller for each set point change. The ARX controller reached the set points very quickly and did not show any overshoots when reaching the new set points.

Table 9–9 Controller Parameter Values for Simulations

Parameter	Value
Minimum output horizon (N_1)	1
Maximum output horizon (N_2)	30
Control horizon (*NU*)	25

Source: Adapted with permission from L. Brescia and R.G. Moreira, Modelling and Control of a Continuous Frying Process: II. Control Development, *Transactions of the Institute of Chemical Engineers*, Vol. 75(C), pp. 12–20, © 1997, Institute of Chemical Engineering.

Table 9–10 Control Weight Sequence for MIMO Control Simulations

Inputs	Control Weight	Outputs	Control Weight
Oil Temperature (°C)	10	color b	1
Submerger Speed (%)	5	oil content	1
Takeout Speed (%)	1		

Source: Adapted with permission from L. Brescia and R.G. Moreira, Modelling and Control of a Continuous Frying Process: II. Control Development, *Transactions of the Institute of Chemical Engineers*, Vol. 75(C), pp. 12–20, © 1997, Institute of Chemical Engineering.

Table 9–11 Constraints on Inputs

Input	Low Limit	High Limit
Oil Temperature (°C)	188	193
Submerger Speed (%)	50	75
Takeout Speed (%)	30	90

Source: Adapted with permission from L. Brescia and R.G. Moreira, Modelling and Control of a Continuous Frying Process: II. Control Development, *Transactions of the Institute of Chemical Engineers*, Vol. 75(C), pp. 12–20, © 1997, Institute of Chemical Engineering.

Table 9–12 Set Points Used in MIMO Controller Tracking Tests

Set Point	Color b	Oil Content (% w.b.)
Set Point 1	61.39	23.77
Set Point 2	59.76	25.53
Set Point 3	61.39	23.77

Source: Adapted with permission from L. Brescia and R.G. Moreira, Modelling and Control of a Continuous Frying Process: II. Control Development, *Transactions of the Institute of Chemical Engineers*, Vol. 75(C), pp. 12–20, © 1997, Institute of Chemical Engineering.

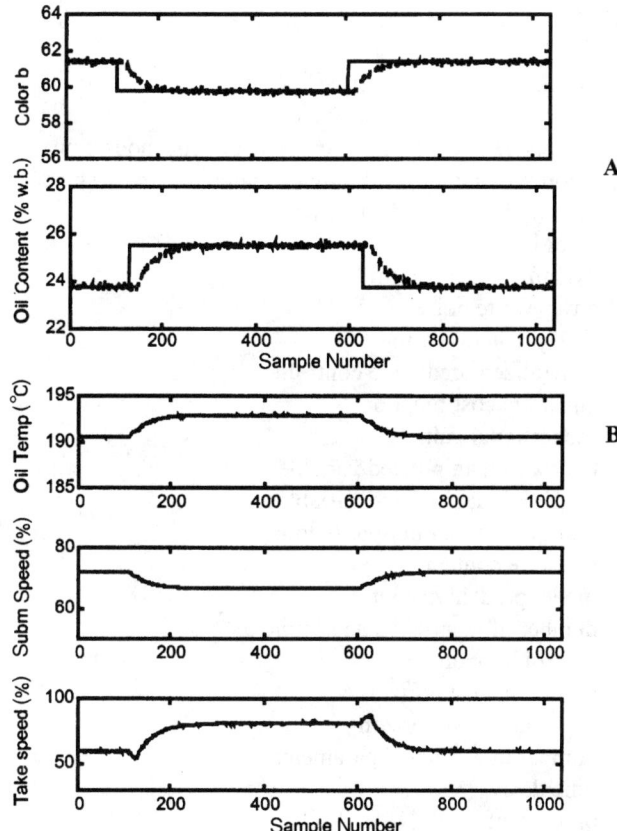

Figure 9–14 Tracking Test for ARX-MIMO Controller. **A**, Step Up and Down Change in Output Set Points—±1.63 in Color b and ±1.76 in Oil Content; **B**, Input Responses. *Source:* Adapted with permission from L. Brescia and R.G. Moreira, Modelling and Control of a Continuous Frying Process: II. Control Development, *Transactions of the Institute of Chemical Engineers*, Vol. 75(C), pp. 12–20, © 1997, Institute of Chemical Engineering.

Table 9–13 Standard Deviations from ARX-MIMO Controller Tracking Tests

Set Point Change	Color b	Oil Content
from 1 to 2	0.33	0.39
from 2 to 3	0.33	0.39

Source: Adapted with permission from L. Brescia and R.G. Moreira, Modelling and Control of a Continuous Frying Process: II. Control Development, *Transactions of the Institute of Chemical Engineers*, Vol. 75(C), pp. 12–20, © 1997, Institute of Chemical Engineering.

NOTATIONS

AIC	Akaike's information criteria
ARX	autoregressive with exogenous input
ARMAX	autoregressive moving average with exogenous input
ARIMAX	autoregressive integral moving average with exogenous input
C'	matrix of controlled variables
CB	color b
DMC	dynamic matrix control
e	error (white noise)
FPE	final prediction error
GPC	generalized predictive controller
J	quadratic cost function
K	proportional gain
K	matrix of manipulated variables
M	matrix of manipulated variables
MIMO	multiple-input, multiple-output
MC	moisture content
MPC	model predictive control
N	number of recorded data samples
NU	control horizon
N_1	minimum cost horizon
N_2	maximum cost horizon
n	number of estimated parameters
nk	time delay
OC	oil content
OT	oil temperature
PQA	product quality attributes
PID	proportional integral derivative
PRBS	pseudorandom binary sequence
RGA	relative gain array
SS	submerger speed
SSG	steady-state gain
TCS	takeout conveyor speed
T_i	integral time constant
t	time
u	input
V	loss function
X	exogenous input
y	input

Δ	difference operator
λ	control weighting sequence
ν	relative gain

TERMINOLOGIES

Adaptive control system: A system that has the ability to self-adjust or self-modify under unpredictable changes in input or environmental conditions.

Closed-loop control systems: A system in which the output signal has a direct effect on the control action; that is, closed-loop control systems are feedback control systems.

Deterministic disturbance: A variable that can be described exactly in analytical form.

Disturbance: A signal that tends to adversely affect the value of the output of a system.

DMC: Dynamic matrix controller. A model predictive control (MPC) technique based on optimization of a quadratic objective function involving the error between the set point and the predicted outputs. The design method is based on a particular type of discrete-time model (a convolution model).

Feedback control: An operation that, in the presence of disturbances, tends to reduce the difference between the output of a system and the reference input (set point) and that does so on the basis of this difference.

Feedback control systems: A system that tends to maintain a prescribed relationship between the output and the reference input by direct comparison while using a difference as a means of control.

Feedforward control: An operation that controls undesirable effects of measurable disturbances by approximately compensating for them before they materialize.

GPC: Generalized predictive control. A generalized form of DMC.

IMC: Internal model controller. The IMC design method is based on an assumed process model and relates the controller settings to the model parameters in a straightforward manner.

Input: Variables that cause action in a system.

MPC: Model predictive controller. In the MPC approach, the process model is used to predict future outputs over a long time period.

Open-loop control systems: Control systems in which the output has no effect on the control action; i.e., in an open-loop control system, the output is neither measured nor fed back for comparison with the input.

Output: Observed variable that is a result of the applied output.

PID controller: Proportional-integral-derivative controller.

Plant: A piece of equipment performing a particular operation.

Process: Any operation to be controlled, e.g., frying process.

Smith Predictor: The Smith predictor is often referred as model-based controller as the control strategy uses the model of the process parameters directly. The Smith predictor for time delay compensation is a feedback controller that has been modified by feeding back the controller output into the controller input through a transfer function

Stochastic disturbance: A signal that cannot be exactly described nor predicted.

System: A combination of components that acts together and performs a certain objective.

REFERENCES

Brescia, L., and Moreira, R.G. 1997a. Modeling and control of a continuous frying process: I. Dynamic analysis and system identification. *Trans Chem Eng*, 75(c): 3–11.

Brescia, L., and Moreira, R.G. 1997b. Modeling and control of a continuous frying process: II. Control development. *Trans Chem Eng*, 75(c):12–20.

Bristol, E.H. 1966. On a new measure of interactions for multivariable process control. *IEEE Trans Auto Control*, AC-11, 133.

Bullock, D.C. 1995. *Modeling of a continuous food process with neural networks.* Masters thesis. College Station, TX: Texas A&M University.

Clarke, D.W.; Mohtadi, C.; and Tuffs, P.S. 1987a. Generalized predictive control, Part I. The basic algorithm. *Automatica*, 23(2):137–148.

Clarke, D.W.; Mohtadi, C.; and Tuffs, P.S. 1987b. Generalized predictive control, Part II. Extensions and interpretations. *Automatica*, 23(2):149–160.

Clarke, D.W., and Mohtadi, C. 1988. Properties of generalized predictive control. In *IFAC proceedings series*, No. 15, ed. Rolf Isermann. Tarrytown, NY: Pergamon, 65–76.

Garcia, C., and Morshedi, A. 1986. Quadratic programming solution of dynamic matrix control (QDMC). *Chem Eng Commun*, 46:73–87.

Haarshma, G. 1994. Project report: development of a dynamic matrix controller for a frying process. Netherlands: Dept. of Agricultural Engineering and Physics, Wagneningen University.

IFT Research Committee. 1993. America's food research needs: into the 21st century. *Food Technol*, 47(3S):1S–40S.

Isermann, R. 1991. *Digital control systems*, Vol II. New York: Springer-Verlag.

Ljung, L. 1987. *System identification: theory for the user.* Englewood Cliffs, NJ: Prentice Hall.

Ljung, L. 1991. *System identification toolbox.* Natick, MA: MathWorks.

Morari, M.; Garcia, C.; and Prett, D. 1988. *Model predictive control: theory and practice.* Proceedings of IFAC Model Based Process Control, 1–12. Atlanta, GA: IFAC.

Moreira, R.G., and Barrufet, M.A. 1996. Spatial distribution of oil after deep-fat frying of tortilla chips from a stochastic model. *J Food Eng*, 27:279–290.

Schonauer, S. 1995. *Product quality adaptive control system for a food extruder.* PhD diss. College Station, TX: Texas A&M University.

Schonauer, S., and Moreira, R.G. 1995. Development of a fixed-GPC controller for a food extruder based on product quality attributes: I. System identification. *Trans Inst Chem Eng*, Part C. 73 (C4):189–199.

Seborg, D. E.; Edgar, T. F.; and Mellichamp, D.A. 1989. *Process dynamics and control*. New York: John Wiley & Sons.

Soderström, T., and Stoica, P. 1989. *System identification*. Englewood Cliffs, NJ: Prentice Hall.

CHAPTER 10

Low-Fat Tortilla Chips

As shown in Chapter 1, tortilla chips account for approximately 21% of the total snacks produced ($3.36 billion) in the U.S. The structure, composition, and processing aspects of tortilla chips are described in Chapter 2.

The traditional method used to process corn into tortillas was developed by the Aztec Indians in the central region of Mexico. The process of transforming corn into masa is called *nixtamalization*. Nixtamalization typically involves boiling, quenching, and steeping of corn in a lime solution (CaO, calcium oxide). The liquor obtained after steeping is discarded, and the cooked corn (nixtamal) is washed to remove the excess alkali and loose pericarp. The nixtamal is then stone-ground to produce masa. Modern commercial plants still use many of the same principles from the original process.

Snack foods are one of the most significant groups of deep-fat fried foods. In these products, low fat content and texture are desired quality attributes. Low-fat snack products will be the driving force of the snack food industry during the next five years (Chapter 1). Sales of new fat-free tortilla chips are increasing, but because these chips are baked rather than fried, they have different taste and textural properties than do fried chips and have engendered mixed consumer reactions. A lower-fat fried version would likely meet with greater consumer acceptance. Ideally, this product should have enough fat to impart the desirable flavor and texture characteristics of typical fried chips. This chapter presents the drying of tortilla chips using impingement drying techniques as an alternative to produce low-fat products.

STRUCTURAL DESCRIPTION OF THE TORTILLA CHIP PROCESS

To understand the quality changes that occur during processing of tortilla chips, it is necessary to analyze the structural and molecular changes that starch undergoes during the process.

During heat processing, starch goes through a series of structural transformations. During baking, for example, starch gelatinizes over a wide range of temperatures from 65°C to 110°C (Hoseney, 1994). *Gelatinization* is typically defined as the physicochemical phenomenon of swelling of starch granules as they imbibe water at temperature sufficient to destroy the birefringence of the granules. The process occurs as the starch/liquid system is heated above a characteristic *gelatinization temperature*. Below this temperature, birefringence of starch granules is preserved. The mechanism of starch gelatinization is described in Figure 10–1 (Remsen and Clark, 1978). Gelatinization is both a physical and a chemical process. The physical process is hydration and swelling of granules, with leaching of amylose and amylopectin molecules into the solution. The chemical process is water breaking intermolecular hydrogen bonds and replacing them with water-polysaccharide hydrogen bonds. Gelatinization results in a loose matrix of granules and long chain molecules. The net effect is an increase in viscosity.

Gomez et al. (1992) analyzed the changes in the starch during tortilla chip processing and related these changes to product quality. Color Plate 10 illustrates the changes that occur in the composition of masa, tortilla, and tortilla chips during the process. The following is the summary of their observations.

During cooking and steeping of corn, hydration and softening of the endosperm and germ, partial starch gelatinization, and disintegration of the pericarp occur (Gomez et al., 1987). Quenching is a common practice used by processors of corn tortilla chip masa to reduce the temperature of the solution before steeping. This is important to prevent excess water uptake after cooking. Masa with a lower moisture content produces chips with reduced oil content (Serna-Saldivar et al., 1990).

Masa grinding disrupts the grain structure, thus releasing starch granules from the endosperm cells and dispersing cellular components and starch polymers. Gomez et al. (1992) defined *masa* as a network composed of two phases (Color Plate 10A): (1) a continuous phase consisting of solubilized starch polymers and (2) a dispersed phase consisting of uncooked and swollen starch granules, cell fragments, and lipids supported by the continuous phase. The combination of swollen and partially gelatinized starch granules helps tortilla shaping during kneading, water vapor retention, and puffing during baking. Additional gelatinization (<5%) caused by shear, heat and water contributes to the adhesion characteristic of the dough. During cooling of the masa, amylose retrogradation occurs. Starch *retrogradation* is defined as the change in starch gels, such as chain aggregation and/or recrystallization, occurring immediately during cooling (Ring, 1985).

The masa is exposed to high temperature during baking (320–420°C) for 20–45 seconds to form the tortilla. Most of the starch crystallinity is lost during baking. Some retrogradation and interaction can also occur during this period. Due to the intense heat, baking causes microstructure changes in starch birefringence and

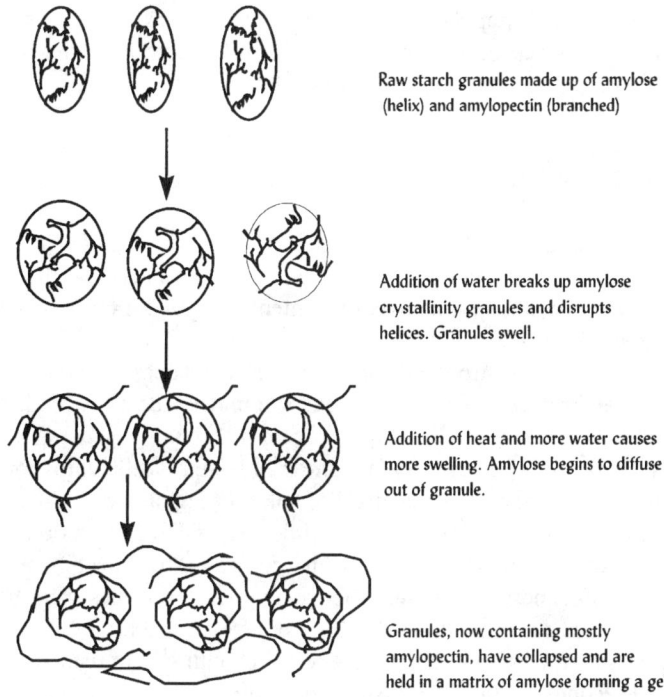

Raw starch granules made up of amylose (helix) and amylopectin (branched)

Addition of water breaks up amylose crystallinity granules and disrupts helices. Granules swell.

Addition of heat and more water causes more swelling. Amylose begins to diffuse out of granule.

Granules, now containing mostly amylopectin, have collapsed and are held in a matrix of amylose forming a gel.

Figure 10–1 Mechanisms of Starch Gelatinization. *Source:* Adapted with permission from C.H. Remsen and J.P. Clark, A Viscosity Model for a Cooking Dough, *Journal of Food Process Engineering*, Vol. 2, pp. 39–49, 1978, © Food and Nutrition Press, Inc.

physical appearance of corn components. Baking (Color Plate 10B) results in faster water evaporation and severe starch gelatinization (>40%). The starch granules at the tortilla surface are partially gelatinized and dehydrated, compared with those at the tortilla core.

During frying (Color Plate 10C), additional starch gelatinization occurs during the first 10–15 seconds of frying, before most of the water is evaporated. Interaction between amylose and lipid produced during frying reduces the potential for starch dispersion. Rapid water removal from the tortilla surface decreases the volume of gelatinized starch, which enhances starch retrogradation during frying.

An alternative method of producing tortilla chips involves production of NCF as an intermediate product. The NCF is made by drying and grinding masa into flour. This flour is sieved and blended from streams of different particle size to produce NCF with optimum properties for various applications (Serna-Saldivar et al., 1990).

Products processed from NCF are more expensive than those made from freshly ground masa, stale faster, and are blander in flavor and poorer in texture (Gomez et al., 1987). However, popularity of NCF continues to increase because it is more convenient and ensures more consistent quality and uniformity of the product than does the traditional method (Serna-Saldivar et al., 1990).

SUN-DRIED TORTILLA CHIPS

In the Mexican food market, there are several brands of *tostadas* (round tortilla chips) that are solar dried before frying (e.g., Zam's Tostadas, Saltillo, Coahuila, Mexico). These tostadas have a lower oil content (about 12–14% w.b.) than do the tortilla chips (22–28% w.b.), have good flavor, and are crunchy. Their structure revealed by scanning electron microscopy consists of a compact appearance, with fewer pores and air cells than the products found in the U.S. market (Lujan-Acosta, 1996).

Lujan-Acosta and Moreira (1997a) sun-dried baked tortilla chips for 180 seconds prior to frying to analyze the effect of solar drying on the product characteristics. Tortilla chips were prepared by mixing nixtamalized masa flour and water to produce a dough with 55% w.b. moisture content. After sheeting and forming the masa to a thickness of 1.5 mm, the triangle-shaped pieces were baked in a three-tier gas oven at temperatures ranging from 190°C to 343°C for 45 seconds to a moisture content of about 35% w.b. The baked chips were then sun dried and fried for 30 seconds at 200°C in fresh soybean oil.

During solar drying, moisture content of baked tortilla pieces is reduced from 34.3% to 7.4% (w.b.). The evaporation rate is higher during the first 45 minutes of sun drying and becomes fairly constant as drying continues. Equilibrium moisture content is reached after 120 minutes of drying. Solar drying time affects the final oil content of the fried chips. This effect is higher during the first 40 minutes of sun-drying time, becoming fairly constant as drying continues (Figure 10–2). During frying, the oil content varied from 31% (for the samples with no sun drying) to around 13% w.b. (for the samples sun dried for 180 seconds).

In the sun-drying process, water is lost slowly, minimizing the cell wall internal stresses. This allows for the material to shrink down to an almost solid core (14% from its initial dimensions) (Figure 10–3), resulting in substantial structural changes. The product structure shows no cracks or holes (Figure 10–4B), unlike the structures of the conventionally prepared tortilla chips (Figure 10–4A). The compact internal microstructure formed during the sun-drying process causes the chip to puff during frying, resulting in a crunchier product with less oil content. The microstructure of the sun-dried chips after frying shows bigger puffed pores (Figure 10–4C).

The chips become more resistant to fracture (harder) as sun-drying time increases, reaching a maximum (32.5 N) at 60 minutes of drying (Figure 10–5). After that time, fracturability (the force required to fracture the product) decreases

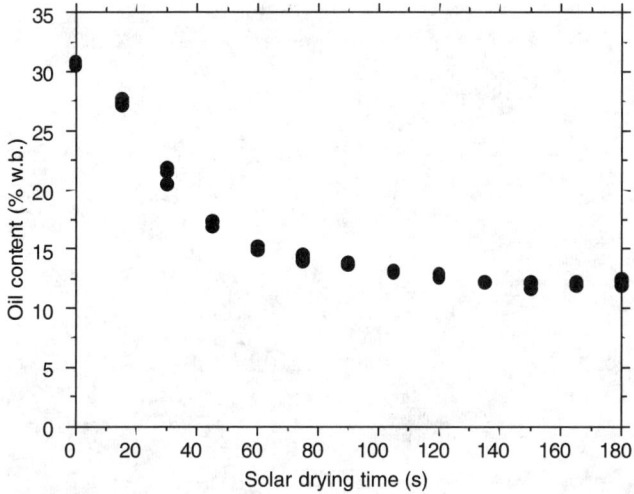

Figure 10–2 The Relationship between Final Oil Content and Sundry Time for Sun-Dried Tortilla Chips Fried in Fresh Soybean Oil at 200°C for 30 s. *Source:* Adapted with permission from F.J. Lujan-Acosta and R.G. Moreira, Effects of Different Drying Processes on Oil Absorption and Microstructure of Tortilla Chips, *Cereal Chemistry*, Vol. 74, No. 3, pp. 216–223. © 1997, American Association of Cereal Chemists.

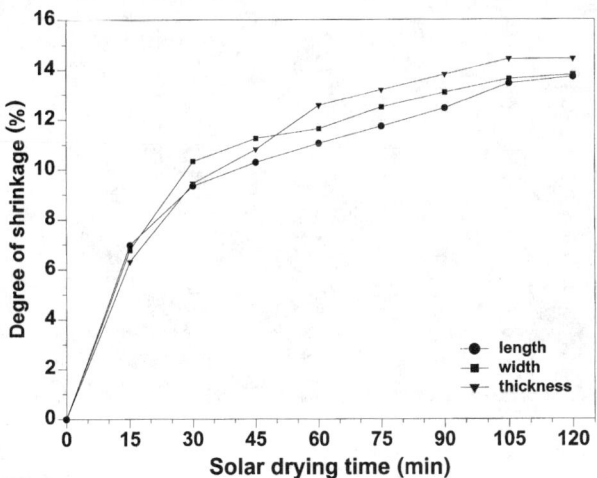

Figure 10–3 Degree of Shrinkage in Dimension of Tortilla Chips as a Function of Sun-Drying Time. *Source:* Adapted with permission from F.J. Lujan-Acosta and R.G. Moreira, Effects of Different Drying Processes on Oil Absorption and Microstructure of Tortilla Chips, *Cereal Chemistry*, Vol. 74, No. 3, pp. 216–223. © 1997, American Association of Cereal Chemists.

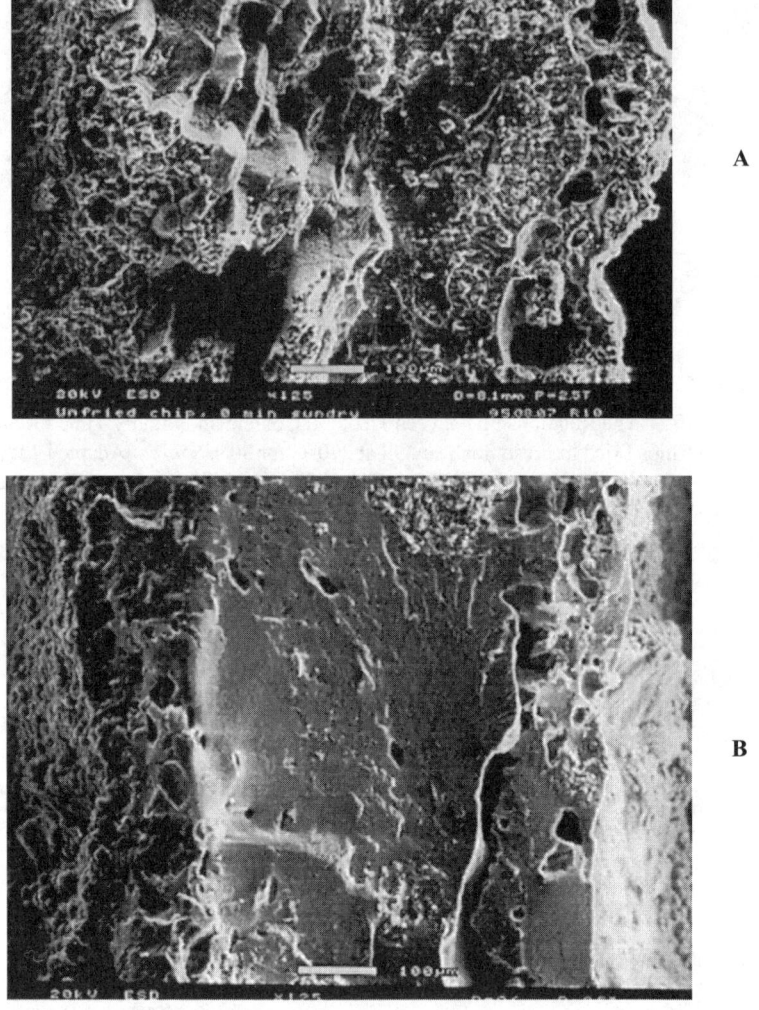

continues

Figure 10–4 Cross-Sections from ESEM of: **A**, Only Baked Tortilla; **B**, Baked and Sun-Dried Tortilla; **C**, Fried Tortilla Chips. *Source:* Adapted with permission from F.J. Lujan-Acosta and R.G. Moreira, Effects of Different Drying Processes on Oil Absorption and Microstructure of Tortilla Chips, *Cereal Chemistry*, Vol. 74, No. 3, pp. 216–223. © 1997, American Association of Cereal Chemists.

Figure 10–4 continued

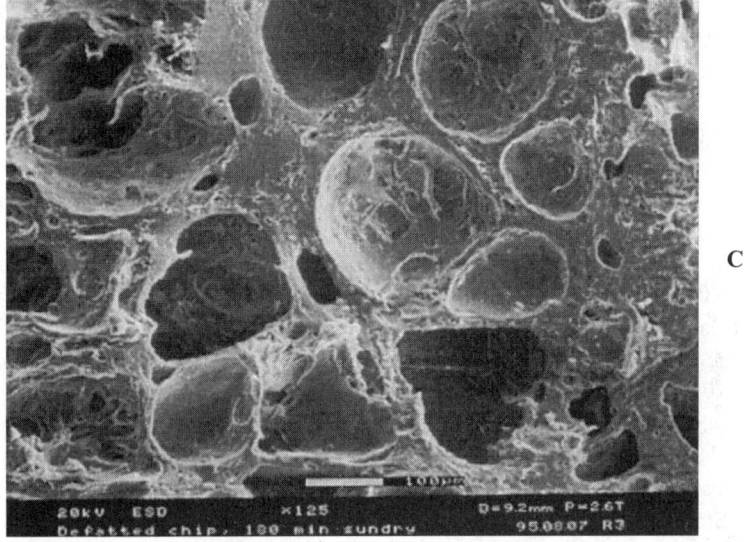

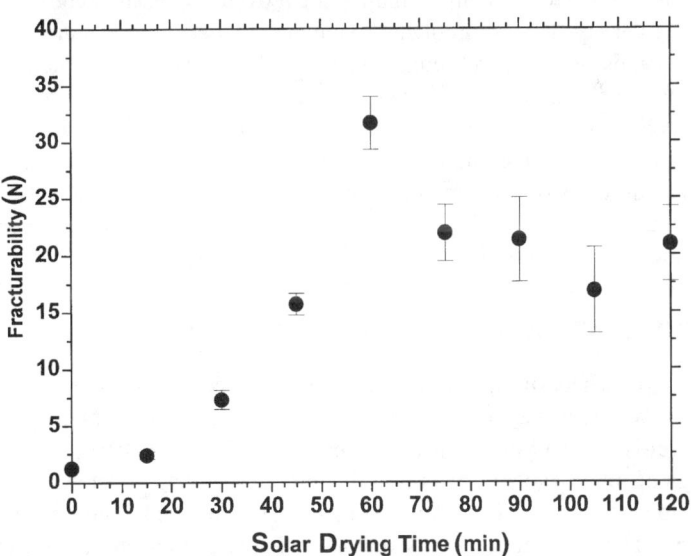

Figure 10–5 Fracturability of Tortilla Chips as a Function of Sun-Drying Time. *Source:* Adapted with permission from F.J. Lujan-Acosta and R.G. Moreira, Effects of Different Drying Processes on Oil Absorption and Microstructure of Tortilla Chips, *Cereal Chemistry*, Vol. 74, No. 3, pp. 216–223. © 1997, American Association of Cereal Chemists.

and remains constant toward the end of drying, and the chips become more fragile and brittle (about 22 N).

Sensory panelists (Table 10–1) found no significant ($P < 0.05$) difference in the attributes of brittleness, friability, flavor, and overall acceptability among the standard chips, sun-dried chips (fried), and commercial chips (El Galindo). The sun-dried tortilla chips had significantly ($P < 0.05$) better scores than did the Baked Tostitos in these attributes. Sun-dried tortilla chips had significantly ($P < 0.05$) lower color scores than all the other samples evaluated. There were no significant ($P < 0.05$) differences for any of the attributes scored among the different treatments of sun-dried chips. The oil content of the sun-dried chips was significantly ($P < 0.05$) lower (14.5–17.8%) than that of the standard (26.3%) and commercial tortilla chips (21.3%). According to the texture analysis, the fracturability values of the sun-dried tortilla chips are higher (8.6–9.4 N) than for the other samples (5.3–7.9 N); therefore they are crunchier.

Even with the oil content reduction, sun-dried tortilla chips, conventionally prepared chips, and commercial chips received the same scores in the sensory evaluation at a consumer level in the attributes of texture, flavor, and overall acceptability. Sun-dried tortilla chips also had better scores than commercial baked chips in those attributes. However, they had lower scores in the attribute of color (lighter).

In the food industry, hot air jet impingement ovens are used to cook, bake, dry, and heat a variety of food products. Examples of food products include pizzas, muffins, breads, chicken, and cereals. The heat transfer coefficients in these ovens are higher and air temperatures lower than in conventional ovens. The food products are cooked in less time and retain more moisture (Marcroft et al., 1997). The impingement drying technique was evaluated by Lujan-Acosta and Moreira (1997b) as a means of reducing oil content of chips during frying. Based on the results obtained in solar drying, impingement drying techniques were used to produce chips with structure similar to the solar-dried product.

IMPINGEMENT DRYING OF TORTILLA CHIPS

Impingement jets of various configurations are commonly used in industrial drying operations as they permit rapid drying rates. Controlled air temperature and air flow rate can result in a high-quality product (Mujundar, 1986).

Design of impingement dryers is both simplified and aggravated by the excessive number of variables or parameters that can be specified or chosen arbitrarily. The difficult task is to choose a system that will give optimal energy consumption, throughput, and product quality. Although significant advances are being made in understanding transport phenomena under jets, the design remains empirical. Impingement can be used to dry foods rapidly, but high heat transfer rates may degrade heat-sensitive products. Therefore, an impingement dryer for foods should

Table 10–1 Sensory Scores and Physical Properties of Different Types of Tortilla Chips

Sample	Brittleness[a,1]	Fryability[a,2]	Color[a]	Flavor[a]	Acceptability[a]	Fracturability[b] (N)	Oil Content[c] (% w.b.)
Conventionally prepared	7.4 ± 1.2	6.5 ± 1.2	6.2 ± 2.2	6.1 ± 1.4	6.0 ± 1.4	7.91 ± 1.62	26.30 ± 0.57
Sun-dried 90 min	7.8 ± 1.2	7.2 ± 1.2	4.5 ± 1.7	6.4 ± 1.8	6.1 ± 1.6	9.39 ± 1.59	17.82 ± 0.38
Sun-dried 120 min	8.0 ± 1.1	7.2 ± 1.5	4.9 ± 1.8	5.7 ± 1.9	6.2 ± 1.5	8.64 ± 0.54	14.50 ± 0.19
El Galindo	6.8 ± 1.6	6.4 ± 1.6	6.4 ± 2.1	5.1 ± 1.8	5.3 ± 2.0	6.17 ± 1.08	21.26 ± 0.17
Baked Tostitos	6.0 ± 2.2	5.7 ± 2.0	6.2 ± 2.2	4.1 ± 2.1	5.1 ± 1.7	5.33 ± 0.44	3.12 ± 0.05

[a]Mean and standard deviation of 25 samples; [b]Mean and standard deviation of 30 measured samples; [c]Mean and standard deviation of 3 measured samples; [1]first bite characteristics, i.e., the toughness or force to break—normally excessive force is not desired; [2]how the chips break apart in the mouth—chips with more fryability quickly break into small pieces that dissolve and are swallowed readily.

Note: Scale is a 9-point hedonic scale; 9 is the highest score.

Source: Adapted with permission from F.J. Lujan-Acosta and R.G. Moreira, Effects of Different Drying Processes on Oil Absorption and Microstructure of Tortilla Chips, *Cereal Chemistry*, Vol. 74, No. 3, pp. 216–223. © 1997, American Association of Cereal Chemists.

be operated based only on experimentally determined drying rate data (Mujundar, 1986).

Impingement Dryers

Due to the high nozzle pressure drop and high recycle ratio needed to achieve reasonable thermal efficiencies, in addition to their complex fabrication, impingement dryers are recommended only if a major fraction of the moisture to be removed is unbound (Mujundar, 1986). Typically, the temperature and jet velocity may range from 100°C to 350°C and 10 to 100 m/second, respectively.

An impingement dryer consists of a single gas jet or an array of such jets, impinging normally on a surface. Figure 10–6 shows an example of a symmetrical exhaust air flow. There is a great variety of nozzles that can be used, and selection of the nozzle geometry and multinozzle configuration have important bearing on the initial and operating costs and on product quality.

Surface Impingement

A single gas jet impinging normally on a surface may be used to achieve enhanced coefficients for convective heating, cooling, or drying. As shown in Figure 10–7, air jets are usually discharged into a quiescent ambient from a round nozzle of diameter D. Typically, the jet is characterized by a uniform velocity profile at the nozzle exit. However, with increasing distance from the exit, momentum exchange between the jet and the ambient causes the free boundary of the jet to broaden and the potential core to contract. Downstream of the potential core, the velocity profile is not uniform over the entire jet cross-section, and the maximum (center) velocity decreases with increasing distance from the nozzle exit.

Convective Heat Transfer Coefficient in Impingement Drying (h)

Heat transfer from the air exiting the impinging nozzle to the tortilla piece is calculated by assuming that the air jet exits its nozzle with a uniform velocity (V_a), temperature (T_a), and species concentration $C_{A,e}$. Thermal and compositional equilibrium with the ambient is assumed ($T_a = T_{oo}$, $C_{A,a} = C_{A,\infty}$), whereas convection heat and/or mass transfer may occur at the impingement surface of uniform temperature ($T_s \neq T_a$, $C_{A,s} \neq C_{A,e}$). Newton's law of cooling and its mass transfer analog may then be expressed as:

$$q' = h(T_s - T_a) \qquad [1]$$

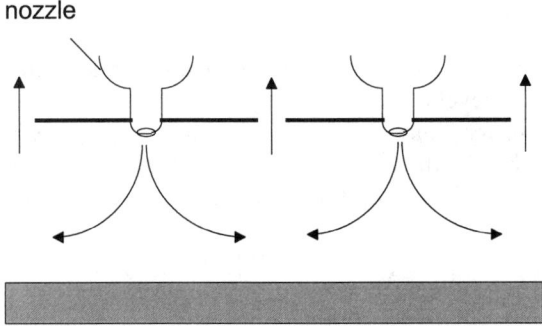

Figure 10–6 Symmetric Arrangement for Exhaust of Spent Jets

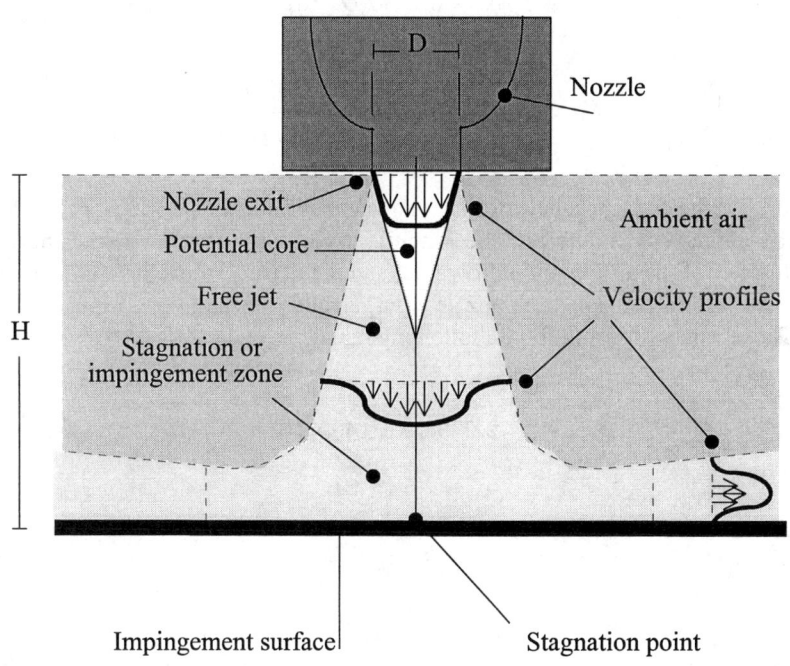

Figure 10–7 Surface Impingement of a Single Round Jet

$$N'_A = h_m (C_{A,s} - C_{A,a})$$ [2]

where q' is the heat flux and N'_A the mass flux. Conditions are assumed to be independent of the level of turbulence at the nozzle exit, and the surface is assumed to be stationary. Convection correlations for a variety of external flow conditions are available. Martin (1977) recommends the following correlation to calculate the average convective heat transfer coefficient for a single round nozzle:

$$\frac{Nu}{Pr^{0.42}} = G\left(\frac{r}{D}, \frac{H}{D}\right) F_1 (Re)$$ [3]

$$F_1 = 2 Re^{1/2} (1 + 0.005 Re^{0.55})^{1/2}$$ [4]

$$G = \frac{D}{r}\left(\frac{1 - 1.1 D / r}{1 + 0.1(H / D - 6) D / r}\right)$$ [5]

$$Nu = \frac{hD}{k_a}; \quad PR = \frac{v_a}{a_a}; \quad Re = \frac{4 \dot{m}}{\pi D v_a \rho_a}$$ [6]

where h is the average convective heat transfer coefficient; $k_a, v_a, a_a, \dot{m}, \rho_a$ the thermal conductivity of the air, the kinematic viscosity of the air, the thermal diffusivity of the air, the mass flow rate, and the density of the air; D the nozzle diameter; H the distance from nozzle to target; and r the target radius.

Those values are valid for the following ranges:

$$2,000 \le Re \le 400,000$$
$$2 \le \frac{H}{D} \le 1$$
$$2.5 \le \frac{r}{D} \le 7.5$$ [7]

The correlation may not be used if the jets emanate from a sharp-edged orifice instead of a bell-shaped nozzle. The orifice is strongly affected by a flow concentration phenomenon that alters convection heat or mass transfer (Martin, 1977). A majority of laboratory studies has used contoured entry nozzles, although this type of nozzle is impractical for industrial use. However, for studying the effects of impingement transport phenomena of other design variables, it provides uniform nozzle exit conditions with low nozzle exit turbulence.

Lujan-Acosta and Moreira (1997b) used impingement air drying techniques to dehydrate tortilla chips before frying to produce low-fat tortilla chips. The chips were prepared as shown in Figure 10–8. Distilled water (100 mL) was mixed with nixtamalized corn flour (100 g) for 9 minutes on a KitchenAid mixer (Model K5SS, St. Joseph, Michigan). The resulting masa was fed through a sheeter/former (model Tortilladora Manual, Gonzalez, Guadalupe, NL, Mexico), then weighed before exposure to drying air. The sheeter was adjusted to give a 1.0-mm chip thickness and a radius of 50.8 mm. The disk-shaped masa samples were then placed between a pair of wire meshes (1 wire/cm) and subjected to drying under the impinging hot air. The weight and diameter of the sample were measured at 2-minute intervals until the end of drying.

The compressed air used in the jets is supplied by a dedicated air compressor system equipped with a heatless air dryer that provides very dry (dew point = –40°C) compressed air at room temperature. Figure 10-9 shows a schematic of the dryer.

Tortillas chips dry significantly faster with increasing drying air temperature and slightly faster by increasing the convective heat transfer coefficient (Figure

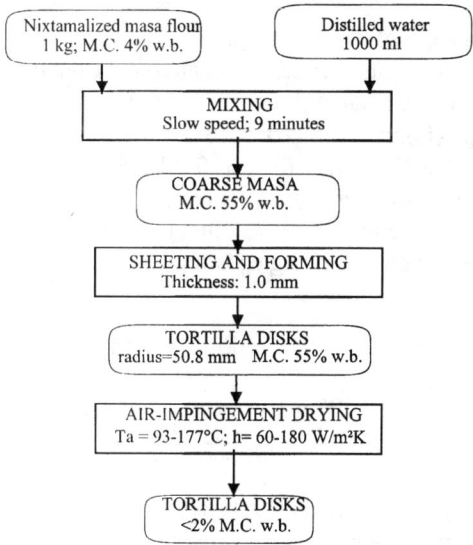

Figure 10–8 Tortilla Preparation by Air Impingement Drying

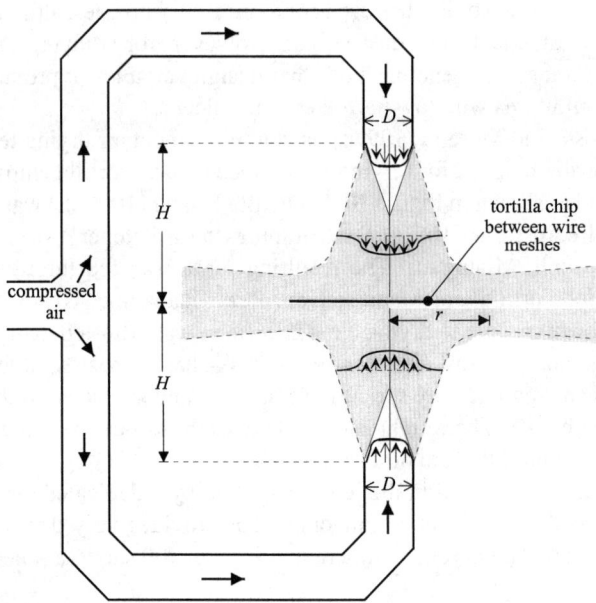

Figure 10–9 Schematic of the Impingement Dryer. D = 0.0254 m; r = 0.508; H = 0.1524 m.

10–10A, B). Shrinkage of the tortilla pieces with dehydration is similar for all conditions. The pieces shrink between 10% and 14% from their initial diameter at the equilibrium moisture content (Figure 10–10). Air temperature has no effect on degree of shrinkage (Figure 10–11A), whereas lower convective heat transfer coefficient allows higher shrinkage (Figure 10–11B).

The relationship between moisture content, $M(t)$, the drying rate constant, k_2 (see Chapter 6), the air temperature, T_{abs}, and the convective heat transfer coefficient was described by the following equation:

$$M(t) = 1.1975 \ \exp(-k_2 t) + Me$$
$$k_2 = 0.206 \ \exp\left(\frac{-1685}{T_{abs}}\right) \exp(0.004h) \qquad \text{[8]}$$

The texture analysis parameters of products from different drying treatments (Table 10–2) show a positive relationship between air temperature and fracturability, and a weak relationship between the convective heat transfer coeffi-

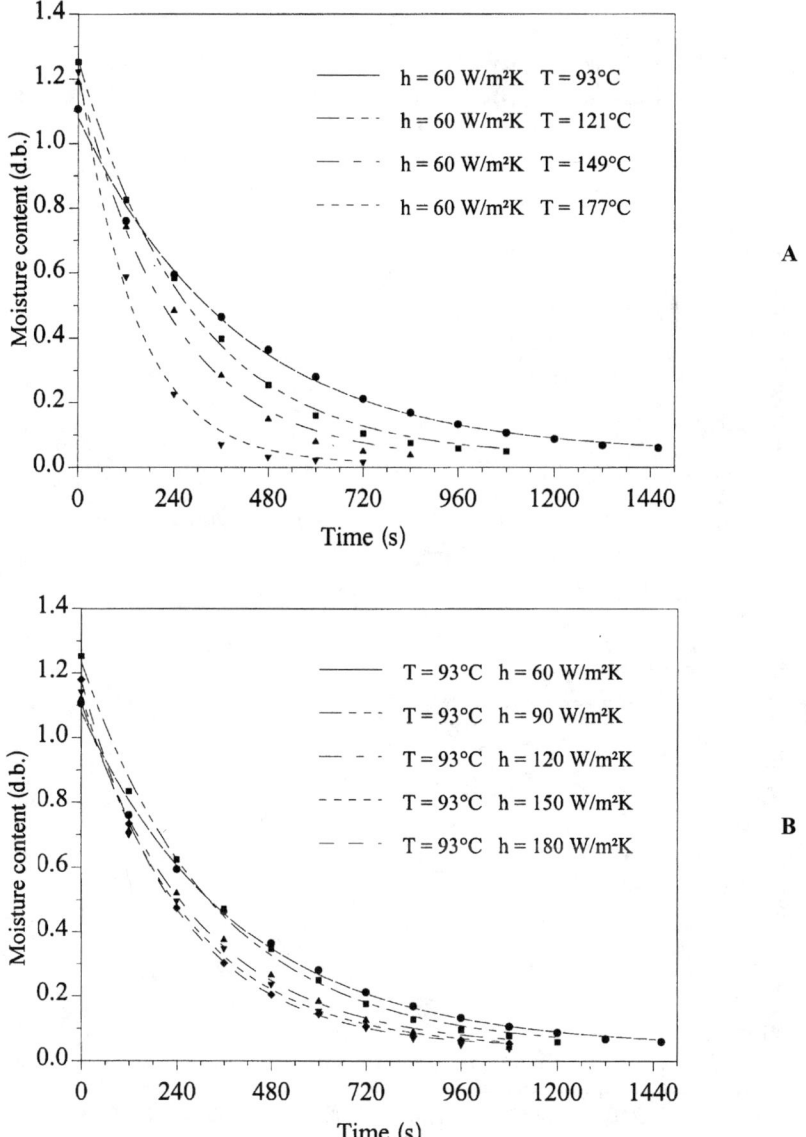

Figure 10–10 Effect of **A**, Temperature and **B**, Convective Heat Transfer Coefficient on Moisture Content during Drying. *Source:* Reprinted from *Lebensmittel-Wissenschaft und Technologie*, Vol. 30, No. 8, F.J. Lujan-Acosta and R.G. Moreira, Reduction of Oil in Tortilla Chips Using Impingement Drying, pp. 834–840, © 1997, by permission of the publisher Academic Press.

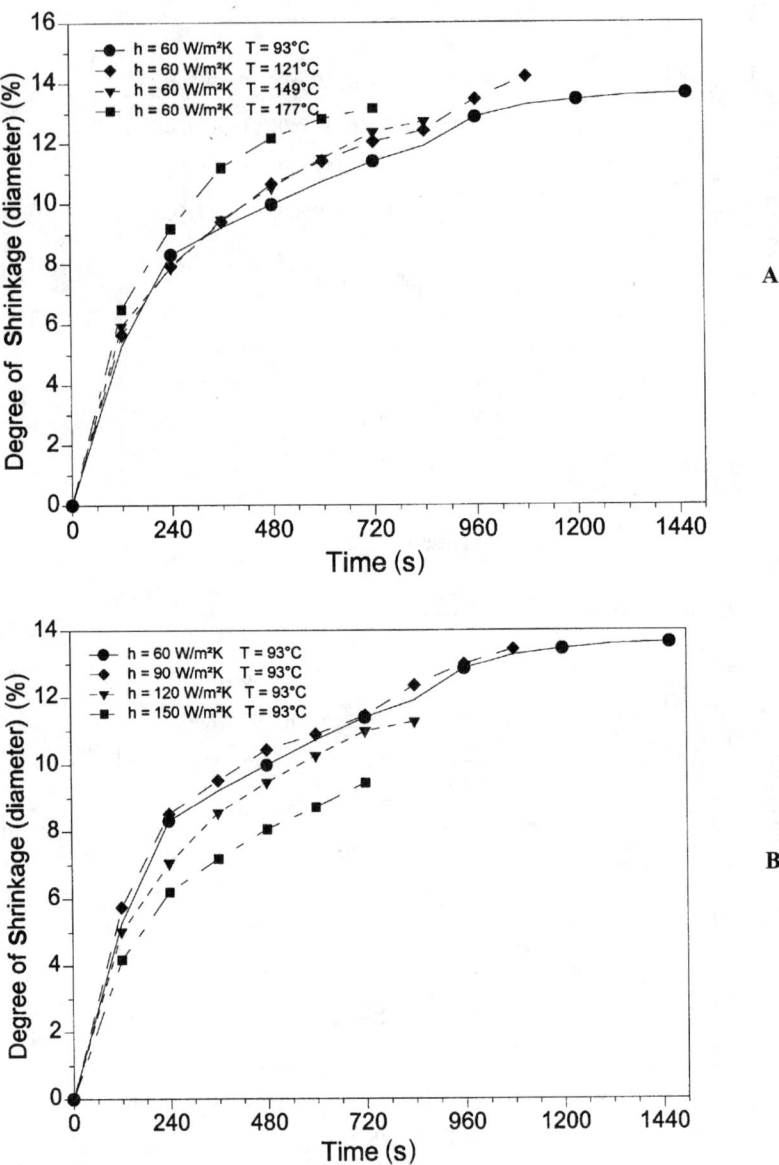

Figure 10–11 Effect of **A**, Temperature and **B**, Convective Heat Transfer Coefficient on the Chip's Shrinkage during Drying. *Source:* Reprinted from *Lebensmittel-Wissenschaft und Technologie*, Vol. 30, No. 8, F.J. Lujan-Acosta and R.G. Moreira, Reduction of Oil in Tortilla Chips Using Impingement Drying, pp. 834–840, © 1997, by permission of the publisher Academic Press.

Table 10–2 Fracturability of Tortilla Chips as a Function of Impingement Drying Parameters

Drying Treatment		
Temperature (°C)	h^a (W/m² K)	Fracturability (N)^b
93	60	6.8 ± 1.6
93	90	5.8 ± 1.0
93	120	4.4 ± 0.8
93	150	5.6 ± 0.9
93	180	4.8 ± 0.8
121	60	5.6 ± 0.8
121	90	4.4 ± 0.6
121	120	4.2 ± 0.7
121	150	4.7 ± 0.4
149	60	6.2 ± 0.7
149	90	4.5 ± 0.8
149	120	4.8 ± 0.7
177	60	5.9 ± 0.3
177	90	4.7 ± 0.7

(a)Heat transfer coefficient; (b)Mean and std. of 10 samples.

Source: Data from F.J. Lujan-Acosta, Production of Low-Fat Tortilla Chips Using Alternative Methods of Drying Before Frying, Masters thesis, 1996, Texas A&M University.

cient and the fracturability. The dried tortilla's microstructure becomes less porous, and the texture becomes harder and more uniform as tortilla pieces are dried at higher temperatures (from 93°C to 177°C), but the texture is less uniform as the convective heat transfer coefficient increases from 60 to 180 W/m² K. Figure 10–12 shows the cross-section from ESEM of tortilla samples at different convective heat transfer coefficients and temperatures.

In conclusion, these results showed that the microstructure that is formed during the air impingement drying is mainly affected by the air temperature. A smoother and tighter microstructure is obtained by drying the chip at an air temperature of 177°C. Based on these results, impingement air drying was used to reduce the moisture content after baking the tortilla in a three-stage, moving-tier, gas-fired oven for 45 seconds. Immediately after drying, the samples were fried for 20 seconds at 200°C.

During the process, the moisture content is reduced from 55% (w.b.) to 50% (w.b.) by baking, and further reduced to 14% (w.b.) by air impingement drying to produce a chip with a final oil content of 14% (w.b.) after frying. The tortilla

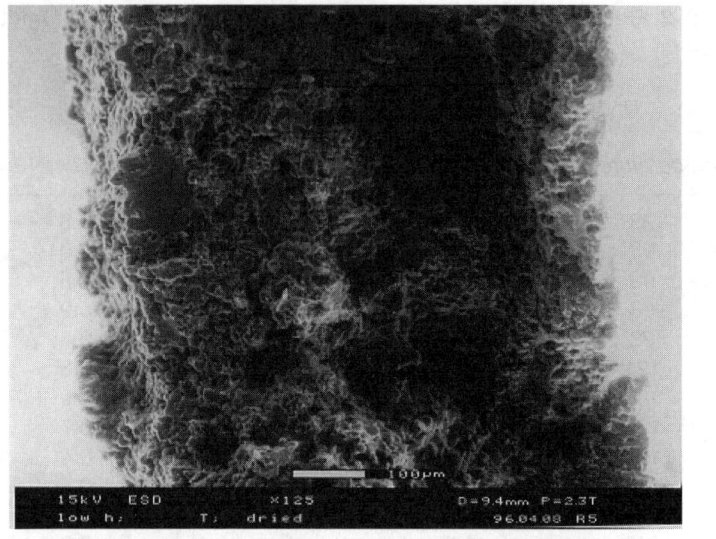

A

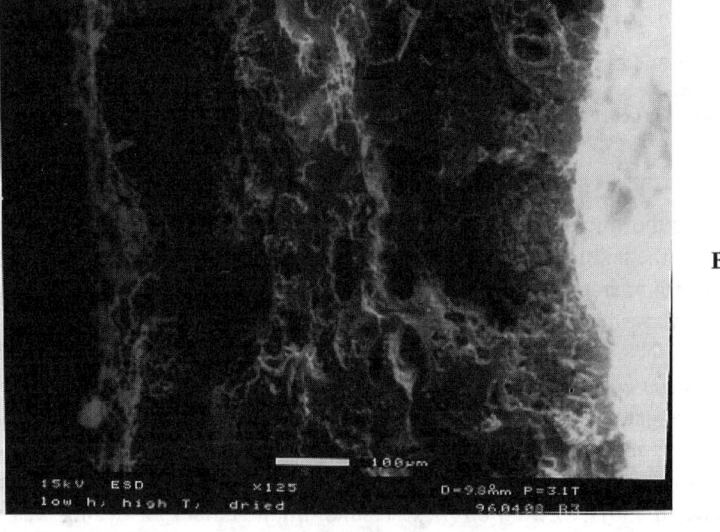

B

continues

Figure 10–12 Cross Sections from ESEM of Tortilla Samples. **A**, h = 60 W/m^2 K and T = 93°C; **B**, h = 60 W/m^2 K and T = 177°C; **C**, h = 180 W/m^2 K and T = 93°C. *Source:* Reprinted from *Lebensmittel-Wissenschaft und Technologie*, Vol. 30, No. 8, F.J. Lujan-Acosta and R.G. Moreira, Reduction of Oil in Tortilla Chips Using Impingement Drying, pp. 834–840, © 1997, by permission of the publisher Academic Press.

Figure 10–12 continued

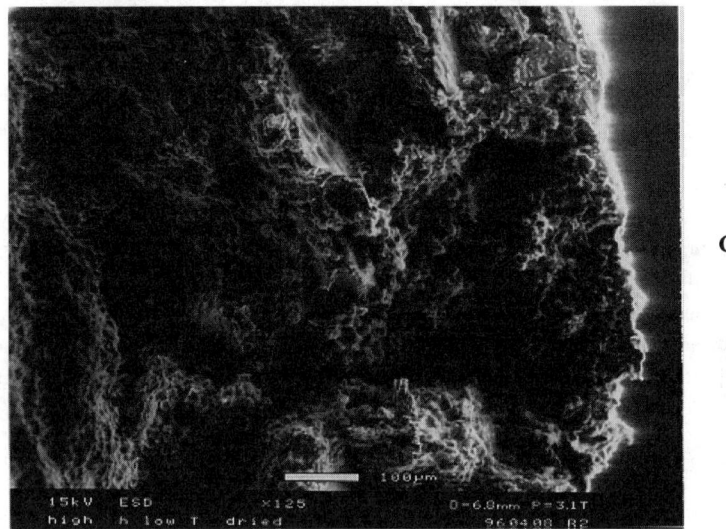

C

pieces shrink from their original diameter by 8.18% during the impingement dry-ing process and puff during frying (10%) in thickness, retaining the same diameter (Color Plate 11). The appearance, texture, and flavor (measured subjectively) are very similar to the tortilla chips obtained using sun drying. However, the impinge-ment-dried chips are slightly harder and less porous. Further studies on changing some of the processing parameters suggested in this work are recommended to improve tortilla chip palatability.

SUPERHEATED STEAM IMPINGEMENT DRYING

The use of superheated steam in impingement drying, may be considered to improve the texture of tortilla chips because steam may cause changes in the tex-ture of the product during drying, thus producing a crispier product after frying than air-dried products.

Yoshida and Hyodo (1966) demonstrated that superheated steam can provide an excellent medium for drying food products. Compared with dry air, super-heated steam is cleaner, provides higher evaporation rate, and there is less oxida-tion in food, thus reducing the loss in nutritional value during the drying process.

The idea of using superheated steam instead of heated air for drying is quite old, but the application of superheated steam drying commercially has taken place only in recent years. The main references to superheated steam drying in the food industry are in the process of animal feed from products such as brewer's waste grains, grass, corn, wheat, baggasse from sugar cane, and sugar beet pulp (Topin and Tadrist, 1997). Spent sugar beets dried with superheated steam are whiter, more porous, and more digestible for cows than air dried products (Bonazzi et al., 1996). Yoshida and Hyodo (1966) dried potato chips using superheated steam at 240°C and found that the product had a better appearance and color than the samples dried with dry air under the same conditions.

Li et al. (1998) used superheated steam impingement drying to dehydrate tortilla chips. The steam was heated by a heater controlled by a PID. The drying experiments were conducted with steam impinging on the top and bottom surfaces of the sample from the two nozzles (Figure 10–13). The diameter of the nozzles (D) was about 15 mm, and the distance from the exit of nozzle to the sample (H) was about 90 mm (Figure 10–9), resulting in a ratio of 6.0 to assure the highest heat

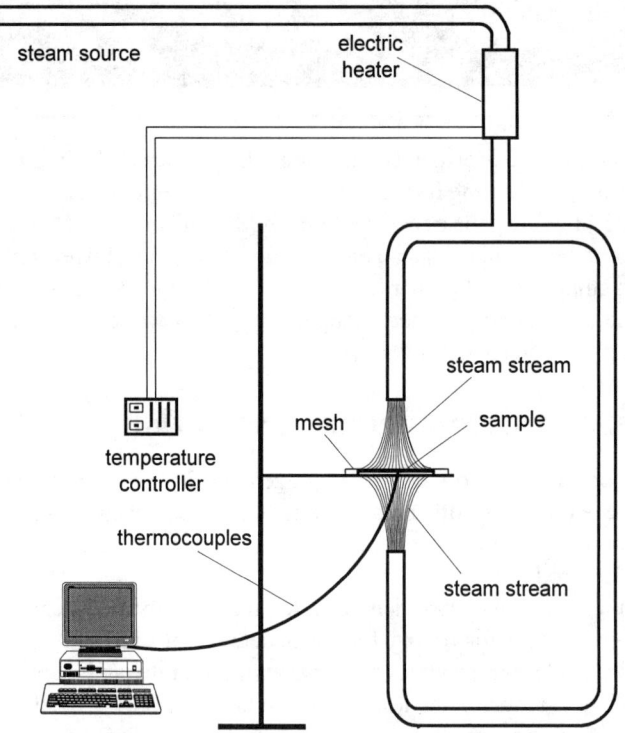

Figure 10–13 Schematic of the Steam Impingement Drying Experiment Set-Up. *Source:* Reprinted from Ref. 99–115 by courtesy of Marcel Dekker, Inc.

transfer efficiency under corresponding conditions (Downs and James, 1987). Tortilla chips were prepared the same way as described in the dry air impingement drying section. Temperature at the product center was measured continuously on line using a data acquisition system. The average moisture content was measured every 1 minute by weighing the product. Steam temperature was varied as 115°C, 130°C, and 145°C. Convective heat transfer coefficient varied from 100 W/m² K, 130 W/m² K, and 160 W/m² K.

Temperature History and Drying Rate

From the temperature history, measured at the center of the chips (Figure 10–14), it can be seen that there is a short sensible heating period at the beginning of drying, whereas the temperature of the sample increases to the boiling temperature of water (100°C). Then the sample temperature stays at the boiling temperature for several minutes, depending on the location of the thermocouple along the sample radius. During this period all the heat is used to evaporate water from the sample. When there is no free water left, the sample temperature begins to increase up to the impinging steam temperature. The higher the steam temperature, the less time is required to evaporate the free water (Figure 10–14A). The results also indicate that the heat transfer coefficient has a similar effect (within the range of study) on sample temperature, compared with steam temperature. However, the effect of steam temperature is more pronounced (Figure 10–14B).

Figure 10–15 shows the effects of the steam temperature and the heat transfer coefficient on moisture ratio profiles of tortilla chips, respectively. As shown in Figure 10–15A the tortilla chip reaches a lower equilibrium moisture content when the higher steam temperature is used (145°C). Due to a small range of heat transfer coefficients considered in this study, its effect on the drying rate is negligible (Figure 10–15B).

The relationship between moisture content, $M(t)$, the drying rate constant, k_2 (see Chapter 6), and the air temperature, T_{abs}, was described by the following equation:

$$M(t') = 1.1799 \ \exp(-k_2 t') + Me$$
$$k_2 = 62.25 \ \exp\left(\frac{-2264.42}{T_{abs}}\right) \qquad [9]$$

Texture

For all the drying conditions, the relationship between the modulus of deformation and the product moisture content is similar. During drying, the decrease in

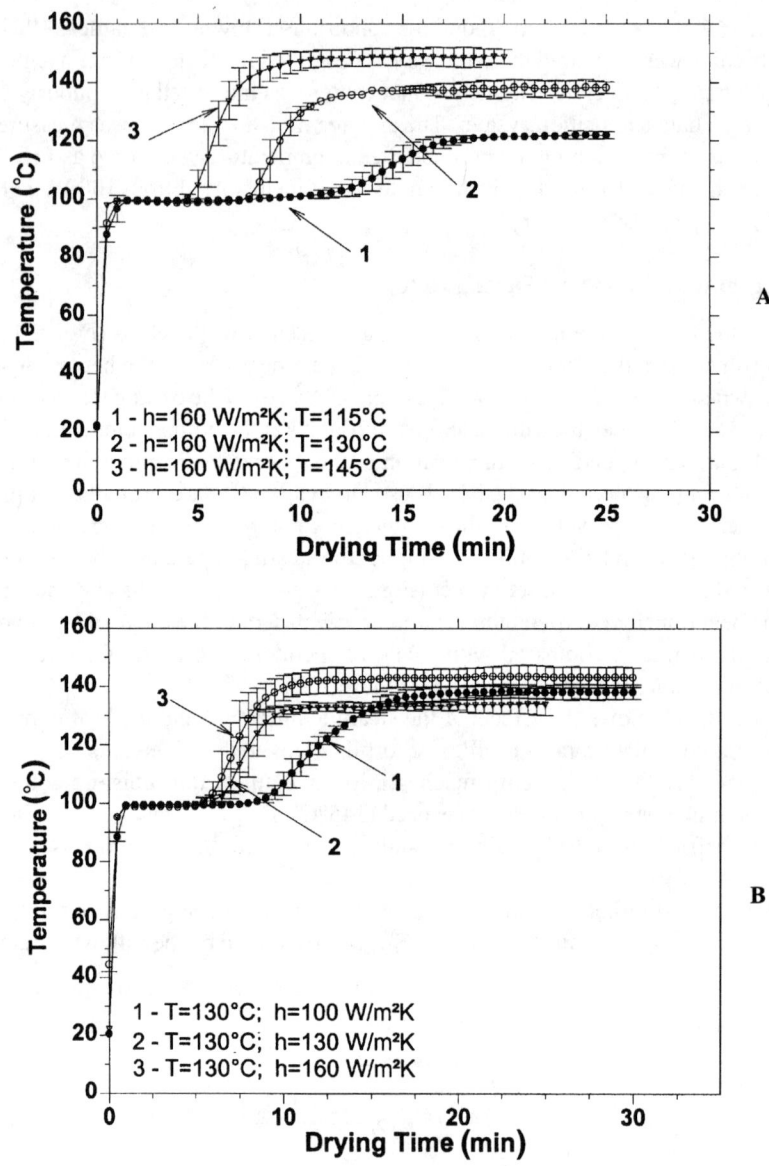

Figure 10–14 Effect of **A**, Steam Temperature and **B**, Heat Transfer Coefficient on Product Temperature during Drying. *Source:* Adapted from Ref. 99–115 by courtesy of Marcel Dekker, Inc.

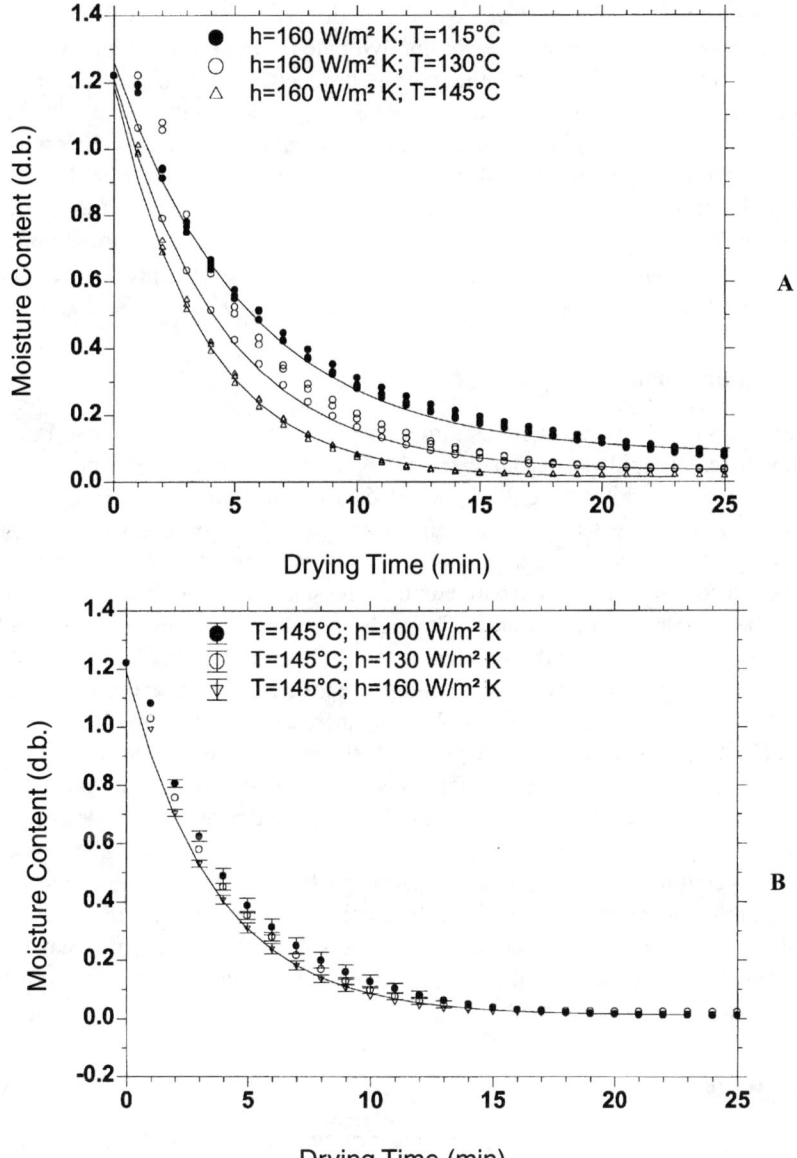

Figure 10–15 Effect of **A**, Steam Temperature and **B**, Heat Transfer Coefficient on Product Drying Rate. *Source:* Reprinted from Ref. 99–115 by courtesy of Marcel Dekker, Inc.

moisture content causes an increase of the modulus of deformation at first, reaching the maximum value, then decreasing with the further moisture removal (Figure 10–16A). As moisture is removed, the product becomes harder, reaching a maximum value at about 15% w.b. moisture, then turns brittle. The maximum values of the modulus of deformation are reached at lower moisture values when a higher steam temperature is used. These moisture values coincide with the transition point of the Tg curve (glass transition temperature) of the sample (not shown). A higher steam temperature corresponded to higher modulus of deformation. The effect of heat transfer coefficient on the chip texture is not very pronounced (Figure 10–16B).

Pasting Properties

Samples dried under different steam temperatures and heat transfer coefficients show different rapid viscous analyzer (RVA) pasting properties. The pasting viscosity developed reflects the ability of particles to absorb water and the capacity of starch to swell as the slurry is heated from 50°C to 95°C in excess water. Samples with high slopes hydrate and develop viscosity more rapidly, reaching the peak viscosity sooner and at lower temperature. The starch is gelatinized more readily than the samples with low slopes. The highest peaks and slopes are observed by the pasting properties of the undried samples (Figures 10–17A,B). The pasting viscosity of samples dried under a higher steam temperature and a higher heat transfer coefficient (145°C and 160 W/m² K) increases at a lower pasting temperature (50°C) and reaches a higher peak and slopes than the samples dried at low steam temperature and heat transfer coefficient. Samples with less fully gelatinized starch have a lower amount of starch retrogradation and, thus, a lower glass transition temperature. Therefore, the pasting viscosity increases at a lower pasting temperature. Moreover, as less gelatinized starch absorbs more water; thus, higher initial and peak viscosities can be reached. Therefore, the samples dried at higher steam temperature and higher heat transfer coefficient seem to have less fully gelatinized starch than do the samples dried at lower steam temperature and lower heat transfer coefficient.

Shrinkage

Shrinkage of the tortilla chips during drying is predominantly governed by moisture loss. The drying conditions also affected the volumetric shrinkage as a different food structure is produced. Figure 10–18A reveals that a higher steam temperature results in a lower degree of volumetric shrinkage, whereas the heat transfer coefficient has a negligible effect on the shrinkage (Figure 10–18B). When higher steam temperature is used, more and larger pores are developed,

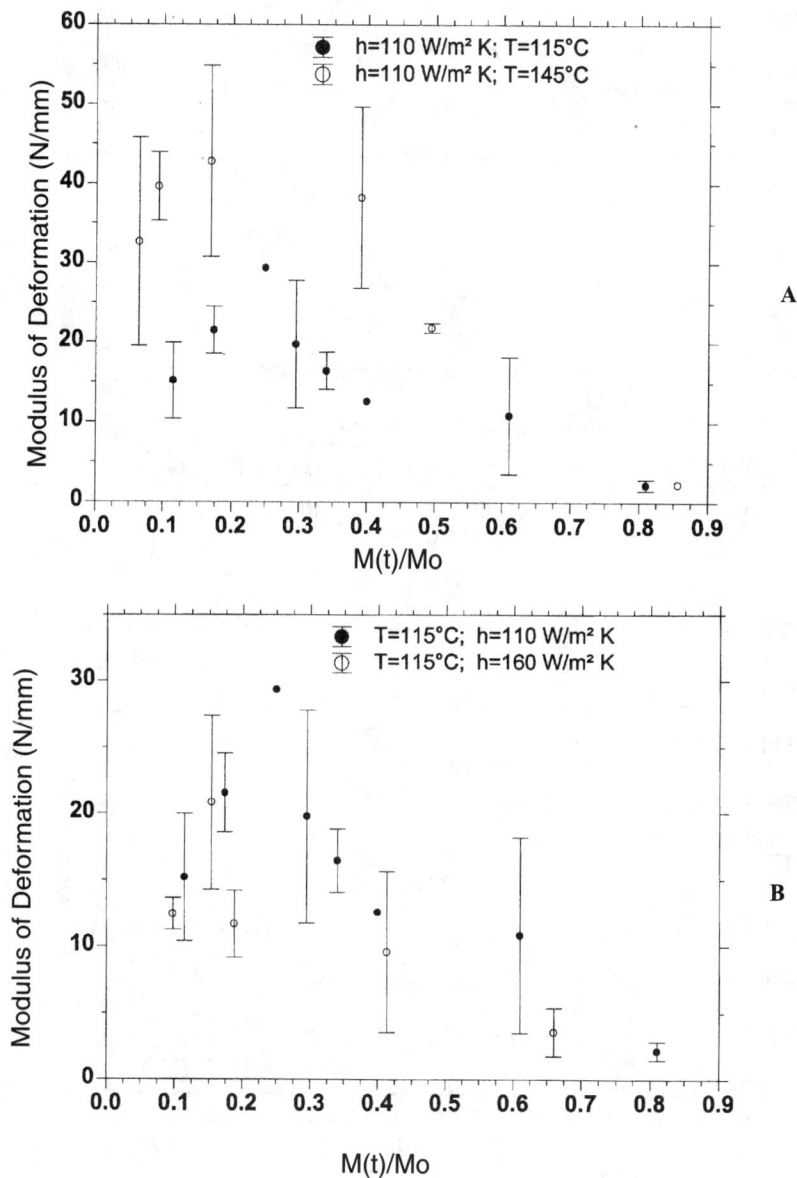

Figure 10–16 Effect of Steam Temperature on Modulus of Deformation and Convective Heat Transfer Coefficient. **A**, Temperature Effect; **B**, Heat Transfer Coefficient Effect. *Source:* Reprinted from Ref. 99–115 by courtesy of Marcel Dekker, Inc.

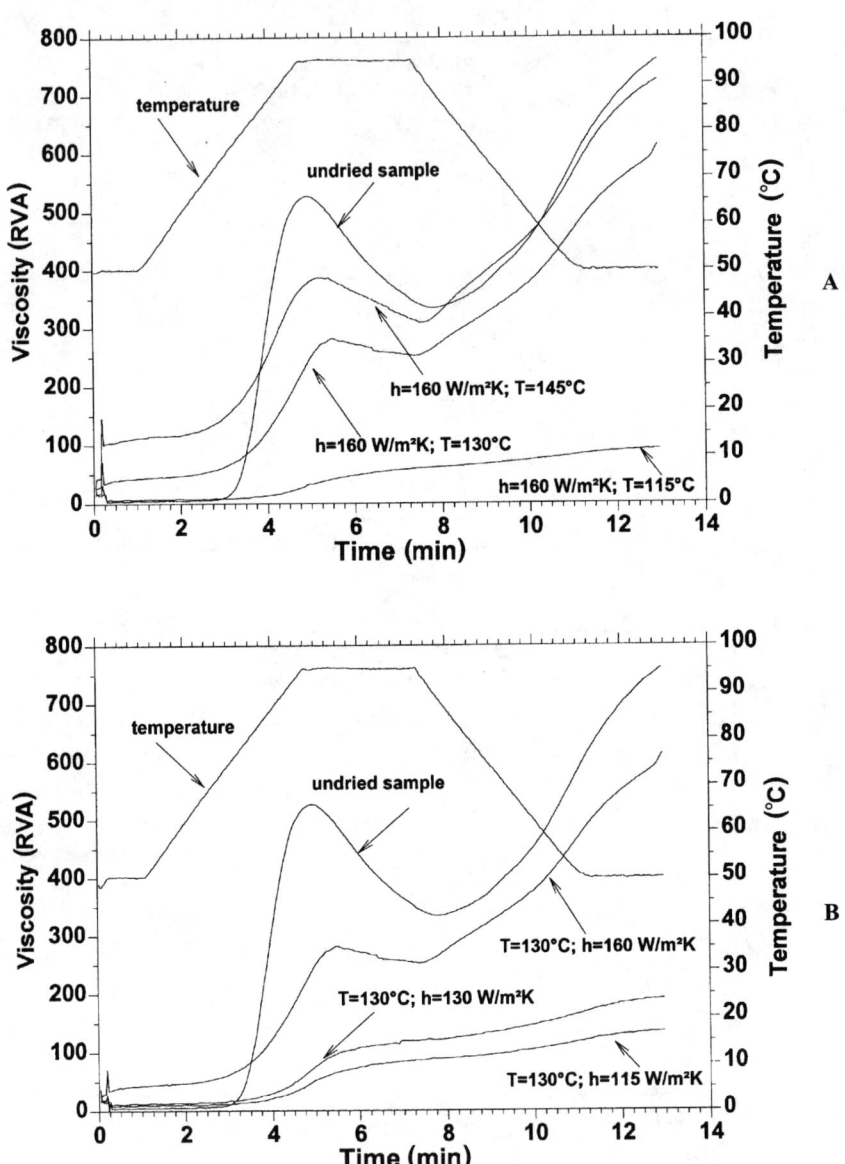

Figure 10–17 Effect of **A**, Steam Temperature and **B**, Heat Transfer Coefficient on Product Pasting Properties. *Source:* Reprinted from Ref. 99–115 by courtesy of Marcel Dekker, Inc.

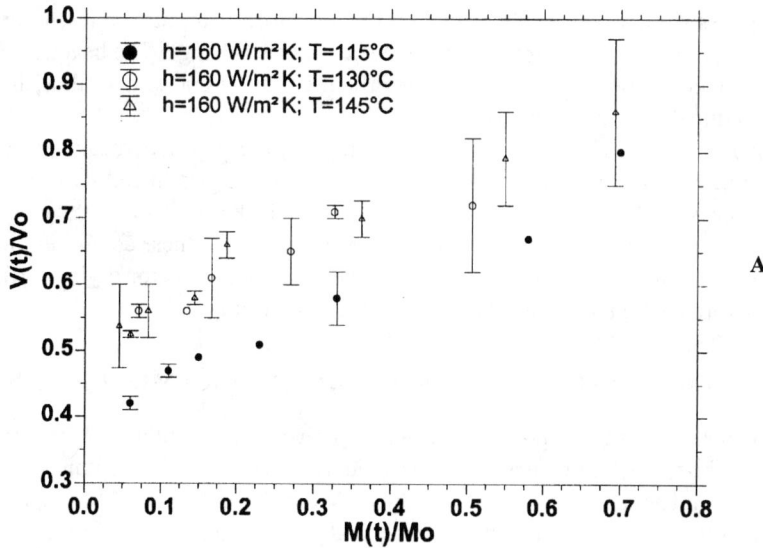

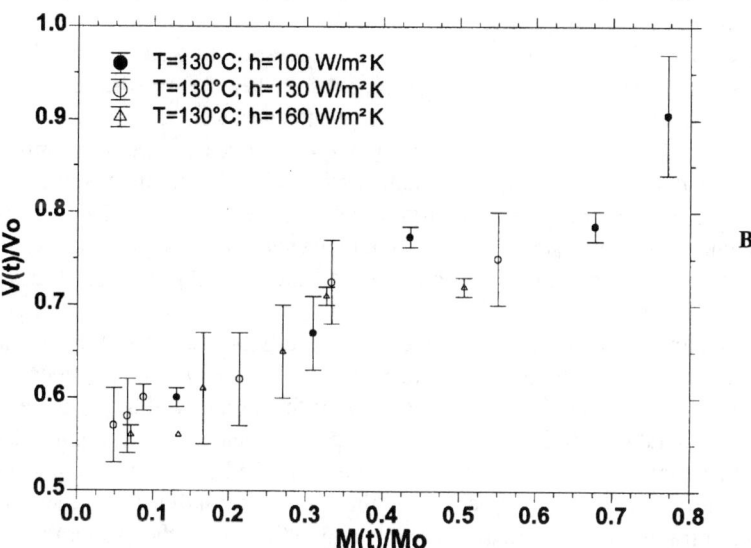

Figure 10–18 Effect of **A**, Steam Temperature and **B**, Heat Transfer Coefficient on Product Shrinkage. *Source:* Adapted from Ref. 99–115 by courtesy of Marcel Dekker, Inc.

causing less shrinkage. The elevated steam temperature also causes the sample to become harder, thus the increased resistance to volume change. The heat transfer coefficient has little effect on the microstructure and hardness, and, thus, little effect on the degree of shrinkage.

Based on the results obtained under sun drying, the best superheated steam-dried product would have a smooth microstructure (more gelatinized starch) with small pore size distribution. It is expected to shrink the most during drying to expand and produce a crunchier texture during frying. Under these conditions, low airflow rate and low temperature would be the best conditions for drying tortilla chips before frying using superheated impingement drying.

SUPERHEATED STEAM VERSUS DRY AIR IMPINGEMENT DRYING

Superheated steam has proven to be an attractive drying medium for materials that are not temperature sensitive. The reader is referred to Beeby and Potter (1992) for a detailed analysis on the subject. Wenzel and White (1951) reported that granular products dried faster in superheated steam than in dry air under the same temperature and flow rate conditions. When drying with superheated steam, the water removed from the product during the process becomes a part of the drying medium, whereas in air drying, the moist air must be replaced by fresh air heated to the desired temperature.

Chu et al. (1953) showed that the rate of evaporation of water in superheated steam is significantly higher than in dry air, except when the superheated steam is relatively close to the saturation temperature. When the steam is near its saturation temperature, the evaporation rate is close to zero, whereas evaporation still will occur when dry air is at the saturation temperature of the liquid. In a second study, Chu et al. (1959) reported that the drying rate will be improved when dry air is mixed with superheated steam. These phenomena lead to the existence of a critical temperature at which the evaporation rates of water into superheated steam and dry air are equal.

Yoshida and Hyodo (1963) showed experimentally that the curves for evaporation of air and steam would intersect at a point called the *inversion temperature*. Above that point, water evaporation increases as the humidity of the air increases, and the evaporation rate is the highest for pure superheated steam. Figure 10–19 shows that evaporation curves, at constant humidity and starting at the dew point of air, would intersect at the same point as the curve of air and steam intersected. If the concentration of steam in air increases above this point, the evaporation rate increases; however, below this point, as steam concentration increases, the evaporation rate decreases.

Any conventional drying process involves the transfer of heat and mass between the drying medium and the wet product and can be characterized by two or more distinct periods:

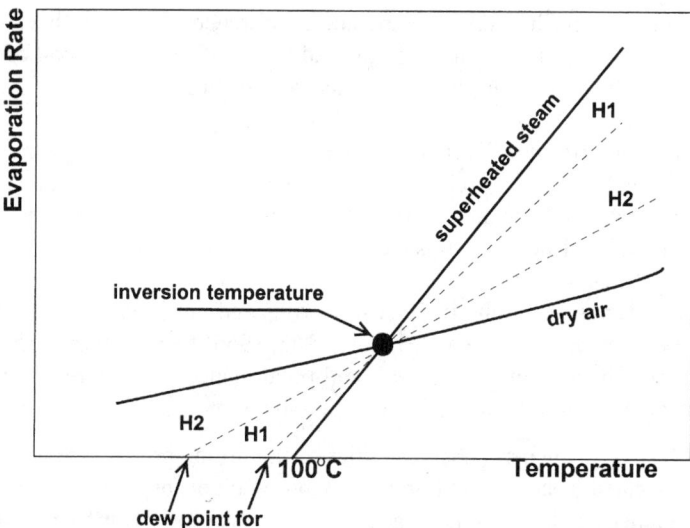

Figure 10–19 Effect of Change of Temperature on Evaporation Rate in Air, Humid Air, and Superheated Steam

- *Initial Heating:* During the initial heating period, the wet product absorbs heat from the drying medium and its initial temperature is increased to the temperature at which moisture begins to evaporate from the material (drying begins). For air drying, this temperature is normally close to the wet bulb temperature; for superheated steam it is close to the boiling temperature of water. In the case of superheated steam drying, the initial heating period involves transfer of a large quantity of heat from the steam, and, if the degree of superheat is not high, condensation will occur at the surface of the cold material. The quantity of condensed steam depends on the thermal diffusivity and moisture content of the product being dried and on the degree of superheating of the drying steam.
- *Constant-Rate Drying:* The constant-drying rate period is observed in products for which the internal resistance to moisture transport is much less than the external resistance to water vapor removal from the product surfaces. The product behaves as if a thin water layer covers the surface. For drying in air, the rate processes include convection of heat from the hot bulk air to the colder surface of the product through an air boundary layer (film) and diffusion of moisture from the surface through this boundary layer to the air bulk. The driving forces for heat and mass transfers are the temperature and vapor pressure differences between the drying medium and the product surface, respectively. In drying with superheated steam, heat is convected similarly

through a steam film, due to temperature differences. However, because the surrounding drying medium is composed solely of steam, removal of moisture from the surface occurs not by means of diffusion mass transfer but by bulk flow.

- **Falling-Rate Drying**: When the surface of the product is no longer moist, the drying rate begins to decrease, and the process enters the falling-rate period. During this period, the internal resistance to moisture transport is greater than the external resistance. In this case, diffusion is the controlling mechanism for mass transport and conduction for heat transport. Therefore, the drying rate is determined by the product properties and not just by the liquid and vapor properties. Yoshida and Hyodo (1963, 1966) found that the surfaces of synthetic fibers (cellulose acetate) and potato chips were more porous and permeable to vapor flow than when dried in air drying.

In conclusion, the general characteristics of the drying process are not altered when superheated steam is used as the drying medium instead of dry air. The major differences include (1) the condensation on the product surface during the heating period and (2) elevated surface temperature during the constant-rate period. The drying rate in superheated steam medium can be faster if (1) the amount of condensation at the product surface is low and (2) the superheated steam temperature is above the inversion temperature. Resistance to moisture movement within the product appears to be reduced for superheated steam drying, thus increasing drying rates during the falling-rate periods.

Li et al. (1998) dried tortilla chips with dry air impingement drying and compared them to the chips dried under superheated steam impingement drying. The results are presented below.

Drying Rate

Figure 10–20A shows the drying characteristics of tortilla chips under superheated and dry air impingement drying at $T = 115°C$ and $h = 160$ W/m^2 K. The drying rate of tortilla chips at these conditions was similar for the superheated steam and dry air. The reason was the amount of condensation that occurred at the product's surface during the heating period, thus increasing the moisture level. This required more drying to be accomplished so that the condensed water could be reevaporated.

For temperatures of 130°C (Figure 10–21B) and 145°C (Figure 10–20B), the drying rate was higher for tortilla chips dried in superheated steam than in dry air. The effect of airflow rate is shown in Figures 10–21A and 10–21B. Again, tortilla chips dried faster in superheated steam than in dry air because the material's surface was softer than with air drying. Thus, resistance to moisture movement within the product is reduced.

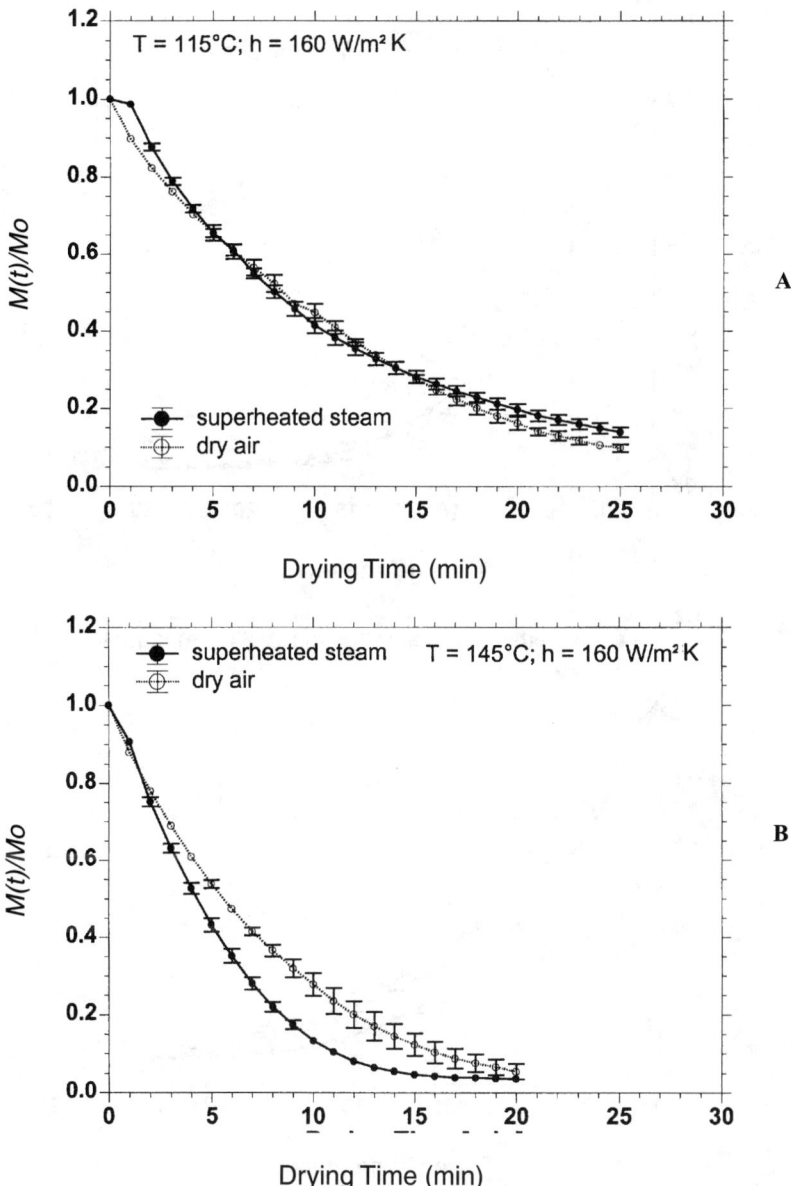

Figure 10–20 Effect of Temperature on the Drying Rate of Tortilla Chips Dried under Superheated Steam and Dry Air. **A**, 115°C; **B**, 145°C. *Source:* Reprinted from Ref. 99–115 by courtesy of Marcel Dekker, Inc.

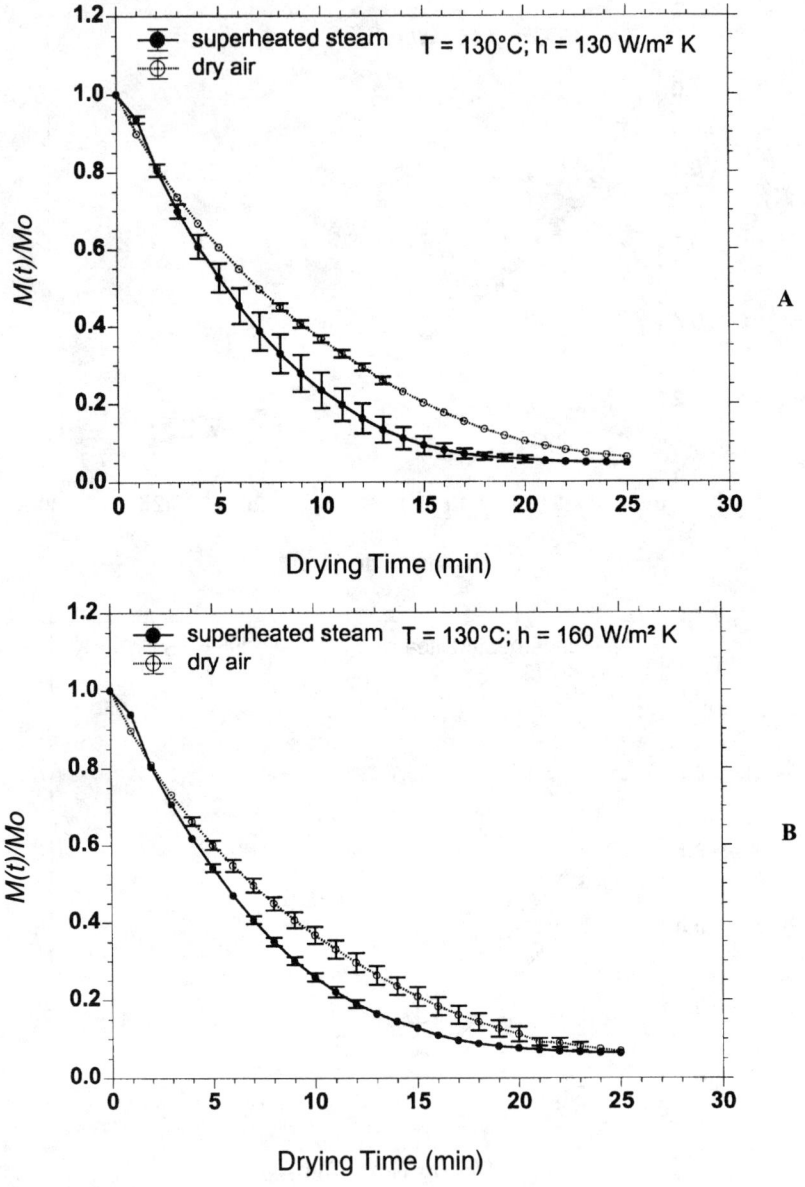

Figure 10–21 Effect of Convective Heat Transfer Coefficient on the Drying Rate of Tortilla Chips Dried under Superheated Steam and Dry Air. **A**, 130 W/m² K; **B**, 160 W/m² K.

Product Temperature

The effect of drying air (or steam) temperature on the temperature history of tortilla chips dried under superheated steam and dry air is presented in Figures 10–22A and 10–22B, respectively. These figures clearly show that superheated steam is a much more efficient drying technique for tortilla chips; it takes less time to evaporate the free water from the chips dried under superheated steam than with dry air. The effect of temperature is the same for steam and dry air—at a higher temperature, a shorter time is required to evaporate the water. As illustrated in Figure 10–23, the heat transfer coefficient has a similar effect on sample temperature.

Texture

The product becomes harder when fried in dry air than in steam when dried at lower temperature and higher convective heat transfer coefficient (Figure 10–24). In this case, the maximum value of modulus of deformation was reached at higher moisture content for the product dried in steam than for the product dried in dry air. When the tortillas were dried at higher temperature and convective heat transfer coefficient, superheated steam caused the material to become more porous (Figure 10–25) than the air dried samples. The surfaces of the air-dried samples were also harder than the steam-dried samples. In the case of air drying, the temperature of the product surface is that of the wet bulb temperature. However, in the case of superheated steam drying, the temperature of the product surface is at the boiling temperature of the water. Further, additional heat caused by the steam surrounding the product causes the material to boil violently. Vapor inside the material is diffused to the outside, thus resulting in a porous final product.

Pasting Properties

Samples dried in the same temperatures and heat transfer coefficients under superheated steam and dry air show different RVA pasting properties (Figures 10–26 A,B,C). Figures 10–26A and 26B show that tortilla chips dried in superheated steam medium have more fully gelatinized starch particles than those dried in dry air at the temperature range of 115°C–130°C and convective heat transfer coefficient of 130–160 W/m² K. For temperatures above 145°C, more fully gelatinized starch granules were found in the air dried samples than in the superheated steam samples. This could be due to the faster drying rate, which resulted in less water available for gelatinization during drying.

The results presented here demonstrate that superheated impingement drying techniques could be an alternative to produce a better-quality product. The tech-

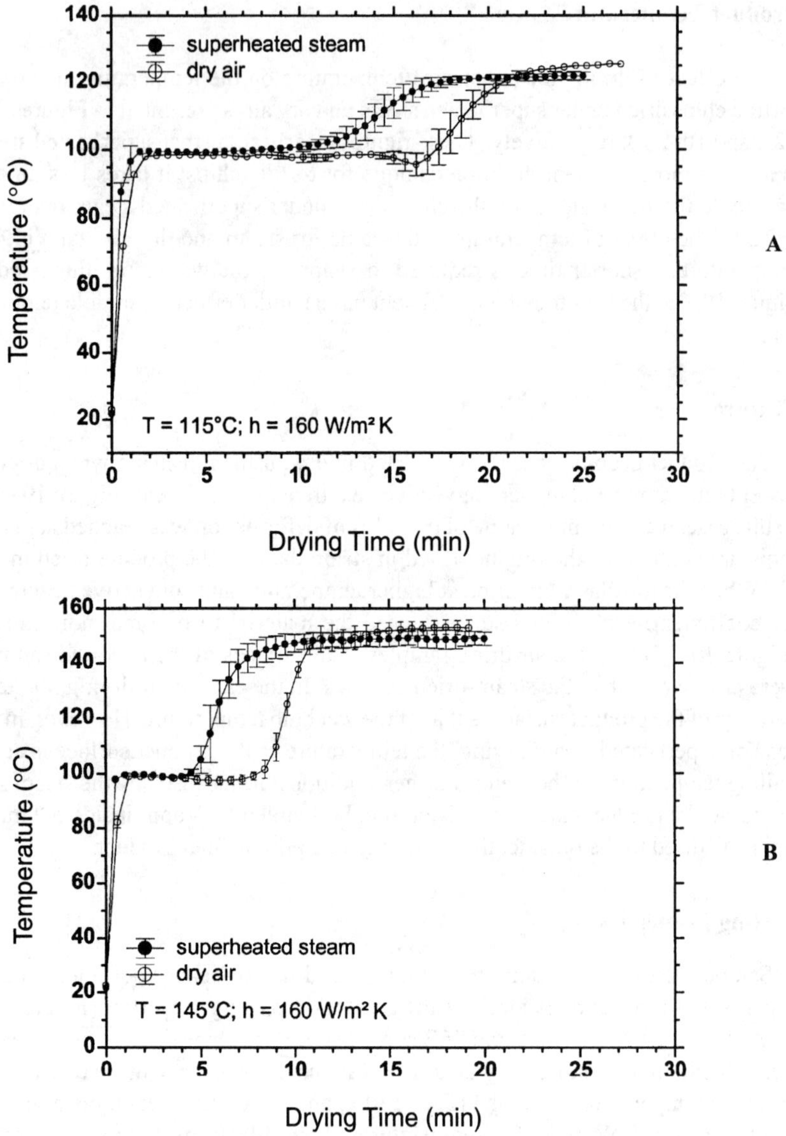

Figure 10–22 Effect of Drying Temperature on the Center Temperature of Tortilla Chips Dried under Superheated Steam and Dry Air. **A**, 115°C; **B**, 145°C.

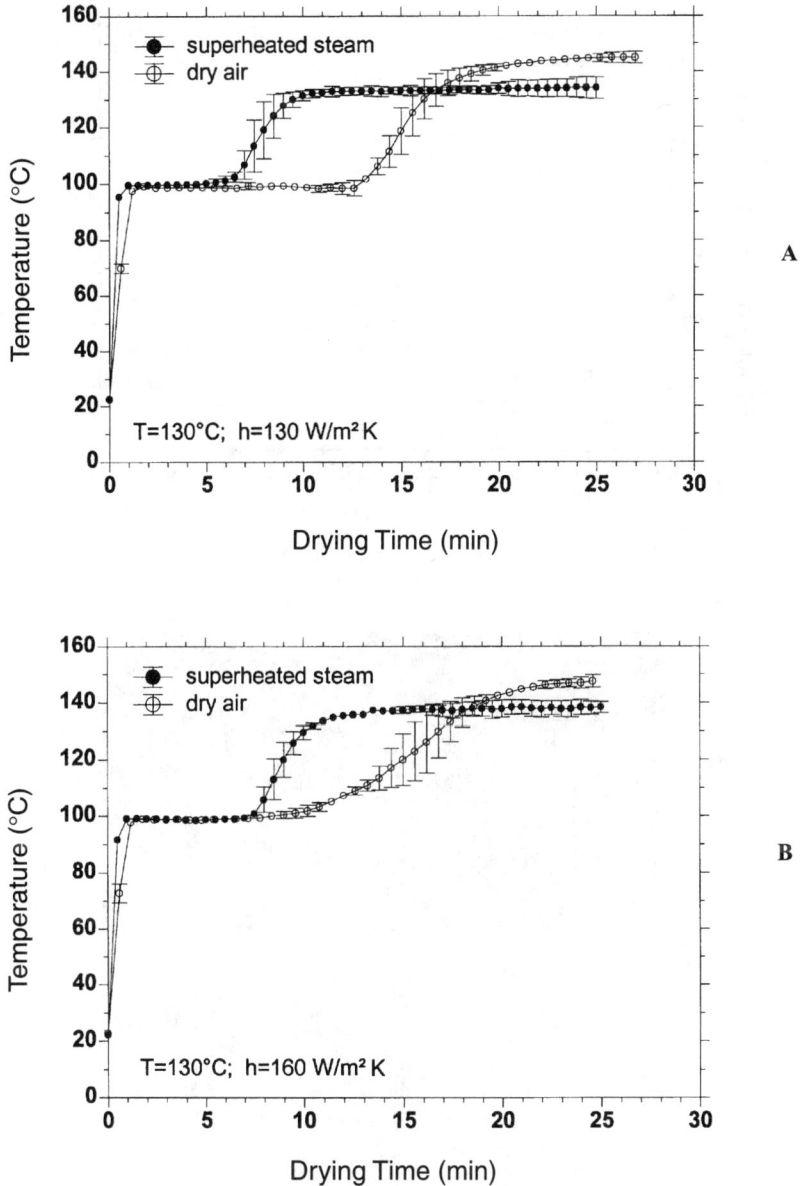

Figure 10–23 Effect of Convective Heat Transfer Coefficient on the Temperature of Tortilla Chips Dried under Superheated Steam and Dry Air. **A**, 130 W/m^2 K; **B**, 160 W/m^2 K.

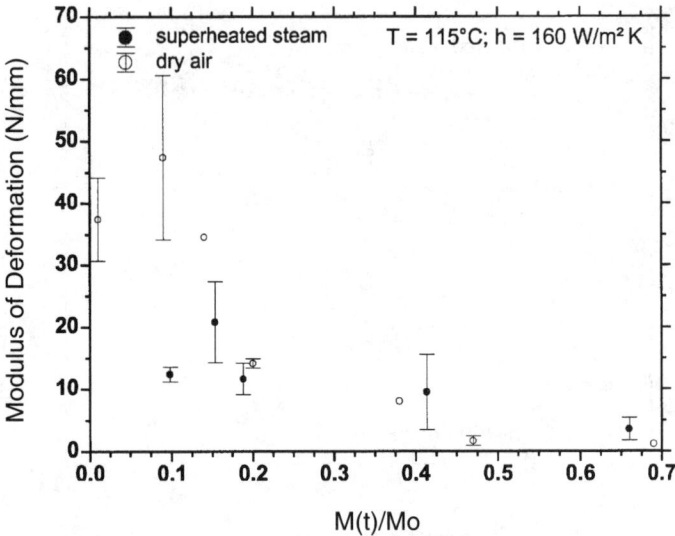

Figure 10–24 Effect of Drying Temperature on Modulus of Deformation

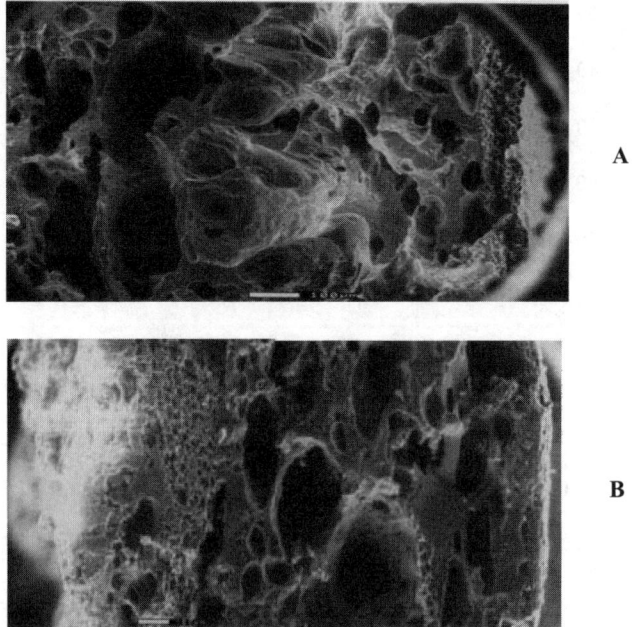

Figure 10–25 Cross-Sections from ESEM of Tortilla Samples Dried at T = 145°C and h = 160 W/m² K with **A**, Superheated Steam and **B**, Dry Air.

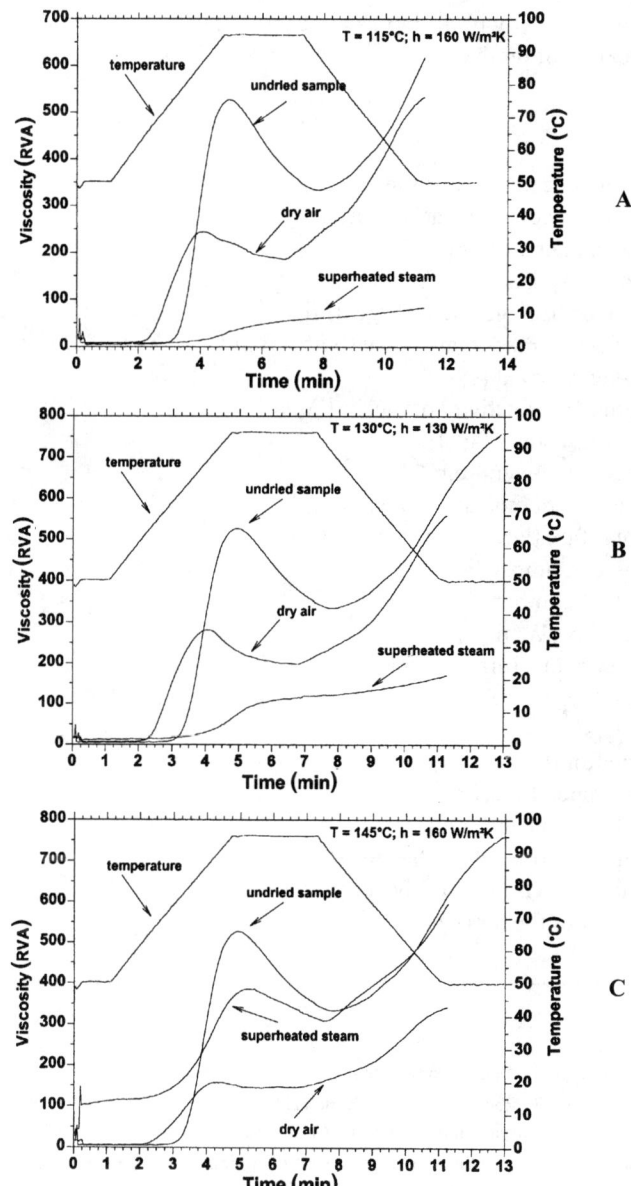

Figure 10–26 Effect of Temperature and Heat Transfer Coefficient on the Pasting Properties of Tortilla Dried under Superheated Steam and Dry Air. **A**, T = 115°C; h = 160 W/m² K; **B**, T = 130° **C**; h = 130 W/m² K; C, T = 145°C; h = 160 W/m² K. *Source:* Adapted from Ref. 99–115 by courtesy of Marcel Dekker, Inc.

nique should be tested with frying as a means of reducing final oil content and improving texture of product.

NOTATIONS

$C_{A,a}$ molar concentration at the air [kmol/m^3]
$C_{A,s}$ molar concentration at the surface [kmol/m^3]
D   nozzle diameter [m]
h   convective heat transfer coefficient [W/m^2 K]
H   distance from nozzle to target [m]
h_m convective mass transfer coefficient [m/s]
k_2 drying constant [s]
k_a thermal conductivity of the air [W/m K]
$\dot{m}$ mass flow rate [kg/s]
M   moisture content [d.b.]
Me   equilibrium moisture content [d.b.]
N'_A molar flux [kmol/s m^3]
Nu   Nusselt number
Pr   Prandtl number
q' heat flux [W/m^2]
r   target radius [m]
Re   Reynolds number
t   time [s]
t' time [min]
T_a air temperature [C]
T_s surface temperature [C]
α_a thermal diffusivity of the air [m^2/s]
ν_a kinematic viscosity of the air [m^2/s]
ρ_a density of the air [kg/m^3]

REFERENCES

Beeby, C., and Potter, O.E. 1992. Steam drying. *Drying '92*, 41–58.

Bonazzi, C.; Dumoulin, E.; Raoultwack, A.L.; Berk, Z.; Bimbenet, J.J.; Coutois, F.; Trystram, G.; and Vasseur, J. 1996. Food drying and dewatering. *Drying Technol*, 14(9):2135–2170.

Chu, J.C.; Lane, A.M.; and Conklin, D. 1953. Evaporation of liquids into their superheated vapors. *Ind Eng Chem*, 38:86–87.

Chu, J.C.; Fenelt, S.; Hoerrner, W.; and Lin, M.S. 1959. Drying with superheated steam-air mixtures. *Ind Eng Chem*, 51:275–280.

Downs, S.J., and James, E.H. 1987. Jet impingement heat transfer—a literature survey. ASME paper No. 87-H-35, National Heat Transfer Conference, Pittsburgh, Pennsylvania.

Gomez, M.H.; Lee, J.K.; McDonough, C.M.; Waniska, R.D.; and Rooney, L.W. 1992. Corn starch changes during tortilla and tortilla chip processing. *Cereal Chem,* 69:275–279.

Gomez, M.H.; Rooney, L.W.; Waniska, R.D.; and Pflugfelder, R.L. 1987. Dry corn masa flours for tortilla and snack foods. *Cereal Foods World,* 32(5):372–377.

Hoseney, R.C. 1994. *Principles of cereal science and technology,* 2nd ed. St. Paul, MN: American Association of Cereal Chemists.

Li, Y.B.; Yamsaengsung, R.; Moreira, R.G.; and Seyed-Yagoobi, J. 1998. *Superheated steam impingement drying of tortilla chips.* IFT Annual meeting. June 20–24. Atlanta, GA.

Lujan-Acosta, F.J. 1996. Production of low-fat tortilla chips using alternative methods of drying before frying. Masters thesis. College Station, TX: Texas A&M University.

Lujan-Acosta, F.J., and Moreira, R.G. 1997a. Effects of different drying processes on oil absorption and microstructure of tortilla chips. *Cereal Chem,* 74(3):216–223.

Lujan-Acosta, F.J., and Moreira, R.G. 1997b. Reduction of oil in tortilla chips using impingement drying. *Lebensm-Wiss U.-Technol, Food Sci Technol,* 30(8):834–840.

Marcroft, H.; Chandrasekaran, M.; and Karwe, M.V. 1997. *Temperature and velocity fields in a hot air jet impingement oven.* IFT Annual Meeting. Orlando, FL.

Martin, H. 1977. Heat and mass transfer between impinging gas jets and solid surfaces. *Adv Heat Transfer,* 13:1–66.

Mujundar, A.S. 1986. Impingement drying. In *Handbook of industrial drying.* New York: Marcel Dekker.

Remsen, C.H., and Clark, J.P. 1978. A viscosity model for a cooking dough. *J Food Proc Eng,* 2:39–49.

Ring, S.G. 1985. Some studies on starch gelation. *Starke,* 37:80–83.

Serna-Saldivar, S.O.; Gomez, M.H.; and Rooney, L.W. 1990. The chemistry, technology and nutritional value of alkaline-cooked corn products. In *Advances in Cereal Science and Technology,* Vol. X. Minneapolis, MN: AACC.

Topin, F., and Tadrist, L. 1997. Analysis of transport phenomena during the convective drying in superheated steam. *Drying Technol,* 15(9):2239–2261.

Wenzel, L., and White, R.R. 1951. Drying granular solids in superheated steam. *Ind Eng Chem,* 43:1829–1937.

Yoshida, T., and Hyodo, T. 1963. Evaporation of water in air, humid air, and superheated steam. *Ind Eng Process Des Dev,* 9(2):207–214.

Yoshida, T., and Hyodo, T. 1966. Superheated vapor speeds drying of foods. *Food Eng,* 38:86–87.

CHAPTER 11

Packaging Fried Snacks

Packaging is a key aspect for successful marketing of a food product. In the particular case of fried snacks (such as potato or tortilla chips), flexible packaging is most commonly used as aluminum foil or laminated bags, although spiral-wound cans are also used. These fried products have specific requirements to preserve their freshness when they reach the consumer. Therefore, it is important to understand the mechanisms responsible for product failure (deterioration, loss of shelf life) as they are related to the packaging material. This chapter introduces the basics of packaging materials, methodology, and requirements for proper packaging of selected fried snacks.

PURPOSE OF PACKAGING

A good package should contain, quantify, and advertise its contents. Most of all, a package should provide adequate protection to its contents against harmful environmental factors. Food items may deteriorate by three mechanisms—physical, biochemical, and microbiologic—which act independently or simultaneously. Almost all biologic systems experience all three forms of deterioration, but one mechanism frequently terminates the usefulness of the product before the others become serious. These factors can be natural in origin, such as temperature, humidity, oxygen, light, and contamination; or manmade, such as mechanical damage, vibration, shock, and stress.

The protective function of a package consists of providing a barrier between its contents and the immediate environment. When unpackaged, a dry product will absorb moisture, reducing its quality properties and/or shelf life; an oxygen-sensitive product should similarly be shielded against contact with the air. Conversely, with products that have a special aroma, measures are required to prevent the escape of flavors. A mechanically fragile product is in danger during handling and storage because of excessive external mechanical stresses; here, the function of the package is to absorb at least part of these stresses.

PACKAGING REQUIREMENTS FOR FRIED SNACKS

Fried snacks constitute more than 67% of the total snack market, with potato chips representing 32.9% and tortilla chips 21% (Chapter 1). The snack market is served by a variety of packaging material constructions, depending largely on the package contents, the shelf life required, and the desired package appearance. For example, potato chips generally use a construction of coextruded oriented polypropylene (OPP) extrusion or adhesive laminated to metallized OPP. Where extended shelf life may be required, high barrier metallized films can be utilized (Robertson, 1993). Corn tortilla chips are packed in polypropylene bags and corn chips in aluminum foil bags, due to their higher fat content. These bags are then put into a transit carton by hand or automatically for transportation to a warehouse or point of sale.

Packaging cannot improve or completely preserve the quality of the packaged products. Some deteriorative mechanism will continue to operate over a period of time at a rate governed by product characteristics and environmental conditions. Packaging can only delay quality loss by regulated time-related factors and by its inherent barrier properties. Specific sensitivities of products to biologic and biotic deterioration become critical considerations in packaging (Robertson, 1993).

Deterioration Mechanisms

There are two major modes of deterioration of fried snack foods: (1) development of fat rancidity and (2) loss of crispness.

Rancidity

Common to all fried snack foods is fat that is used as a processing agent to dehydrate the product (as in the case of potato chips), to puff it (as in the case of some extruded products), and to develop characteristic flavors. As a consequence, the end of shelf life of many fried snack foods is closely related to the development of rancidity by the fat.

All fats are subject to deterioration by oxidative rancidity, which leads to the formation of objectionable odors and flavors. Oxidative rancidity results in food spoilage associated with fat deterioration, i.e., the presence of pungent or acrid odors, and this is very important with respect to food acceptability (Robertson, 1993).

The susceptibility of fried snacks to oxidative rancidity depends on the type of fat used and the number of unsaturated bonds in the fatty acid moiety. Analysis of volatile compounds in fresh and aged potato chips and unused fresh and aged frying oils showed that oxidation of oils was mainly responsible for volatile compound changes in potato chips during storage (Chapter 3). To minimize the development of such rancidity, the product must be protected from oxygen, light, and trace quantities of metal ions. The packaging material, if a light barrier, may reduce the rate of rancidity development, as it decreases the catalytic effect of light on the oxidation reaction. Moisture content may affect the rate, and there can be an induction period after exposure of the fat to oxygen before rancidity is noticed. Diffusion into the fat can reduce the rate of development of rancidity, making thin fat layers most vulnerable (Hine, 1987).

Crispness

Crispness is a salient textural characteristic for fried snack foods, and its loss due to absorption of moisture is a major cause of product rejection by softening the starch/protein matrix, which alters the mechanical strength of the product (Robertson, 1993). A water content limit at which the textural quality of a snack food product becomes organoleptically unacceptable is typically 3–3.5%; however, the water content limits are strongly dependent on the method used for moisture determination. Potato chips, for example, are packaged at a moisture content of 1–1.5%. When these chips reach a 4–5% moisture content, they are considered stale. A more reliable approach to establishing moisture conditions for textural appearance is as a function of water activity, a_w. The critical a_w for potato chips is 0.40, with the same value having been reported for corn chips. Another study has shown that potato chips become organoleptically unacceptable at 0.51 a_w (Katz and Labuza, 1981).

The moisture content of a moisture-sensitive product and the relative humidity (or water activity) with which it is in equilibrium are linked by a characteristic curve for the product (moisture absorption isotherm) (Figure 11–1). Representative data for a specific product should be available for proper determination of quality losses due to moisture absorption.

Low-Fat Fried Products

Low-fat fried snacks typically contain high levels of moisture and require high-performance packaging to maintain freshness and palatability. This requires that the packaging film have a higher barrier than as used for the regular-fat version. These products also have a different flavor and aroma profile than full-fat products because there is less fat for the flavors and aromas to dissolve into. Thus, off-flavor and -aroma chemicals, which diffuse into or are generated in the package, remain at higher concentrations in the headspace of the bag and can be easily smelled. Flavors added to the product escape faster into the headspace of the bag

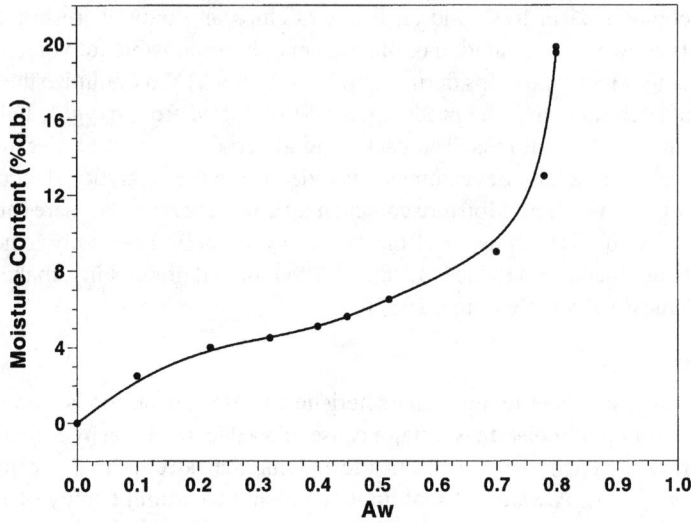

Figure 11–1 Moisture Sorption Isotherm for Potato Chips at 20°C. *Source:* Reprinted with permission from E.E. Katz and T.P. Labuza, Effect of Water Activity on the Sensory Crispness and Mechanical Deformation of Snack Food Products, *Journal of Food Science*, Vol. 46, p. 405, © 1981, Institute of Food Technologists.

---

and can then be scalped by the bag or lost by diffusion through the packaging materials (Taoukis et al., 1997).

Metallized film by itself cannot guarantee a good aroma because the aromas usually interact in the substrate film and somehow disrupt the interface of the metal, degrading the barrier. Therefore, films in these applications should be chosen for their chemical resistance to the flavor and aroma chemicals used in the products or to which the products are exposed. The resistance should be evaluated for each case to obtain optimum results. Several existing films, both metallized and unmetallized, as well as several new developmental films, can be combined in several configurations to enhance the overall resistance of the package to chemical challenges from inside or outside the bag. An example is the increased use of metallized OPP with its benefits of high barrier protection for light, oxygen and moisture, and high graphic appearance (Scherer, 1998).

BARRIER MATERIALS AVAILABLE

Retailers used to dispense potato chips in bulk from cracker barrels or glass display cases. In 1926, Laura Scudder developed the first potato chip bag by ironing sheets of waxed paper into bags. The bags were filled by hand, ironed shut, and delivered to retailers. In 1933, the Dixie Wax Paper Company of Dallas, Texas introduced the first "preprint" waxed glassine bag, called *Dixie's Fresheen*. This allowed potato chips to stay fresh longer. Today, a Saran coating provides superior protection from fat, water vapor, odors, and gases (Hanlon, 1994).

Currently, a range of barrier materials used as primary packages is available. Selection of the best product for a given application requires careful consideration of many factors. Foils provide excellent barriers but preclude product visibility. A wide variety of plastics has been developed in the form of films or rigid containers, and one important property of these materials is their permeability, particularly when used in films. This chapter briefly describes the performance characteristics of such materials when used in food packaging.

Lamination consists of combining two or more materials with adhesive. Before beginning to choose the components of a lamination, it is good practice to make sure there is no single film that will do the job. Next consider whether a *coated film* will do the job. The addition of a coating is less costly as a coating is generally thinner than the lightest film that could be used for the same purpose. If it is decided that a lamination is necessary, every effort should be made to keep the number of plies to a minimum. As a general rule the most protective ply should be nearest to the product. Oriented polypropylene/cellophane is commonly used for snack foods. Usually, it is a combination of paper, foil, and polyethylene (PE)— pouch paper/PE/foil/PE (out to inside). A typical laminate for chips and snacks is polyvinylidene (PVDC)-coated PP/PVDC-coated PP (Fellows, 1996).

A material up to 0.25 mm in thickness is known as a *film*. When two or more films are extruded at the same time and combined, they are known as a *composite film* (Hanlon, 1994). Films are increasingly called upon to protect their contents from oxygen, moisture, light, or any combination of the three to extend shelf life. These packaging films have become much more important in snack food companies' successes with a product, especially as most companies want their products to hit a consumer's mouth with as much freshness and appeal as it did when it was first packaged. To meet consumer demand for freshness on the first bite, these processors have had to find ways to package products so that flavor is maintained while still having excellent graphics, high barrier, and cost-effectiveness.

Blown-film coextrusions are thinner and suitable for high-speed form-fill-seal and pouch equipment. Typically, a three-layer coextrusion has an outside presentation layer, which has a high gloss and printability, a middle bulk layer that provides stiffness, strength, and split resistance, and an inner layer that is suitable for heat sealing. These films have good barrier properties and are more cost-effective than wax-coated paper or laminates. They are also used for snack foods (Fellows, 1996).

Snack companies will likely always look for a packaging film that protects the product and shelf life while not being too expensive. At the same time, films must be handled at a high rate of speed by the converter. Converters' need for high-speed films is driving the push for thinner films. Food manufacturers want reduced film costs, but they do not want to give up the protection that thicker films provide. In response, suppliers are creating improved barrier films with thicknesses from 0.55 mils to 0.70 mils (1 mil = 0.00254 cm) (Scherer, 1998).

Barrier Performance

Gas Barrier

In the case of the oxygen barrier, the primary function is that of excluding the gas from the package content. For example, providing a potato chip package with a Saran barrier (PVDC copolymer), while leaving normal atmosphere in the package, serves no purpose. The oxygen present inside the package will react with the contents and will ultimately produce rancidity. Because the bulk density of potato chips is typically 0.056 g/mL, they have a very large headspace volume per unit weight of product. If the product is packaged at atmospheric oxygen concentration, the headspace oxygen is sufficient to cause oxygen uptake in excess of 3 mL O_2 (STP)/g (Robertson, 1993).

Similarly, use of inert gas packaging on potato chips in a light transparent barrier material would severely restrict the benefits available from such a costly package as Saran. Even minute amounts of oxygen that would penetrate the barrier would set off rancidity in the presence of light. Evacuation (removal of air) may take place to a greater or lesser extent by design. The normal food package requires a vacuum level of 711 mm Hg. Extremely high vacuum levels, however, demanding an almost absolute 0% of residual oxygen, are rare. Normal vacuum packaging provides for about 0.5% residual oxygen at best. It has been shown that inert gas packaging would result in a very significant increase in the storage life of potato chips, provided that the headspace oxygen concentrations attained are below 1% and the package permeability to oxygen is very low (Quast and Karel, 1972).

Nitrogen is the gas most frequently employed in the gas flushing operation. This gas is inexpensive, readily available, and, above all, inert, causing no change in the packaged product. Product sensitivity and shelf life expectancy must be the guiding factors in the selection of a gas flush system (Paradis, 1993).

Moisture Barrier

The packaging material may be required to maintain the moisture content at either a low or high level. Fried snacks maintain their quality, due to the absence of moisture. It is, however, most urgent to exclude even minute amounts of moisture, as these would lead to the rapid deterioration of the food item thus exposed.

The packaging must exclude the migration of even traces of moisture over the expected storage life of the product. A properly functioning package would be designed to maintain the moisture content at the desired level.

The oxygen transmission rate (OTR) and water vapor transmission rate (WVTR) formulas are especially applicable for flexible films or their composites. Materials differ widely in their permeabilities toward oxygen, carbon dioxide, and water vapor (Tables 11–1 and 11–2).

Aluminum Foil

In fried snack packaging applications, foil is one of the choices available for a high-barrier, flexible packaging material. However, foil packages are very expensive and, furthermore, preclude visibility of product. These two shortcomings must be considered prior to the selection of this type of packaging material. The transmission rate of both oxygen and moisture can be drastically reduced with the aid of foil. However, it is erroneous to assume that the use of foil in the packaging material will reduce the oxygen and moisture transmission rate to the ideal zero. The need for foil will be determined by the shelf life expectancy of the product and by its maximum permissible moisture loss or gain, as well as oxygen susceptibility. Very often, however, data on maximum permissible moisture loss or gain are not available (see section on permeability).

Foil (more than 0.015 mm thick) is impermeable to moisture, gases, light, and microorganisms (Fellows, 1996). Because of both cost and flexibility considerations, the thinnest foil material suitable is employed for the desired application. Very thin gauges, such as 0.00762 mm or 0.0089 mm of foil (7.5–8.75 μm) have minute pinholes and, thus, are not absolute barriers. For example, 8.9-μm foil has a WVTR of up to 0.3 mL/m^2-day at 38°C and 100% relative humidity (RH) (Robertson, 1993).

Metallized Films

If properly selected, metallized films can extend product shelf life with a lightweight and colorful packaging system that adds shelf appeal. However, to ensure that the protection of the metallized film is appropriate for the product, some information on the product's modes of failure must be known for the product in its unpackaged form, as well as in its current packaging. Also, knowledge of the product's distribution history is helpful in choosing between the various substrates used to carry the aluminum layer, e.g., OPP, biaxially oriented polyester (OPET), biaxially oriented nylon (OPA), and oriented polyethylene (OHD). The composition of a product plays a major role in determining its fit. The protection profile of all metallized films, in order of importance is:

Table 11–1 Comparative Properties of Packaging Films

Material[a]	Tensile Strength (kPa)	Elongation (%)	Impact Strength (Kg-cm)	Tear Strength (g/mil)	Heat Seal Range (°C)
MECHANICAL					
Aluminum foil	0	16	1–5	4–20	121–163
ABS	62,055–82,740	10–60	2–8	2–15	177–232[b]
Acrylonitrile	65,552	5	30–90	15–25	88–204
Cellulose Acetate	48,265–82,740	15–50	4–6	50–150	177–260[b]
Ionomer	20,685–34,475	350–450	high	20–40	204–221
Capron 6	68,950–124,110	250–500	25–30	13–80	135–204[b]
Polycarbonate	117,215	110	1–3	15–300	135–154[b]
Polyester (uncoated)	117,215	70–130	7–11	100–400	121–177
HDPE	20,685–68,950	5–400	1–3	40–330	163–204[b]
LDPE	6,895–24,133	225–600	1–5	4–20	121–163[b]
PP (unoriented)	20,685–41,370	200–500	40		149–191
Polystyrene (oriented)	62,055–82,740	10–60	12–20		93–177[b]
Polyurethane	48,265–62,055	300–700		high	138–149[b]
PVC	13,790–131,005	5–500	12–20	varies	
SAN	75,845		10–15		
Saran	55,160–137,900	40–80	5–15	10–20	
Other Films					
Cellophane (polycoated)	48,265–124,110	15–25	good	2–10	110–149
Fluorohalocarbon	34,475–68,950	50–400	good	10–40	177–204
Polyvinyl alcohol	34,475–62,055	400		300–500	191–254

CHEMICAL	Oxygen Barrier[c]	Water Vapor Trans. Rate[d]
Aluminum foil	0	0
ABS		5
Acrylonitrile	5.52	very high
Cellulose acetate	12,411–21,375	22–30
Ionomer	24,133–51,713	very high
Capron 6	207–759	100
Polycarbonate	32,062	15
Polyester (uncoated)	359–897	5–10
HDPE	3,585–23,443	18
LDPE	26,890–89,635	8–10
PP (unoriented)	8,964–44,128	100
Polystyrene (oriented)	17,927–53,092	40
Polyurethane	very high	8
PVC	531–51,713	high
SAN	high	1.5–5
Saran	55–179	
Other Films		
Cellophane (polycoated)	varies	18
Fluorohalocarbon	very low	0.4-1.0
Polyvinyl alcohol	very low	none

[a]1 mil thickness; [b]unsupported film may not be sealable; [c]cc/m²/24 h/1 atm/38°C @ 0% RH; [d]g/24 h/m² @ 38°C, 90% RH. ABS, acrylonitrile-butadiene-styrene; SAN, styrene-acrylonitrile.

Source: Data from M.J. Lewis, *Physical Properties of Foods and Food Processing Systems,* © 1996, Woodhead Publishing Limited.

Table 11–2 Permeability of Some Gases and Vapors through a Variety of Film Materials

Film Material	Permeability ($mL/m^2 \cdot MPa \cdot day$)			
	Nitrogen (30°C)	Oxygen (30°C)	Carbon Dioxide (30°C)	Water Vapor (25°C, 90% RH)
Poly(vinylidene chloride) (Saran)	0.7	0.35	1.9	94
Polychlorotrifluoroethylene	0.2	0.66	4.8	19
Polyester (Mylar A)	0.33	1.47	10	8,700
Polyamide (Nylon 6)	0.67	2.5	10	47,000
Poly(vinyl chloride) (unplasticized)	2.7	8	6.7	10,000
Cellulose acetate (P912)	19	52	450	500,000
Polyethylene (ρ = 0.45–0.960 g/cm³)	18	71	230	860
Polyethylene (ρ =0.922 g/cm³)	120	360	2,300	5,300
Polystyrene	19	73	590	80,000
Polypropylene (ρ =0.910 g/cm³)	—	150	610	480

1. *Light barrier.* Most films are manufactured with an optical density of 2, which represents 1% light transmission through the film.
2. *Moisture barrier* (WVTR). All films supply a permeability to moisture vapor at a rate between 0.005 and 0.02 g-m/100 in²/24 hours.
3. *Oxygen barrier* (OTR). There is a wide variance in oxygen barrier, depending on substrate, ranging from 0.001 to 10 cc/100 in²/24 hours.
4. *Chemical (or aroma) barrier.* Variable, depending on film substrate, as well as permeating chemical.
5. *Appearance.* If the metallized film is used only for visual appeal and package decoration, the optical density should be in the range of 1.6, or 4% light transmission (Scherer, 1996).

If the product contains unsaturated oils, either naturally or from the cooking process (as is the case of fried foods), the product can benefit from the light protection of metallized film. Eliminating light as the primary cause of product degradation focuses attention on moisture gain or loss as a cause of failure. Metallized films in the form of sealable laminations offer a significant WVTR advantage over the previous glassine/polymer laminations and have essentially replaced them, producing a significant increase in product shelf life in relation to staling, or moisture level changes (Scherer, 1996). A demand for metallized films with high moisture barrier, as well as higher oxygen barrier, has resulted from previous packaging of snacks with OPP or polyesterterepthtalate (PET) films, which allowed secondary oxidation mechanisms to dominate (thus producing rancidity).

Selection of the most appropriate film is determined by a variety of fitness-for-use criteria and cost considerations. These films span the range of high-performance metallized OPP with an OTR of 1–2 cc/100 in²/24 hours to metallized OPA and PET with an OTR of 0.01-0.5 cc/100 in²/24 hours. In general, oxygen levels of 2 cc/100 in²/24 hours or lower are desirable for many products, although some applications require true high oxygen barrier at 0.5 cc/100 in²/24 hours or lower. The majority of metallized OPP films have an average OTR of 4–6 cc/100 in²/24 hours. In the 0.05 and lower range, there are metallized PET and OPA films, along with a new developmental OPP film available (Scherer, 1996).

Rigid Packaging

Composite cans made of paperboard (pure virgin sulfate stock suitable for food packaging, more than 0.01 mm thick) or combinations of paperboard with plastic films or metal foils coated on both sides with adhesive are used for packaging of high-fat dry products. These packages are economic and practical, recyclable, water-resistant, and convenient. Barrier materials can be laminated to or coated onto paper or foil to improve product shelf life. The material used for this liner is determined by the characteristics of the product, its shelf life requirements, and the design of the closure. There are three major types of bodies for composite cans: convolute-wound, spiral-wound, and lap-seam (Figure 11–2). The most commonly used is the spiral-wound type because it is the most economic (Hanlon, 1994). End pieces can be made from aluminum up to 0.368 mm thick or plastic.

U.S. Food and Drug Administration Regulations

The materials used for packaging foods must meet certain standards as required by law and by the regulations issued by the FDA to supplement and define the intent of the law. They must be cleared by the FDA for direct food contact. The package design should also meet any legislative requirements concerning labeling of foods.

PACKAGE SELECTION

Efforts on package design may be justified by the need to revitalize an image, minimize cost, become more competitive, improve the product shelf life, or a combination of all these. Good packaging includes eye appeal, convenience in size, and shape and ease of opening. A package should also retain odors or prevent odor pickup.

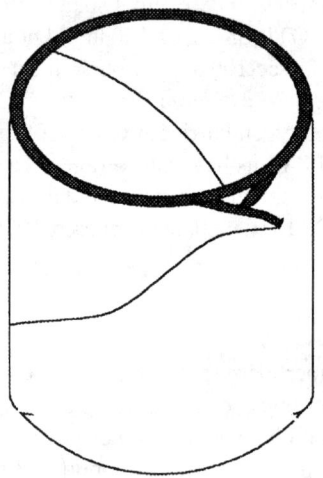

Figure 11–2 Schematic of a Composite Can

The steps to follow for proper selection of a package for fried foods are:

1. Determine failure mode of product.
2. Establish distribution and storage conditions.
3. Estimate product shelf life.
4. Select package design concept (shape, size, etc.).
5. Conduct cost analysis.
6. Consider printability.
7. Determine convenience of package.

From the modes of deterioration, it is clear that a satisfactory package for a fried snack such as potato, tortilla, and corn chips should provide a good barrier to oxygen, light, and moisture. Limited information is available about the effects of packaging materials on the stability of fried snacks during ambient storage. Because these products are frequently displayed for sale under fluorescent light, flexible packages are usually pigmented or (occasionally) placed inside paperboard cartons. The use of metallized films has become widespread in recent years, and although they are reasonably efficient light barriers, they do permit some light to penetrate into the package (Robertson, 1993). Fried snacks are typically packaged in multilayer structures (Figure 11–3), although spiral-wound, paperboard cans lined with aluminum foil or a barrier polymer and sealed under vacuum with an LDPE/foil end are used for some specialty products.

Studies on potato chips showed that, when packaged in pouches constructed of high-density polyethylene (HDPE), low-density polyethylene (LDPE), LDPE-

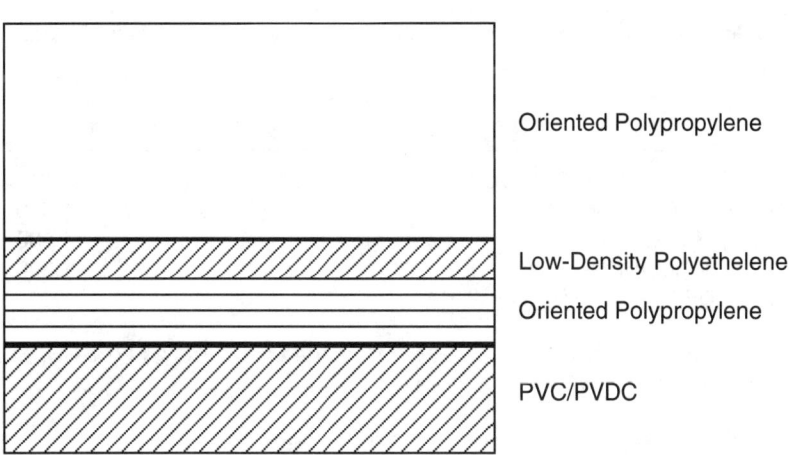

Oriented Polypropylene

Low-Density Polyethelene

Oriented Polypropylene

PVC/PVDC

Figure 11–3 Typical Pouch Lamination

coated regenerated cellulose film (RCF) or polyvinyl chloride (PVC)/ polyvinylidene chloride (PVDC) copolymer-coated RCF and stored at 27°C and 65% RH, the chips had a shelf life of 15 days. None of these films offered any protection from light, and the conditions of lighting were not reported (Robertson, 1993). Potato chips packaged in PVC/PVDC copolymer-coated PP were reported as being more sensitive to water vapor than to oxygen, and, when stored at 55–65% RH, became unsalable after 8–10 weeks because of loss of crispness. Potato chips packaged in PP/aluminum foil pouches required about 27 weeks to become unsalable because rancid flavors developed (Robertson, 1993).

Potato chips packaged in OPP/LDPE/PVC, HDPE/ethylene vinyl acetate (EVA) copolymer and an ultraviolet light-absorbing compound, or HDPE/EVA copolymer and a titanium dioxide light barrier developed distinct oxidized flavors within 7 days when stored at 21°C, 55% RH, and under 140–230 ft candles of continuous fluorescent light. Potato chips stored under the same conditions but packaged in HDPE and titanium dioxide, and a brown light-absorbing pigment construction or an aluminum foil/LDPE construction were stable throughout 10 weeks of storage (Booth, 1990). Oxygen-barrier film characteristics did not influence the oxidative stability of the air-packaged potato chips. The barrier properties of the various films used are presented in Table 11–3.

Hernandez (1997) describes a typical snack food package as a structure composed of extruded or laminated films of PET/ethylene vinyl alcohol (EVOH)/ EVA Nylon/EVA/Nylon/Ionomer OPP/EVA/EVOH/EVA. Table 11–4 shows typical packaging materials used in snack packaging.

Table 11–3 Barrier Properties of Selected Packaging Materials for Potato Chips

Film Construction	WVTR $(cm^3/m^2\ day)^a$	OTR $(cm^3/m^2\ day)^b$
OPP-LDPE-OPP-PVC/PVDC	3.5	6
HDPE-HDPE + UVab-EVA	3	1,500
HDPE-HDPE + TiO$_2$-EVA	3	1,500
HDPE + TiO$_2$-HDPE + Br^c-HDPE	3	1,500
OPP-LDPE-Foil-LDPE-HDPE-EVA	0	0

[a] Measured at 37.8°C and 90% RH; [b] Measured at 22.8°C and 0% RH; [c] Brown-pigmented light barrier.

Table 11–4 Materials Used in Snack Foods Packaging

Material	Properties
LDPE	Good resistance to acids, alkali, and water. Poor resistance to grease and oil.
EAA	Similar barrier properties to LDPE. Superior mechanical properties.
Surlyn	Combined with LDPE, PVDC, and aluminum foil to form heat seal layer. Tough and flexible. Great visual appearance.
Acrylic-Coated PP	Good machinability, attractive appearance, cost-effectiveness. Metallized OPP provides excellent protection for oxygen-sensitive foods as low-fat chips.
Saran (PVDC)	Barrier material to moisture, gases, flavors, and odors.
EVOH	Outstanding O_2 and odor barrier properties. High resistance to oils.

Source: Data from R.J. Hernandez, Food Packaging Materials, Barrier Properties, and Selection, in *Handbook of Food Engineering Practice*, K.J. Valentas, E. Rotstein, and R.P. Singh, eds., © 1997, CRC Press.

In the case of fried snacks, 100% nitrogen is recommended. At least two package design elements must be simultaneously present for a successful nitrogen flushing system: (a) quality package seals (films without folds, puckers, or leaks) and (b) gas-flushable packages (films with acceptable oxygen, aroma, and moisture barrier properties). The actual nitrogen cost is typically one-tenth the film upgrade cost (Paradis, 1993). A high-barrier film with good seals at 21°C and 50% relative humidity will maintain the oxygen levels under 4% through 50 days and under 8% through 100 days.

PRODUCT SHELF LIFE

Foods can be classified according to the amount of protection required. For fried snacks (moisture and oxygen sensitive), shelf life depends largely on the water and/or oxygen permeability of the package, which can be defined as the time to reach a critical humidity (corresponding to a maximum moisture content) or extent of oxidation. If one of these limits is reached before the other, the product is overprotected with respect to the other. Because good barrier materials are expensive, this overprotection is unnecessarily costly (Karel, 1975).

In the case of fried snacks, the following is required from a package for 1 year shelf life at 25°C: maximum oxygen gain of 5–15 ppm, maximum water gain or loss of 5%, and high fat resistance (Salame, 1974). When the shelf life of a product is known under specific conditions and a projection is required at different conditions, the same type of calculation may be conducted. An estimate is needed of the average temperature in which the package will find itself during transit (Taoukis et al., 1997).

Prediction of product shelf life requires knowledge of specific information, such as:

1. The mechanism of product deterioration
2. The factors affecting the rate of deterioration
3. The expected quality of the food in the package
4. The desirable shape and size of the package
5. The environmental conditions during distribution and storage
6. The barrier properties of the packaging materials

This approach allows estimation of the time before the product becomes unacceptable. It is precisely the definition of what is acceptable that is crucial to this process and it depends on a true understanding of item 1. This sort of analysis is good for evaluation of film-based packaging materials, which provide various degrees of protection (Robertson, 1993). Taoukis et al. (1997) present an in-depth review of shelf life prediction and applications into food packaging.

Concept of Permeability

This is one of the most important properties in food packaging. It concerns a mass transfer effect appropriately known as *permeation* and associated with a partial pressure differential of a gas or vapor between the two sides of a film. This is defined as the *flux* (rate per unit area) of the permeant through the material. (It should be noted that the term *permeation* refers to transfer on a molecular scale through microscopic holes but not to flow through cracks, flaws, and poor seals due to a total pressure gradient.) The permeability of a material depends on its chemical structure and morphology (degree of crystallinity, nature, and size of crystallites), the nature and size of the permeant, and temperature (Hernandez, 1997).

The permeability coefficient of a material, P, is a constant for a specific permeant at a given temperature. The most common units for gases is cm^3 (standard temperature and pressure conditions)-mm/m^2-day-atm and, in the case of water vapor, the quantity is given in grams, and the partial pressure differential is expressed as the difference in relative humidity. The expression for the steady-state permeation of a gas or vapor through a film can be written as:

$$\frac{\delta w}{\delta t} = \frac{P}{X} \cdot A \cdot (p_1 - p_2)$$

[1]

where P/X is the permeance (the permeability constant P divided by the thickness of the film X), A is the surface area of the package, p_1 and p_2 the partial pressures of water vapor outside and inside the package, and $\delta w/\delta t$ the rate of gas or vapor transport across the film. Integration of the above equation with appropriate boundary conditions yields:

$$\frac{Q}{t} = \frac{P}{X} \cdot A(\Delta p) \qquad [2]$$

with P/X called the *permeability* or *permeance*. As a guide, it is possible to assume that if the thickness of the barrier film is doubled, the transmission rate of a gas or vapor is halved (Robertson, 1993).

The significance of optimal permeabilities is best illustrated by the example of a film laminate. If a single film has high oxygen permeability and low water permeability, it must be very thick to prevent oxygen transmission. However, the thickness and oxygen transmission can be greatly reduced if a second film with high water permeability and low oxygen permeability is laminated to the first film (Karel, 1975). This lamination can significantly reduce the cost of the package.

For a typical fried snack package, several layers of material may be used. Multilayer materials can be considered as a number of membranes in series (Figure 11–4). In the case of three layers, Equation 2 becomes:

$$P_T = \frac{X_T}{(X_1/P_1)+(X_2/P_2)+(X_3/P_3)} \qquad [3]$$

where X_T is the total thickness $(X_1 + X_2 + X_3)$ and P_T the permeability coefficient of the multilayer material. Eq.(3) assumes that the permeability coefficients P_1, P_2, and P_3 are independent of pressure. Otherwise, differing permeability coefficients will be obtained, depending on the position of the layers (Labuza, 1982).

Quast and associates (1972) initially developed a mathematic procedure to describe the changes in quality of potato chips during storage. Optimal conditions were calculated for a specified set of conditions:

1. Initial environment in the package
2. Product weight
3. Storage conditions
4. Film area
5. Film thickness

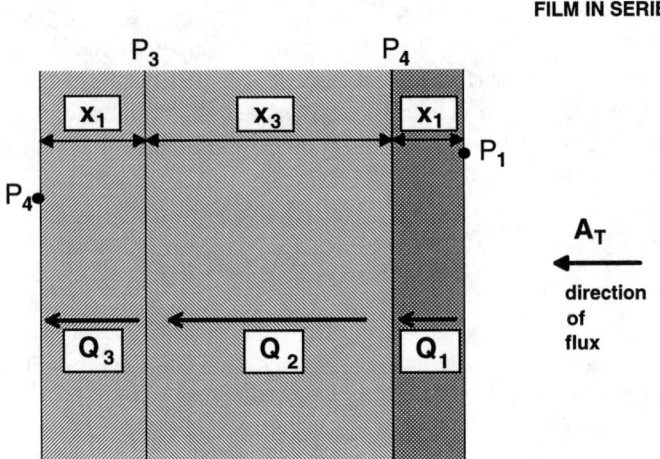

Figure 11–4 Schematic Representation of Permeation through Several Materials in Series

---

The uniqueness of their approach is that it takes into account the fact that fried snacks deteriorate simultaneously by loss of crispness and development of rancidity. Because both deterioration mechanisms interact, the representative equations must be solved simultaneously. The interaction arises as alteration in humidity in the package alters the rate of oxidation. The technique produced three differential equations (Quast et al., 1972):

(1) Change in Oxygen Partial Pressure

$$\frac{d\left(\dfrac{VO_2}{V}\right)}{dt} = \frac{d\left(\dfrac{PO_2}{P}\right)}{dt} \qquad [4]$$

$$\frac{d(VO_2)}{dt} = \frac{T \cdot A \cdot KO_2}{TO \cdot X}(PO_{20} - PO_2) \qquad [5]$$

where VO_2 is the volume of oxygen, V the total headspace volume, t the time, PO_2 the oxygen partial pressure inside the package, T the temperature in K, TO is the reference temperature (273 K), X is the film thickness, KO_2 is the oxygen permeability, and PO_{20} is the outside oxygen partial pressure.

(2) Extent of Oxidation

$$RATE = \frac{d(EXT)}{dt} = \left[EXT + \frac{P_1 + P_2 \cdot EXT}{\sqrt{RH}} \right] \left[\frac{PO_2}{P_3 + P_4 * PO_2} \right] \qquad [6]$$

$$EXT = \frac{VO_{2t} - VO_{2p}}{w} \cdot \frac{TO}{T} \cdot 1000 \qquad [7]$$

where *RATE* is the rate of oxidation, *EXT* the extent of oxidation, P_1 *and* P_4 empirical constants, *RH* the equilibrium relative humidity, VO_{2t} the volume of oxygen transferred into the package from time 0 to time *t*, VO_{2p} the actual volume of oxygen inside the package, and *w* the product weight.

(3) Relative Humidity Changes During Storage (related to the rate of moisture transfer through the packaging film):

$$\frac{d(m)}{dt} = \frac{A \cdot KW \cdot PWS}{X \cdot w}(a_o - a_i) \qquad [8]$$

where *m* is the product moisture content (g/g solids), *A* is the area of package film (m^2), *KW* the water vapor permeability (g.mil/m^2.hr.mm Hg), *PWS* the pressure of saturated water vapor, a_o the water activity outside the package, and a_i the water activity inside the package.

Measurement

Three different methods can be used for measuring the permeability of a barrier film to gases or vapors (Miltz, 1992).

1. *Absolute Pressure Method:* This method for determining steady-state rates of gas transmission through barrier films is carried out according to the American Society for Testing and Materials standard test method ASTM D-1434 (ASTM, 1990). In this method, a film is mounted between two chambers of a gas transmission cell. The cell and sample are sealed so that the only path for gas to move from one chamber to the other is through the film. The tested gas then permeates from the high-level chamber, and the intake of the low-level one is measured by withdrawing samples at suitable

time intervals and analyzing them (e.g., by means of a gas chromatograph). This intake is then plotted as a function of time. At the beginning of the test, transient conditions prevail, during which the permeant dissolves in the film and diffuses through it. After a while, steady-state conditions set in with a constant flux of the permeant at which the permeability coefficient is determined.

2. *Isostatic Method:* The gas or vapor used for the measurement is present either as a pure gas or in conjunction with other gases or air in a cell chamber on one side of the film. The partial pressure of the permeating gas must be known. The gas that has permeated into the second or low-concentration cell chamber is conveyed by a carrier gas to a suitable sensor. The total pressure on both sides of the film is constant (generally at 1 atm), as the gas currents in both chambers have contact with the external environment.

3. *Quasi-Isostatic Method:* The gas or vapor used for measurement is present as a pure gas or in conjunction with another gas or gases in the high concentration chamber of the permeability cell. The partial pressure of the permeating gas must be known. The gas or vapor flows through one cell chamber at a total pressure of 1 atm. Concurrently, gas or vapor permeates the test sample and is accumulated in the lower concentration chamber of the cell, which is closed off from the atmosphere. At predetermined time intervals, aliquots are withdrawn from the lower chamber for analysis using gas chromatography. The concentration changes of the permeated gas accumulated in the lower concentration chamber is plotted as a function of run time, and the slope is related to the sample's permeability.

Water vapor permeability is commonly measured using the standard gravimetric method described by ASTM E96-66 (ASTM, 1990). In this method, a desiccant maintaining low water vapor pressure is sealed in an aluminum cup by the test film. The cup is placed in a chamber maintained at constant temperature and relative humidity, and the gain in weight is plotted as a function of time. The desiccant must maintain a low vapor pressure of water, regardless of water sorption. Calcium sulfate, magnesium prechlorate, and calcium chloride have been used for this purpose. Another method to measure the water vapor transmission is the measurement of the relative humidity in a chamber separated from a source of water vapor by the test sample.

Permeability of organic compounds, flavors, and aroma is described by Hernandez et al. (1986).

FUTURE TRENDS

A new concept introduced into snack packaging is the use of holograms, which can be valuable with the launch of a new product or the reintroduction of an old favorite. A hologram has a more desired appeal as it uses a visual effect to attract the eye. Using this type of packaging coincides with the increased use of metallized films, as the basis of a hologram begins there (Scherer, 1988).

Another growing trend is nitrogen gas flushing with cold sealing. The technology for this process is currently being developed so that potato chips, for example, could have their packaging gas flushed with nitrogen to block out oxygen, then sealed with cold seal. However, unless the package is recloseable, this gas flushing will not be as convenient for consumers who desire to capture the first-taste freshness every time they open the package. Also, cold sealing adhesives currently are not strong enough to withstand nitrogen gas flushing.

Current high-barrier flexible packaging uses aluminum metallized films/foils or specialized polymer films, such as PVDC, to limit oxygen and water vapor permeation. It is anticipated that new transparent, clear, low-cost, high-barrier packaging systems will be developed that will be environmentally benign, transparent, and able to maintain high-barrier properties after significant mechanical manipulation. This new technology will allow new packaging designs that can benefit consumers, distributors, and manufacturers of foods worldwide by enhancing the quality of the distributed foods through improved shelf life and reduced costs. This product, which is completely recyclable and contains no environmentally unfriendly materials, could replace tens of millions of pounds of more expensive PVDC/PET, as well as other lower-performance products (Scherer, 1988).

NOTATIONS

EAA   Ethylene acrylic acid
EVA   Ethylene vinyl acetate
EVOH   Ethylene vinyl alcohol
EXT   Extent of oxidation
HDPE   High-density polyethylene
KO_2 Oygen permeability (cc STP mil.m^2.h.atm)
LDPE   Low-density polyethylene
OPP   Oriented polypropylene
OTR   Oxygen transmission rate
PET   Polyesterterepthtalate
P_1-P_4 Empirical constants, Equation 6
PO_2 Oxygen partial pressure inside the package [atm]
PO_{2O} Outside oxygen partial pressure [atm]
PP   Polypropylene
PVC   Polyvinyl chloride
PVDC   Polyvinylidene chloride (Saran)
RATE   Rate of oxidation

RCF   Regenerated cellulose film
RH   Equilibrium relative humidity
Surlyn   DuPont's trade name for ionomers
t   Time [hour]
T   Temperature [K]
TO   Reference temperature [273 K]
V   Total headspace volume [cm³]
VO_2 Volume of oxygen [cm³]
VO_{2p} Actual volume of oxygen inside the package
VO_{2t} Volume of oxygen transferred into the package from time 0 to time t [cm³]
w   Product weight [g]
WVTR   Water vapor transmission rate [gm/100 in³/24 hr]
X   Film thickness [mil]

REFERENCES

American Society for Testing and Materials. 1990. Methods ASTM D-1434 (*Gas transmission rate of plastic film and sheeting* and ASTM E96-66 (*Water vapor transmission of material in sheet form*).

Booth, R.G. 1990. *Snack Food*. New York: Van Nostrand Reinhold.

Fellows, P.J. 1996. *Food processing technology, principles and practice*. Cambridge, England: Woodhead Publishing.

Hanlon J.F. 1994. *Handbook of package engineering*. New York: McGraw-Hill.

Hernandez, R.J. 1997. Food packaging materials, barrier properties, and selection. In *Handbook of food engineering practice,* ed. Valentas, Rotstein, and Singh. Boca Raton, FL: CRC Press.

Hernandez, R.J.; Giacin, J.R; and Baner, A.L. 1986. The evaluation of the aroma barrier properties of polymer films. *J Plastic Film Sheeting*, 2 (July):187–211.

Hine, D.J. 1987. Shelf-life prediction. In *Modern processing, packaging and distribution systems for food,* ed. Pain. New York: AVI.

Karel, M. 1975. Protective packaging of foods. In *Principles of food science. II. Physical principles of food preservation,* ed. Karel, Fennema, and Lund. New York: Marcel Dekker.

Katz, E.E., and Labuza, T.P. 1981. Effect of water activity on the sensory crispness and mechanical deformation of snack food products. *J Food Sci*, 46:403.

Labuza, T.P. 1982. *Shelf-life dating of foods*. Westport, CT: Food and Nutrition Press, Inc.

Miltz, J. 1992. Food packaging. In *Handbook of food engineering,* ed. Heldman and Lund. New York: Marcel Dekker.

Paradis, A. 1993. Nitrogen in total quality for snack food. *INFORM* 4(12):1378–1386.

Quast, D.G., and Karel, M. 1972. Computer simulation of storage life of foods undergoing spoilage by two interacting mechanisms. *J Food Sci*. 37:679–682.

Quast, D.G.; Karel, M.; and Rand, W.M. 1972. Development of a mathematical model for oxidation of potato chips as a function of oxygen pressure, extent of oxidation and equilibrium relative humidity. *J Food Sci*, 37:673–678.

Robertson, G.L. 1993. *Food packaging*. New York: Marcel Dekker.

Salame, M. 1974. The use of low permeation thermoplastics in food and beverage packaging. In *Permeability of plastic films and coatings*, ed. Hopfenberg. New York: Plenum Press.

Scherer, C. 1998. The future of films. *Baking and Snack*, 20(2):81–86.

Taoukis, P.S.; Labuza, T.P.; and Saguy, I.S. 1997. Kinetics of food deterioration and shelf-life prediction. In *Handbook of food engineering practice,* ed. Valentas, Rotstein, and Singh. Boca Raton, FL: CRC Press.

Index